Telecommunications Politics

Ownership and Control
of the Information Highway
in Developing Countries

TELECOMMUNICATIONS

A Series of Volumes Edited
by Christopher H. Sterling

Telecommunications Politics

Ownership and Control of the Information Highway in Developing Countries

Edited by

Bella Mody
Johannes M. Bauer
Michigan State University

Joseph D. Straubhaar
Brigham Young University

Foreword by
Andrés B. Bande
Ameritech International, Inc.

IEA LAWRENCE ERLBAUM ASSOCIATES, PUBLISHERS
1995   Mahwah, New Jersey

Lawrence Erlbaum Associates, Inc., Publishers
10 Industrial Avenue
Mahwah, New Jersey 07430

Library of Congress Cataloging-in-Publication Data

Telecommunications politics : ownership and control of the information
highway in developing countries / edited by Bella Mody, Johannes M.
Bauer, Joseph D. Straubhaar.
    p.   cm.
Includes bibliographical references and index.
ISBN 0-8058-1752-2. — ISBN 0-8058-1753-0 (pbk.)
1. Telecommunication policy—Developing countries.
2. Privatization—Developing countries. I. Mody, Bella.
II. Bauer, Johannes M. III. Straubhaar, Joseph D.
HE8635.T44 1995
384'.09172'4—dc20   94-47647
                                                            CIP

Books published by Lawrence Erlbaum Associates are printed on acid-free
paper, and their bindings are chosen for strength and durability.

Printed in the United States of America
10 9 8 7 6 5 4 3 2 1

Contents

III   CASE STUDIES

IV   THE ROLE OF REGULATION

Foreword:
Balancing Foreign Investment
and National Development

Andrés B. Bande
Ameritech International, Inc.

I recently spoke at a Michigan State University workshop where early drafts of the following chapters were discussed. It was a pleasure to inaugurate that seminar, and an honor to be invited to share those inaugural comments with a wider readership through this Foreword. I have always been in favor of corporate participation in academic initiatives and believe that such exchanges benefit not only the participants, but the public interest at large.

There is no doubt that we are experiencing an Age of Privatization. However, as is often the case, after a generation or so, the pendulum of public opinion swings strongly in the other direction. The renewed faith in privatization now reverses the trend toward state ownership that began at the turn of the century in Britain, France, Germany, and other industrially advanced nations.

After the Great Depression, which was interpreted internationally as the failure of the marketplace, governments perceived the necessity to intervene to ensure the smooth functioning of the national economy. Interventionist philosophy, along with natural monopoly and national security considerations, came together to justify the national ownership and operation of telecom networks.

More recently, government corruption, the influence of Margaret Thatcher, and the international discrediting of centrally planned economies have contributed to swinging the pendulum in the other direction—renewing faith in private capital. However, we must be careful not to let the pendulum swing *too* far. We must objectively evaluate the nature of private ownership, particularly with foreign participation.

TELECOM MODERNIZATION

It is a well-established fact that telecom infrastructure and service, in the age of information and transnational communication, are linchpins of a healthy, growing economy. Telecommunications is the backbone of business activity, productivity, and trade. And for a developing nation, effective provision of telecom services is a precondition to the emergence of a strong market economy. Today, the vitality of important industries and social activities—such as banking, healthcare, information services, transportation, and education—is extremely dependent on adequate telecom infrastructure and service.

Research has clearly demonstrated a correlation between telephone penetration and gross domestic product (GDP). It has shown the benefits of telecommunications both as an input and as a catalyst to development. At the WorldTel preparatory seminar in New York in 1993, Sam Pitroda, former Secretary in India's Department of Telecommunications and Adviser on Technology Missions to the Prime Minister, estimated that the addition of a single phone line contributes an average of $3,700 to GDP in developing nations.

There is no doubt that extending telecom service into remote areas that have never been serviced before will generate economic growth. Local efficiencies are created, and forward and backward linkages develop over time. Moreover, telecommunications brings about important technological and social change: openness, connectivity, decentralization, accessibility. It brings people together, links like-minded groups, affects interpersonal relationships, bridges distance and time, and integrates cultures.

However, in weighing costs and benefits, one must remember that at some point, in some configurations, *the quality of national life could actually decline with merely more telecommunications.* Obviously, basic service is critical for a minimum standard of living with adequate healthcare, information, and entertainment. Beyond that point, however, telecommunications may make our lives fairly complicated. Was life better before the fax machine? Is it better now than it will be with the advent of personal communication networks (PCNs)? At what point does expediency, accessibility, and openness overtake our lives? After all, life has more to it than efficiency goals. This is a lesson that industrialized countries need to learn from developing nations.

Given that investment and liberalization are important to improving telecom service, which in turn can provide growth to other sectors of a developing economy, how does a nation in the periphery stimulate this investment and liberalization? *What are the options it can exercise?*

INVESTMENT OPTIONS

There are several options a nation can exercise. First, it could try to modernize telecom infrastructure and service on its own through state intervention. France Telecom has, for instance, done a pretty good job of modernizing its infrastructure and service offerings.

But in developing countries, where limited government funds are torn between satisfying basic subsistence and other important social needs, there just may not be enough money available. In addition, government bureaucracy and corruption can pose serious threats to flexibility and innovation, which are two critical components of any modernization plan. Public telephone and telegraph entities (PTTs) have generally been over-staffed and inefficient; bureaucracies in developing nations can be particularly bad. Also, PTTs may not have the technology, management skills, and human resources necessary to institute modernization and competitiveness. And, unless their status as a monopoly changes, they may not have the incentives either.

A second option is bank loans. However, bank loans have been providing less and less conventional equity. Increasingly, they are interested in providing debt-for-equity swaps—another form of foreign investment.

A third option is multilateral lending institutions that can provide funds for projects in telecommunications. However, their pockets are not as deep as they should be. For instance, the World Bank lends less than 2% of its total funds for telecom projects. Moreover, in such lending situations, funds are sometimes tied to later purchases from certain equipment suppliers. These purchases may or may not have anything to do with the host country's particular equipment needs, and could, therefore, inhibit critical technological advancement.

Because the multilateral lending institutions are not providing adequate funding for telecom development, there is definitely the need for a multinational organization—run along commercial lines—that will tap into all financial resources available for telecom development. This organization should work in close cooperation with the ITU and with other international organizations. This is the objective of WorldTel, a new organization that seeks to promote telecommunications growth worldwide, with special emphasis on developing nations.

A fourth option in securing capital is foreign investment.

Foreign Investment Advantages

Foreign investment can be an effective means of securing necessary capital, technology, management, and human resources critical to achieve modernization and liberalization aims. Funds from investors can be used by the government for several important development purposes. First, investment funds can be used to improve telecom infrastructure and service to reach stipulated modernization goals. In developing nations, networks are often underdeveloped and do not extend into remote areas. Venezuela and Mexico are two good examples of countries that used privatization as a means to upgrade and extend the telephone network. Second, investment funds can be used to invest in other industries critical to economic growth, to assist in balancing the government's budget deficit, and to sponsor other important social and welfare programs. Third, foreign investors can transfer technology and management skills that will allow the host country to operate its telecom entity more efficiently. In addition to introducing technologies not necessarily available in the country and training engineers and other technicians on how to use them, private investors can share new attitudes toward management, customer service, and marketing. By bringing outside investors and new managers into a PTT, privatization can eliminate constrictive government practices, which tend to focus on the needs of the operator rather than on the needs of users. A private investor can also play an important and expedient role in streamlining staff. Finally, increased efficiencies and improved network infrastructure can produce more service revenue that may ultimately benefit government owners. Incomplete calls and pent-up demand from a poor network means lost revenue opportunities.

Disadvantages of Foreign Investment

It should be remembered, however, that foreign investment in telecom privatization is not a magical solution. There may be disadvantages to host nations. First, foreign investment might cut out local investors who are willing to invest their monies in domestic privatization projects. Often in developing nations, local industrialists and entrepreneurs invest their money abroad, and may not necessarily want to bring those funds home. Or they may not be willing or able to work with the longer term repayment plans involved with financing for these investment projects. A second disadvantage is the selling off of the strategic assets of a nation. Undoubtedly, foreign investment raises sensitive questions about national sovereignty and foreign control. As telecom plays a vital role in providing national security, the subject is even more sensitive. Finally, foreign investment introduces speculation into the financing of essential national infrastructure. For example, Bond Corporation sold its stake of CTC Chile only 2 years after purchase.

INVESTOR REQUIREMENTS

Like all investors, Ameritech looks at the macroeconomic environment and the particular business case before making a decision to invest in a privatization. Dimensions of the macro environment include the following:

- Macroeconomic superstructure such as banking, transport, and so on.
- Pattern of economic growth.
- Infrastructure development and investments of the World Bank and other multilaterals.
- Other private investments.
- Current and projected inflation rate.
- Laws governing repatriation of earnings.
- Stability and convertibility of local currency.
- Level of political risk and availability of risk insurance to offset it.

Dimensions of the particular business case include the following:

- Length of startup phase.
- Nonrecourse financing availability for the investment.
- Tax relief incentives.
- Security of market share—ability to migrate to next technology/ service.
- Licensing—sufficient frequency allocation to maintain business.
- Management influence—to have a role in future planning.
- Local staffing issues—skills of the local labor force.
- Length of time before return on investment.

THE BALANCING ACT: GOVERNMENT AND THE PRIVATE INVESTOR

To implement successful privatization, governments undoubtedly face a balancing act. They need foreign capital but must avoid complete foreign control. Their objective might be expressed as having overall control, without necessarily owning or running every bit of the PTT. The key is for a government to take control by using incentives and regulation, or carrots and sticks, if you will, in a reasonable way. In this way, it can gain needed foreign participation and achieve other important development goals. I suggest the following:

Establish Joint Ventures and Selloffs. The government can limit foreign involvement and attempt to build a "level playing field" by forcing outside investors to partner with local competitive entities (government or business) in joint venture agreements or, over time, to sell their stakes to local firms. They can establish equity caps for foreign participants.

Create Incentives or Stipulate Goals. Governments must either create incentives or negotiate to extract concessions for foreign investors to provide universal service and support certain economic development goals. But, they must seek to do this in a way that balances profitable activity for the investor with important national goals. They may, for instance, use certain tax holidays to encourage investors' actions. For example, before privatizing Telmex in 1990, Mexico's government clearly defined the conditions under which the new owners could operate the company. The government structured regulation to provide incentives for improving the quality of service and increasing penetration, and held competitors at bay for a 7-year period until rates could be brought closer to costs of service. The government has been tough; the license establishes specific improvement conditions the company must meet over a 5-year period, with stiff penalties for failure to comply. The government has also indicated that any tariff changes in 1997 will be related to discussions on the quality of service provided.

Set Limits on the Monolopy. Set limits on how long foreign investors can operate as a monopoly, or tie their monopolistic position to perform-ance records or fulfillment of certain concessions.

Establish a Regulatory Authority. Establish an independent regula-tory authority with well-qualified policymakers, who will plan, imple-ment, and oversee the privatization process. Investors view regulation that is fair, consistent, and flexible as a favorable situation, one that provides a clear indication of the industry's future and stable means to arrive there.

CONCLUSION

There is no doubt that the global economy is inextricably linked to rapid and sophisticated information exchange. Countries that are able to acquire the telecommunications infrastructure necessary to access this informa-tion will exercise great power. Telecom-rich nations will reap the full benefits of expanding trade and investment opportunities, whereas na-tions that are stuck with primitive telecommunications infrastructures will become ever more marginal players in the world economy. For this and for other aforementioned reasons, it is critical that developing nations pursue improvements in their telecommunications infrastructure and services. Privatization involving foreign participation provides attractive opportunities for nations in the periphery to secure capital and other needed resources. But it also raises sovereignty, regulatory, and balanced growth issues that must be anticipated and managed properly.

I am basically optimistic that governments in the developing nations will strike the right balance in attracting investment and advancing development policy. I'm optimistic about governments because there is such strong recognition that telecommunications is the backbone of a healthy economy, that any government wishing to improve the economy at all, even perhaps for their own reasons of personal gain, would have a difficult time "selling out" the sector.

I am also fairly optimistic regarding the behavior of foreign investors. From Ameritech's perspective—and many other investors see the situation in the same way—participation in a privatization is a substantial and potentially long-term investment. For such an investment, the most important thing is the nation's stability. We will not participate in a country that is unstable. Naturally, in addition to a tethered environment, we will need to earn a reasonable return on our investment. That is obviously a basic business requirement. However, we are not willing to earn our return in a country where the government is exploiting the industry for its own purposes, and to the detriment of the people. Exploitative policies that are not structured in the public interest will present problems for the government over time. And so, both for moral considerations and for the desire to protect stable investments, we will not work with governments under those conditions.

Although not all private investors see the situation this way, I am confident that many do, and that is progress!

Preface

The reason for this book is a present-day gold rush. Investors, telecom operators, and speculators from saturated telecom markets in industrialized countries are out looking for joint venture partners and opportunities for investments in the less-mined telecom service markets of Asia, Latin America, and the Caribbean. Telecom is a good investment: Financial rates of return are generally 15% (Wellenius & Stern, 1994), but are frequently much higher. Southwest Bell Communication's $1 billion investment in Telemex doubled in value in spite of the 45% loss due to the December 1994 devaluation of the peso.

Our focus is on political aspects of this outward expansion of First World investors searching for new telecom service markets in the periphery of the world capitalist system. Although most chapters in this book address ownership and control of telecommunications in middle-income and low-income countries in Asia, Africa, Latin America, and the Caribbean, many of our observations hold for economies in the European periphery,[1] for

---

[1] In 1990, the former Soviet Republic of Georgia gave an exclusive contract to erect an elaborate international telephone company to Videotel, a seven-person New York firm that had never installed a telephone system. Three years later, the system could handle only six international calls simultaneously. When the Government of Georgia terminated the contract and sued the U.S. firm, the firm responded with a breach-of-contract lawsuit against Deutsche Telekom, claiming it was trying to steal Videotel business. The New York firm's counterattack was led by lawyers that included former U.S. Defense Secretary Caspar Weinberger, a former adviser to President Clinton, and a former State Department official. The U.S. State Department and Treasury Department defended the U.S. firm and pressured the World Bank to withhold a $40 million telecommunications loan to Georgia.

example, Russia and Central and Eastern European countries. We have used the term *developing countries* for lack of a better alternative that would communicate with both telecommunications practitioners and researchers.

TELECOMMUNICATIONS IN DEVELOPING COUNTRIES

There continues to be a considerable disparity in the distribution of the world's 575 million existing main telephone lines between developed and developing countries in the 1990s. The average of 49 lines per 100 inhabitants in the 24 developed member countries of the Organization for Economic Cooperation and Development compared to 3.5 in the rest of the world does not reveal the reality of less than 1 line per 1,000 population in many countries of South Asia and Africa ("Perspectives," 1994). Approximately 4 billion of the world's 5 billion people still do not have a telephone. Given the experience of relative deprivation of food, clothing, and shelter, why should telecommunications service provision matter in developing countries? What is the relative priority of an information distribution system over a food distribution system, water supply, or clothing? The problem for low-income and middle-income countries is that agriculture, industry, and social services are all important and interrelated, and that telecommunications can potentially facilitate the functioning of each one of them. When health, education, agriculture, industry, and other service sectors are efficient and effective, the use of an *enabling technology* such as telecommunications can amplify productivity. Thus, aggressive private firms (domestic and foreign) and government agencies wanting to compete in the increasingly global economy are demanding instantaneous telecommunications of text, audio, and video facilities that can be reconfigured flexibly to suit changes in their needs. The expanding middle class in developing countries wants the same residential and business telecommunications services that its counterparts enjoy in North America, Western Europe, and Japan. Low-income populations and physically disadvantaged individuals in developing countries have never experienced the advantages of telecommunications.

SUPERHIGHWAY?

Because we are appreciative of the potential advantages of telecommunications, we see only disadvantage in resorting to superlatives and exaggerated claims about its miraculous capability across the board. We eschew the hyperbole surrounding electronic networks carrying information around countries and across the globe, be it of the Naisbitt or Orwellian varieties. The scholarly reasons for our skepticism include the following:

1. Much of the rhetoric and the reality is technologically driven: Advances in technology address only technological parts of communication problems, not their economic, social, organizational, administrative, and political aspects. Hence, hardware alone rarely constitutes the total solution. Interactive cable TV trumpeted by the rhetoric of the electronic town hall has been technically possible for decades in the United States, but Warner closed down its experimental QUBE facility in Columbus, Ohio, in the 1970s because they could not figure out how to make it profitable. The supply of interactive hardware capability did not generate demand. Democratic governments bemoaning voter apathy did not deploy this interactive information capability then, so Vice President Al Gore's hope that the national information infrastructure and the global information infrastructure will automatically enhance citizen participation today (and international cooperation tomorrow) is questionable. Teletext and videotex were similarly praised for their technological capability, but there are few large operational systems serving business or residential applications anywhere. Discourse on the "information highway" continues society's preoccupation with means/media and neglects questions relating to ends/goals of human existence (see Foreword by Bande) and the quality of the electronic environment it heralds. Rather than think of networks as conduits or highways, Samarajiva (1994) suggests they may be better conceptualized as electronic environments designed with control capability unsurpassed in pervasiveness, complexity, and sophistication.

2. Different systems of national and local ownership and control differentially limit and/or energize utilization of telecommunications hardware.

3. Promoters' agendas are central to the calculus. The so-called "information superhighway" is being advocated in the United States in the 1990s as a government strategy to boost the economic competitiveness of its domestic industries. The focus is on interactivity to promote electronic commerce through electronic cash. In 1934, universal telecommunication service was the result of a state mandate, too.

4. The information highway and other such proposals driven by economics and technology are not socially and spatially neutral. They are capital-intensive and will therefore tend to be deployed when and where they can provide worthwhile rates of return to the investor. Thus business users, who are able to pay for information processed in particular ways to be delivered to their offices, will have a better chance of being served than rural and residential users with limited purchasing power. Hundreds of millions will not benefit from this information-skills-based borderless economic order.

5. The convergence of computers and communication channels has implications for individuals (privacy, surveillance) and nations (security, sovereignty, access, exclusion). Whereas scholars and activists have paid attention to the effects of broadcasting entertainment and news flows across borders in the 1970s and 1980s (Nordenstreng & Varis, 1973; Varis, 1974), we need to build on the good critical work of Mosco (1982), Schiller (1985), Samarajiva and Shields (1990), Shields and Samarajiva (1990), and Samarajiva (1994) in the telecommunications sector. The basic question needs to be restated in a different metaphor: Who will own and control the new electronic control environment, and in whose interest?

DOES OWNERSHIP MATTER?

That telecommunications will ensure consumption of the goods and services produced by the telecom-energized flexible production economy is unquestionable. Does the nature of ownership (public, private, cooperative, users) make the major difference in how efficiently a nation's telecommunications system is run, and how universally its services are available? Previous research does not support this position (Cook & Kirkpatrick, 1988; Vickers & Yarrow, 1988). There are many exceptions to the popular generalization that state-owned telecom firms perform poorly (e.g., Singapore, France, Botswana, South Korea, Sweden) and privately owned firms perform well (e.g., PLDT in the Philippines, ITT in several Latin American countries before nationalization in the 1960s). Countries with the highest telephone density operate under a variety of ownership and market structure conditions. Recourse to private capital may rectify some part of the problem of inefficiency, but its causes are many and private capital is only one tool. Even the ITU that has been promoting private sector participation in the telecommunications sector found that operating entities perform best when run as profit-driven enterprises, *irrespective of who owns them* (ITU, 1989). A single transparent global network to meet customer needs can be provided through coordinated interconnections of public and private providers of different parts (e.g., local, long distance, international). If ownership changes are all that take place, significant problems in efficiency and equitable coverage could continue due to institutional reasons such as networks of influence and market power (see Melody, chapter 11, this volume). Surely, it is organizational practices that make the difference in efficiency in both public and private enterprises.

If ownership does not determine efficiency, then why is ownership an issue at all? The ownership framework is important in terms of social goals and social governance. Ownership determines who will control our electronically mediated environment capable of unprecedented surveillance. Different forms of ownership lead to different forms of control. Public ownership of telecommunications in developing countries was frequently inevitable because there was no other willing and able private investor. However, public ownership implied that governance and control would ideally be public. Several contributors (see Bauer, Mody, Sinha) discuss why the reality of public ownership has been distinct from this norm. Private ownership has no such norm or standard to which it is accountable: Control explicitly rests with the few who have capital to invest; the public accepts that decisions on how, when, and at what price they can communicate across distances will be based on consolidation and maximization of the interests of mobile private capital. There is an obvious contradiction in attempting to formulate telecom policies for national competitiveness through heavy reliance on liquid extranational capital with a distinct agenda. In some societies, a degree of social and political control may be achieved by government regulation of private activity, but the effectiveness of regulation depends on the political structures, on the political will, on the relative freedom from cooptation by private interests, and a relatively low level of corruption.

So what combination of ideals and investors will own and control telecommunications in developing countries in this phase of global capitalist expansion? Will it be the state, national capital where it exists, foreign institutional investors like the New York City banks that financed the creation of International Telegraph and Telephones in 1904, or private investor-operators like Cable and Wireless and Southwestern Bell?

When nation-states, like South Korea, that have developed a modern efficient telecommunications system without any foreign investment are the exception, what are the surveillance and other implications of foreign direct investment by telecommunications operators? There are no estimates of the economic benefits of surveillance. Some firms (e.g., STET from Italy, Bell Atlantic from the United States) will want to earn revenues through controlling votes on the board of directors while keeping their equity participation to a minimum, whereas others may be willing to risk greater financial resources. In Latin American sales, a single buyer was given controlling interest on the board although ownership of the companies was widely dispersed among domestic investors, foreign investors, employees, and the public. Domestic capital in the Philippines handled this a couple of decades ago, with the political patronage of former President Marcos.[2] In an environment in which large providers of tele-

---

[2] The Ramon Cojuangco family, who indirectly own only 2.7% of the total issued capital of the domestic oligopoly PLDT, actually control 6 of the 11 seats on the board. Filipinos own more than 70% of the stocks, although only 14.8% carry voting rights. A voting majority is actually held by an ineffective cast of U.S. speculators who could change in New York every day.

communications services can select their investment projects on a global basis, political unrest or an unexpected economic downturn may impact the success and the degrees of freedom of a national policy relying on foreign capital. This is illustrated by the recent Mexican currency crisis brought on by speculative capital flight from Mexican financial markets. As an immediate result, the major foreign investors suffered losses in the value of their assets, leading them to consider new investment carefully. But the crisis has also triggered some indirect effects. Telephone rates increased sharply. And the Mexican government may be willing to extend the monopoly privileges of the incumbent carriers for some additional time. Whereas such a policy would clearly stabilize the earnings of TEL-MEX, it may cause welfare losses for Mexican customers. Will the International Telecommunications Union-initiated *Worldtel* (a multinational commercial mechanism to finance developing country systems mentioned in the Foreword by Andrés Bande of Ameritech International) address the problem of the growing dependence on global capital markets that are beyond the control of any single government? Unlike less volatile foreign direct investment in productive activity within a country, financial speculators can move their investments in moments to take advantage of better interest rates elsewhere. Thus, foreign speculators become state-less unelected legislators that can threaten the control of states, *and* their World Bank and IMF handlers.

TELECOMMUNICATIONS POLITICS

Our book elaborates on the position that economics and technology generate options for action; politics chooses between the alternatives. By *politics*, we mean the struggle for power, control, and profits between different contending forces in developing country contexts (Mody, 1983). Contending wielders of economic, political, and technological force in the international and domestic context of particular nation-states influence the nature of goals set for their national telecommunications sector and how they will be achieved. Changing national contexts of power also influence the selection of public or private ownership (or some balance thereof) and the nature and degree of competition (or lack of it). Thus, national decisions in favor of privatization have been changed with changes in the ruling party in many countries (e.g., Greece, Turkey). In other countries, the state has consolidated its own power position through a calculated loss of autonomy by accommodating to some pressures (e.g., from the World Bank, domestic capital, and government users) and sub-jugating others (e.g., its own bureaucracy and unions).

Until the mid-1980s, the state owned and controlled the national tele-communications system in most developing countries. State telecommu-nication entities, particularly in Latin America and parts of Asia, are looking for investment capital and new technology at a time when tele-communications companies in North America and Western Europe are looking for places to invest their monopoly profits. Capital has not flowed exuberantly across the globe since the Rothschilds and the Morgans over 100 years ago. Although the overall share of private finance in telecom-munications investment is still relatively small (World Bank, 1994), it represents a big increase in international financial flows into the telecom-munications sector. Private capital flows have consisted of long-term private capital in the form of both foreign direct investment for production and speculative portfolio flows.

East Asian and Latin American countries have been using both inter-national public offerings and concessions to build telecommunications systems (as part of build-and-transfer schemes). In October 1993, Tele-communications Argentina placed much of its $500 million 7-year bond issues with U.S. and Asian investors; Argentina's Telefónica has also used bond market placements to raise expansion funds. In 1990, Compañía de Teléfonos de Chile (CTC) raised $42 million on the New York Stock Exchange through an issue of equity in the form of American Depository Receipts in the U.S. capital market. India's international telecommunica-tions operator (VSNL) hopes to raise $1 billion through a Euroissue of global depository receipts. In Latin America, foreign direct investment by telecommunications firms and financial institutions has opened an-other route into international equity markets. In 1995, India is licensing private competitive providers of basic services.

Private and statist strategies wax and wane. Inevitable cyclical changes in capitalism will leave the information and telecommunications sector behind in search of new areas of profits, but for now, the politics of telecommunications in the developing world has changed. On the one hand, the context is characterized by competition for state resources, worsening terms of trade for Third World exports, an increasing external debt burden, and internal fiscal deficits. On the other hand, we see increasing demand from foreign and domestic users in industry and government, perception of telecommunications as a revenue-making cash cow, bureaucratic management of the sector by a government department or a public sector enterprise, and poor service. At the political level, there is an international ideological consensus in Washington, Tokyo, London, and other European capitals that delegitimizes the use of state-owned enterprises and pushes governments toward privatization (Hills, 1994). At the domestic level, many societies are also rethinking the role of the state because of internal political divisions and controversies (Migdal, Kohli, & Shue, 1994).

ORGANIZATION OF THE BOOK

The structure of this book reflects our time- and space-specific, contextually situated intellectual perspective.

Part I is an opening historical and conceptual introduction to changes in the organization of telecommunications. Organization and reorganization of a state sector is a political response to a political situation,[3] whether the impetus is debt reduction in Latin America or infrastructure enhancement and competitiveness in East Asia. Joseph Straubhaar presents an overview of the restructuring process and trends from public post, telegraph and telephone operator, the PTT, to private firm. Daniel Headrick presents a historical perspective on this current cycle of the global expansion of capitalism through a case study of the relations between private and public ownership in telegraphy before World War II. His research shows that the salience of different aspects (economic, political, technological) of telecommunications at different points in time lead to different combinations of public and private ownership and control; for example, in times of technological control, telecom entities resort to private capital. When political issues (public service, national security) become salient, the state intervenes. Both private commercial ownership and government ownership of telecommunications must contend with conflicting public and private political interests.

Part II analyzes the major political forces, both external and internal to a country, that have influenced the process of private sector participation in telecommunications. These forces and factors include the expansion of global capital, the World Bank, the International Telecommunications Union, domestic capital, and the nature of the state. The restructuring path recommended by the European Community Green Paper of 1987, the World Bank, and the ITU begins with commercializing and separating operations from government, increasing the participation of private enterprise and capital, restraining monopolies, encouraging competitive supply of services, and shifting government responsibility from ownership and management to policy and regulation.

---

[3]Proponents of privatization often restrict their causal analyses to technological changes and economic demands. They neglect to point out that it is unfavorable terms of trade and protected Northern markets for products from Asia, Africa, Latin America, and the Caribbean that contributed to the Third World external debt crisis and internal fiscal deficits. On the other hand, opponents of privatization frequently fail to deal with the frustrations of working with bureaucracies mired in self-serving corruption or sometimes controlled by a small political elite.

Gwen Urey analyzes the forces of global capitalism that have generated such pressures. Major telecommunications operator-investors who are expanding their international operations in this present wave of private capital expansion include Nippon Telegraph and Telephone, British Telecommunications, Cable and Wireless, France Telecommunications, Deutsche Telecommunications, Telefónica of Spain, STET of Italy, Bell Canada, GTE, AT&T and the regional Bell companies from the United States, and Korea Telecom International and Singapore Telecommunications International from Asia. Foreign institutional investors include New York investment banks returning to global telecommunications, such as J. P. Morgan and Chase Manhattan. Sharmishta Bagchi-Sen and Parthavi Das look at the different patterns of foreign direct investment of the regional Bell operating companies in the United States. Gwen Urey looks at the facilitating role that the World Bank is playing in the increasing privatization of the telecommunications sector in developing countries. Jill Hills (1994) argues that the promotion of privatization by international development banks helps increase exports of available capital from industrialized countries (who hold major shares in the banks), and ensures a seat for its personnel at the highest policy table of every country wanting a loan. The World Bank receives special attention because it has influenced almost all countries considering use of private capital in their telecommunications sectors. Although much less than 5% of its annual budgets have been spent on telecommunications, "the Bank" (as it is popularly known) has financed preparation of restructuring and privatization attempts in many developing countries. In some cases, it has awarded the government a large balance of payment support loans, partly on condition of good progress on privatization (e.g., Uganda). Other multilateral agencies such as the Asian and African Development banks have also been supportive of privatizing telecommunications, as have bilateral aid agencies (particularly USAID and France) and the International Telecommunications Union (ITU). Eduardo Barrera and Meheroo Jussawalla present the different roles of domestic capital in the distinct Latin American and East Asian restructuring scenarios. Bella Mody and Lai-Si Tsui address the reactions of different states in the developing world to the pressure they face to improve the functioning of their telecommunications service providers.

In Part III, Joseph Straubhaar, Patricia McCormick, Johannes Bauer, and Consuelo Campbell present comparative country case studies. The eight selected cases present evidence of the diverse conditions, goals, and processes of the realignment of public and private tasks in the telecommunications industry. As the cases illustrate, the goal of creating economically efficient conditions for the development of telecommunications is frequently superseded by other more mundane goals such as reduction of foreign debt balances or privatization to reduce the influence of the state per se. The reviewed cases suggest, furthermore, that the conditions for success of increased private sector participation are more complicated than frequently assumed. Liberalization, deregulation, and privatization are no panacea for improved sectoral performance. Important issues, such as providing open access conditions to networks and services, fair competitive structures, and service for economically disadvantaged areas and strata of the population, demand continued public sector supervision of the industry. Last but not least, trade-offs become visible, for instance, between the goals of fast versus equitable infrastructure deployment. To counteract the panacea perception of private ownership, we asked Alexandrina Benedicto Wolf and Gerald Sussman to spell out what is known about the performance of the telecommunications system in the Philippines in terms of efficiency and equity, because it is one of the longest running privately owned systems in the developing world. While the state in the Philippines is trying to restructure the private system through the introduction of competition, the moral of the story for promoters of privatization is that sociocultural traditions of the country and power politics do not change overnight.

Part IV is the pragmatic, how-to-cope part of the book for developing country decision makers. Contributors address the issue of regulation from different perspectives. William Melody and Johannes Bauer argue that the issue of ownership is less important than the creation of institutional structures that support efficiency and allow for measures to pursue equity objectives. Pointing to the experience of selected industrialized countries, Bauer illustrates that a wide variety of institutional and regulatory design options exist for telecommunications. He argues that these options cannot be ranked uniquely on a spectrum ranging from more efficient to less efficient as much of the current policy debate seems to suggest, but rather that different approaches will create incentive structures that will lead to characteristic trajectories of telecommunications development. Arguing from a property rights perspective, Nikhil Sinha outlines how to create a stable system of regulation in developing countries, and what the roles of the various actors would be. Harry Trebing provides a critical analysis of the limited workability of competitive forces in telecommunications and suggests various generic approaches of how developing countries should conceptualize and implement regulation. The World Bank and other promoters of private sector participation, like the United States, ITU, and the European Community, stress the importance of regulation, as do public interest activists and lawyers. The promoters of privatization are concerned with protecting private capital from political risks such as nationalization. Hence they insist on an independent regulatory mechanism and transparent rules on capital entry, exit, rates of return, and exclusive licenses. For public-interest advocates, the key issue is how to substitute for the ideal of public control that is lost when "public" utilities and "public" enterprises are privatized. The latter are concerned with the design of a tough-minded regulatory agency with human and financial resources to resist capture by those it is meant to regulate. The poor are not provided with the kind of services for which they are willing to pay. Models of regulatory agencies range from AUSTEL, which represents a proregulation stance, to the laissez-faire approach in New Zealand, which is having problems dealing with the inevitable disputes that accompany burgeoning competition. Neither of the two distinct groups that urge clear regulatory frameworks before privatization commences have been happy with the regimes set up in any of the recently privatized PTTs: Mexico, Malaysia, Venezuela, Peru, Jamaica, Argentina, or Chile.

FUTURE RESEARCH

Many questions could not be addressed in sufficient detail in this volume and demand further research. These issues include, but are not limited to, the ideological or idea structure in which political decision making takes place (Pirie, 1988); the question of the power of the nation-state to meaningfully regulate transnational capital in telecommunications; the medium and long-term sectoral dynamics resulting from the shifting boundary between public and private players in telecommunications; many issues related to the design of functioning and efficient regulatory institutions and instruments that are appropriate for the context of developing countries; more meaningful measures to assess the local, national, and international welfare effects of institutional restructuring; and, last but not least, the impact of the reformed governance regimes on fairness and equity both within and between nations. Most important is control of telecom conceptualized as an environmentwide control mechanism, as electronic space (Samarajiva, 1994): What are the implications of public versus private ownership, of foreign direct investment versus speculation, and foreign versus domestic control of our total environment? Perhaps Africa should feel it is just as well that its telecom penetration is lower than the rest of the world.

A VARIED PERSPECTIVE

Do the three editors take a position? We maintain that telecommunications is not technological apparatus alone, or strictly an economic investment. The domestic and foreign political context of each country will

influence the particular balance of public and private capital in the ownership and control of its restructured telecommunications sector. Each of
us comes to the topic from different national backgrounds (Austria, the
United States, and India), formal preparations (economics, international
relations, psychology), and work experiences. We discovered by chance
that each of us was researching different parts of the same topic simultaneously and independently while we were colleagues in the Department
of Telecommunications at Michigan State University in 1992.

The contributors to this volume (e.g., Meheroo Jussawalla, Gwen Urey)
do not necessarily agree with each other, and the editors made no attempt
to negotiate resolution of differences. We hope the diverse positions in
this volume illuminate the political struggle between the different forces
contending for control of new markets in developing countries.

ACKNOWLEDGMENTS

The collaborative work began by sharing our analyses of different parts
of the situation with each other, our students, and the surrounding Michigan community at a small workshop funded by the Center for Advanced
Studies in International Development (CASID) at Michigan State University. Despite our distinct personal and professional socializations that led
to different positions on privatization, we agreed that privatization is a
lot more complex than its current fans and critics made out. We needed
to make the simple depiction of choices between state central planning
and markets as elaborate as it was. We recognized the need to provide
scholars and developing country policy makers with a more subtle understanding of current telecommunications politics surrounding the hype
of the information highway.

We therefore decided to invite others working on this topic around
the world to join us in producing an edited volume of research on this
topic. A U.S. Department of Education award to the Center for International Business Education and Research (CIBER) at Michigan State University helped support a large conference in May 1993, which enabled
contributors to present early drafts of our chapters to each other and get
the reactions of representatives of the World Bank and industry involved
in investing in developing countries. The inaugural address of the conference, by Andrés Bande, President of Ameritech International, is presented as the Foreword to this collection.

In addition to CIBER, other academic units at Michigan State University
that contributed financially to the May 1993 conference include the Center
for European and Russian Studies, the Office of Academic Computing
and Technology, the Office of the Vice President for Research, the Office

of the Dean of the College of Communication, International Studies and Programs, the Center for Advanced Study of International Development, Center for Latin American and Caribbean Studies, the Asian Studies Center, and the Department of Telecommunications. We are grateful to the many formal and informal reviewers of our chapters; as always, the final product could be further improved.

Former graduate students Kathy Busse, Parthavi Das, and Luiz Guilherme Duarte helped organize the conference. Discussers of draft chapters presented at the conference intentionally included many who would argue with our positions, ranging from our MSU political science colleagues Peter Lewis (now at American University) and Nicholas Van de Walle (now at the Overseas Development Fund), Rohan Samarajiva of Ohio State University, Ahmed Galal and Bjorn Wellenius of the World Bank, Jill Hills of The City University in London, and economist Aydin Cecin of Central Michigan University. Graduate students Mitra Barun Sarkar, Pierre Azoulay, and Ramsey Kemal helped with library research and discussion of revised chapters. Many of the chapters benefitted from editing services financed by the Center for Advanced Studies in International Development at Michigan State University. The credit for production of this book goes to Barbara Wieghaus of Lawrence Erlbaum Associates (LEA). It was a pleasure to work with LEA's cheery, efficient Senior Editor Hollis Heimbouch.

—*Bella Mody*
—*Johannes M. Bauer*
—*Joseph D. Straubhaar*

REFERENCES

Cook, P., & Kirkpatrick, C. (1988). *Privatisation in less developed countries*. Brighton, UK: Wheatsheaf Books.

Hills, J. (1994, July). *Economics as ideology: The World Bank and privatization*. Paper presented at the annual conference of the International Association for Mass Communication Research, Seoul, South Korea.

ITU. (1989). *The changing telecommunications environment: Policy considerations for the members of the ITU*. Report to the Advisory Group on Telecommunications Policy, Geneva.

Migdal, J. S., Kohli, A., & Shue, V. (1994). *State power and social forces*. New York: Cambridge University Press.

Mody, B. (1983). First World technologies in Third World contexts. In E. M. Rogers & F. Balle (Eds.), *Communication technology in the United States and western Europe*. Norwood, NJ: Ablex.

Mosco, V. (1982). *Pushbutton fantasies*. Norwood, NJ: Ablex.

Nordenstreng, K., & Varis, T. (1973). The nonhomogeneity of the nation-state and the international flow of communication. In G. Gerbner, L. P. Gross, & W. H. Melody (Eds.), *Communication technology and social policy* (pp. 393–412). New York: Wiley.

Perspectives: Report on the state of the world's telecommunications. (1994, 1). *ITU Newsletter,* 7.

Pirie, M. (1988). *Privatization.* Hants, UK: Wildwood House.

Samarajiva, R. (1994, March). *Electronic public space: Dystopian and other futures.* Paper presented at the Fourth Computers, Freedom, and Privacy Conference, Chicago, IL.

Samarajiva, R., & Shields, P. (1990). Value issues in telecommunications resource allocation in the Third World. In S. B. Lundstedt (Ed.), *Telecommunications, values, and the public interest.* Norwood, NJ: Ablex.

Schiller, D. (1985, January). The emerging global grid. *Media, Culture and Society (7)*1, 105–125.

Shields, P., & Samarajiva, R. (1990). Telecommunications, rural development and the Maitland report. *Gazette, 46,* 197–217.

Varis, T. (1974). Global traffic in television. *Journal of Communication, 24*(1), 102–109.

Vickers, J., & Yarrow, G. (1988). *Privatization: An economic analysis.* Cambridge: MIT Press.

Wellenius, B., & Stern, P. A. (1994). *Implementing reforms in the telecommunications sector: Background, overview and lessons.* In B. Wellenius & P. A. Stern (Eds.), *Implementing reforms in the telecommunications sector.* Washington, DC: The World Bank.

World Bank. (1994). *World development report.* New York: Oxford University Press.

I

INCREASING PRIVATE SECTOR PARTICIPATION

1

From PTT to Private: Liberalization and Privatization in Eastern Europe and the Third World

Joseph D. Straubhaar
Brigham Young University

In this chapter, we discuss the rise and rationale of Postal, Telephone, and Telegraph (PTT) traditions, patterns of PTT operations, and then the problems that have arisen with the PTT structure that have created pressures for change. We then examine the general patterns to restructuring that countries are taking. First we consider liberalization, opening up selected sectors of the telecommunications area to competition by private firms, which varies widely but is more common and, in some ways, more important, than privatization of the principal telecommunication companies themselves. Additionally, changes in regulation to fit various goals for and approaches to restructuring are discussed. Particularly, we examine *deregulation*, reducing the control of the state over firms in the marketplace; *re-regulation*, separating the operations from regulation, sometimes even the reduction of regulatory powers; and *privatization*, selling or transferring at least part of the ownership of state-owned telecommunications companies or assets to private owners. Finally, although it is a relatively new concept, we offer *corporatization* as an alternative to privatization. Corporatization allows for some telecommunications operators to remain primarily government-owned but nonetheless separated from regulator(s). This process gives management greater financial and managerial autonomy, changes the institutional structure to resemble a corporation, and applies standard coporate laws to the restructured operator(s).

For years, nearly all telecommunications in the Third World, as in the industrialized nations and Eastern Europe as well, were owned and

3

operated by governments in what were typically called Postal, Telephone, and Telegraph administrations (PTTs). In many countries, starting with the United Kingdom and Japan, but spreading rapidly into Eastern Europe, Latin America, and Asia as well, governments have been changing their policies on telecommunications. The first step was usually to separate postal operations from telecommunications, but many have moved further to liberalize, de- or re-regulate, and, in some cases, privatize their telecommunications operations.

THE PTT TRADITION

Before we can discuss how the traditional state telecommunication administration or PTT is changing, we need to understand how the PTT developed in Eastern Europe and the Third World, what functions it served, and why it tends to persist in many countries. There has been a strong economic argument that telecommunications is a natural monopoly, that economies of scale are such that competition did not make sense. Most specifically, it has been argued that the extension of basic telephone services beyond areas that are profitable, notably into rural and poorer areas, is best conducted by a monopoly that will cross-subsidize unprofitable rural and remote services from profitable urban and business services.

Although the point has become controversial, for years regulators and scholars argued that even in the United States, the achievement of universal service was best conducted by a private monopoly under close government regulation (Tunstall, 1986). However, until recently, most experts held that in many developing countries, the state was best able to provide the financial and operational resources required to pursue universal or at least extended service (Hills, 1990).

In many countries, perhaps most, there have been strong cultural and legal tendencies toward monopoly structures dominated by the state. There is a tendency in many countries toward seeing state monopoly, particularly in basic infrastructure activity, as expected, whereas in the United Kingdom and the United States, there has been a traditional distrust of direct government economic activity and a culturally based sense that most activity should stay with the private sector. Unlike the United States, many countries lack an antimonopoly, antitrust legal tradition. In many countries, even those that the United States has had long ties with, such as those in Latin America, there is a tendency to distrust private ownership, particularly in the area of infrastructure. This tendency is now changing but has deep historical and cultural roots.

Correspondingly, many political groups in Latin America and elsewhere question the morality of selling assets created by public fees, taxes, and contributions. There is a sense that the public patrimony achieved through public resources should not be given to private parties for private profit. For instance, in Mexico, even after what many consider a model privatization, center-left groups are now raising questions about whether a visibly profitable telecommunications company (Telmex) should have been kept in public hands rather than permitting private profit, particularly when that profit is repatriated abroad (Barrera, 1991).

Nationalization and the Formation of PTTs

If we look at the history of telecommunications structures in the developing countries, we need to acknowledge that much of the initial structure derived in many countries from colonial regimes. This was particularly true in Africa, the Caribbean, and parts of Asia, where telecommunications systems were built under direct colonial control until 1945. In parts of the world where independence had come earlier but trade and investment was dominated by the United States, such as in the Philippines and most of Latin America, telephone companies were started by foreign companies. A number of companies, such as Cable and Wireless, International Telephone and Telegraph (ITT), and GTE, invested in or created local telephone companies in Latin America, Asia, the Arab world, and Africa. Some of these operations remain in the private sector, such as the GTE (now PLDT) operations in the Philippines, GTE in the Dominican Republic, or the Cable and Wireless operations in Hong Kong, Abu Dhabi, and several Caribbean nations. These represent a group of continuous private operations (see Benedicto-Wolf & Sussman, chap. 9, this volume). A number of local telephone operations were also started by local entrepreneurs in various countries in Latin America.

After gaining formal political independence from colonial regimes following World War II, many new governments in various countries took over colonial systems. The architecture of the early private or colonial systems was frequently designed only for the convenience of colonial administrators and not for increasing domestic business, residential, or even government communications, so national governments tried to take them over and adapt them. This created one major set of government PTTs, particularly in Africa and Asia.

In countries where telecommunications companies were initially private and/or foreign owned, governments began to think about nationalizing telecommunications. Many of these initial private operations were seen as inadequate for supporting national development by governments in the postcolonial era after World War II. The private telecommunications operators often covered only selected urban areas and did not necessarily plan to extend coverage into rural or poorer urban areas. For example, analysis by Benedicto-Wolf and Sussman (chap. 9, this volume) and Sussman (1991) of the Philippine GTE operation showed that it only tended to pursue urban and business routes that were profitable, and, being essentially unregulated, GTE was not required to invest in or pursue rural and universal service as was AT&T in the United States. When operations were started by local entrepreneurs, they were often not very well networked together, either with each other or with local operations of foreign telecommunications firms.

To extend service into uncovered areas or to more efficiently network various local, regional, and national telecommunciations operations together, many national governments tended to nationalize both private or colonial government telecommunications operations. From the 1940s through the 1970s, many locally or foreign-owned private operations were nationalized and colonial networks were taken over by government operations. For example, telecommunications were nationalized for both economic and national security reasons in Brazil in 1965 (Alencastro e Silva, 1990), but in Mexico, they were completely nationalized only in 1976 (Casasuz, 1990). A number of the foreign owners resisted efforts at nationalization. Perhaps the most notorious example was in Chile, where ITT was a principal mover behind efforts to overthrow the Allende regime (1970–1973), which was determined to nationalize ITT's operation there (Sampson, 1973). However, most foreign telecommunications operators were eventually nationalized.

Similarly, governments in Western Europe, Eastern Europe, and the Third World were concerned about the service responsibilities of telecommunications companies. They often felt that private telecommunications firms neglected key areas, such as the need to pursue universal service and extend residential and rural services. It was often felt that private firms would not pursue these kinds of investments because they were not profitable. Concern about the ability or willingness of private firms to extend telephone services to rural and less profitable urban and suburban areas was reinforced by a rising preoccupation with the role of telecommunications in national development and national security.

In the 1960s, a number of governments began to realize that telecommunications was an important aspect of the infrastructure for economic development. Supportingly, studies in the 1970s and 1980s about the relationship of telephone penetration, particularly into rural and residential service areas, and economic growth supported government impressions (Hardy, 1980; Hudson, 1984). In some countries, economic growth and the integration of large national territories for purposes of creating national identity and enhancing national security were seen as overlapping, complementary goals. For instance, under the military governments that ruled from 1964 to 1985, Brazil nationalized, centralized, and consolidated telecommunications operations in the 1960s and invested considerable resources in national infrastructure, particularly microwave and satellite systems (Alencastro e Silva, 1990; Quandt de Oliveira, 1992). Until the 1980s, most international development agencies reinforced government control of infrastructure for development, including telecommunications. The state was seen as a necessary factor in gaining and planning the resources necessary for telecommunications development (Hills, 1993; Saunders, Wellenius, & Warford, 1983).

In addition, Third World and European governments saw PTTs as a source of revenue generation. Perhaps the most common practice in almost all countries with PTT structures was for PTT telecommunications operations, which were usually profitable, to subsidize the PTT postal operations, which were seldom self-sufficient (Saunders et al., 1983). This argument was persuasive, until recently, even in Germany. However, countries, particularly in the Third World, often diverted profits from telecommunications to various parts of the government that seemed to require subsidy. For instance, although the military governments in Brazil (1964–1985) reinvested most telecommunications profits, the post-1985 civilian regimes have been more concerned about other priorities and have seen telecommunications as a source of revenue for them (E. Siqueira, personal communication, October 10, 1992). Saunders et al. (1983) showed that a major problem in expanding telecommunications infrastructure in Third World countries was the tendency of the governments to see telecommunications as a revenue source for other operations and to not reinvest telecommunication revenues in the expansion of telecommunications services.

In addition to being treated as revenue sources or "cash cows," PTTs have sometimes had tariffs set either artificially low or artificially high for political reasons. Politicians in several coutries have seen PTT basic telephony tariffs as an area in which prices could be held low to provide visible evidence of their fights against inflation. In Brazil, for example, tariff values for basic telephone service prices in 1987 represented only 18.5% of their value in 1975 (Lerner, 1988). Both military and civilian regimes had become more concerned about inflation's effect on their popularity than about the need to maintain tariff values to keep up maintainence and reinvestment of profits for expansion of service. South Korea, in contrast, kept rates high throughout the 1970s and 1980s to create internal savings for expansion (Park, 1993), as Brazil had earlier.

Furthermore, in some countries in the Third World, as in Europe, PTTs had a significant status as large employers, whose relationships with national unions was politically important. In fact, in some Third World countries telecommunications was seen as a convenient area for generating jobs to be given for political patronage. In some countries, payrolls are inflated and the number of employees actually inhibits efficiency. Since they fear cuts in employees after privatization, labor unions in both

European and Latin American countries have been among the fiercest opponents of privatization. Privatization has been fought in Mexico, Puerto Rico, and Malaysia, and effectively stopped in Costa Rica because of union pressure (Ambrose, Hennemeyer, & Chapon, 1990). In Brazil, union pressure has been one key element in keeping privatization at the discussion stage (E. Siqueira, personal communication, October 10, 1992).

PTT Patterns

Until quite recently, most PTTs maintained an extensive monopoly over nearly all equipment and services. They controlled access and connections, including restrictions on terminal equipment, restrictions or prohibitions on private satellite ground stations, network terminating equipment, and network facilities. In services, PTTs tended to control basic local and long-distance service, as well as virtually all other services: analog and digital data services, electronic mail, packet switching, value-added services, satellite services, and mobile services. Most PTTs were marked by extensive vertical and horizontal integration of various operations. They were vertically integrated for coordination of all aspects of telecommunication operations: construction, financing, operations, marketing, and so on. There were not only clear-cut horizontal divisions between post, telecommunications, and other areas of responsibility but also extensive integration via financial subsidy, common personnel systems, and the like.

PTT relations with national equipment vendors were often very close. Some PTTs literally had their own hardware and equipment divisions, adding another layer of integration. Most governments required PTTs to purchase or procure terminal, switching, and other equipment from national manufacturers when available. Some governments also used standards-setting procedures to keep imported equipment out, as the United States has consistently accused Japan of doing, for instance. Some governments used such rules as barriers to build up a national switch or terminal equipment manufacturing industry, with the PTT as a reliable market for such equipment. This was most common among the larger Third World countries, such as Brazil, China, and India, but was pursued in smaller countries as well (Mody, 1989a). In Eastern Europe, some countries, such as Czechoslovakia and Poland supplied equipment for other Eastern Bloc countries. Since the collapse of the Eastern Bloc, those countries are trying to sustain these equipment industries without the captive market that the Comecon Bloc used to provide (Bauer & Straubhaar, 1994).

Some countries had close ties with selected foreign firms that had invested in manufacturing operations in their countries, like Ericsson, Siemens, and NEC in Brazil. Others were tied to particular vendors by continuing postcolonial ties, like Alcatel in former French colonies. Others became tied to vendors by virtue of either official or private foreign aid, which was tied to the purchase of equipment from manufacturers in the donor country. Various African countries, which are almost completely dependent on aid and soft loans for telecommunications equipment, have found themselves in this situation.

In both Europe and the Third World, PTTs were often marked by a residential versus commercial or business orientation, in the kinds of infrastructure that they built and the kind of services that they offered. In the Third World and Eastern Europe, basic telephone service is far from universal, particularly outside the rapidly industrializing countries of East Asia—Taiwan, Hong Kong, South Korea, Singapore—where service has been rapidly expanded in the 1980s to levels equivalent to southern Europe (around 40 telephones per 100 people). Some policy actors fear that privatized telecom operators will not worry about universal service. Traditional PTTs have also tended to emphasize basic service over development of new technologies, usually related to business services, for similar reasons.

Particularly in Eastern Europe, but also in China and some other countries, change in the PTTs required a major change in communication policy toward tolerating more direct horizontal communication via telephone and data systems. In some countries, such as in the former Soviet Union, industrial policy often desired a better telecommunication infrastructure for encouraging investment but was at odds with information control policies that feared the political implications of unrestricted access to telephone, electronic mail, and other communications (Dizard & Svensrud, 1987).

PROBLEMS OF THE PTT MODEL AND PRESSURES FOR CHANGE

In a number of countries, the history of the PTT offers examples of both success and failure. The national government monopolies on telecommunication did largely succeed in consolidating and expanding services. In Brazil, for example, Telebrás, the government telecommunications holding company (which includes state-level operating companies and the national long-distance and data company), was seen as very successful from the 1960s through the early 1980s (Quandt de Oliveira, 1992). It expanded rural and residential service, so that nearly all villages had at least one phone, and created new data and business services in major urban areas. Its expansion was largely financed with internal savings and cross-subsidies. However, in the 1980s, accumulating national problems with debt, inflation, and eroding savings hurt the PTT, as did increasing political interference in pricing, after the return of civilian rule in 1985, which cut its ability to finance either maintainence or expansion from internal savings or cross-subsidy (Lerner, 1988; E. Siqueira, personal communication, October 10, 1992).

Studies of PTT behavior became concerned with the inflexibility of public administrations to respond to the crises that hit many PTTs in the 1980s (Saunders et al., 1983). PTT management was often seen as too bureaucratic or too inefficient. In the face of both public residential demand and business demand for new services, PTTs often had limited incentives to innovate. Long waiting lists of residential customers and long waiting periods for business services characterized problems of responsiveness in a number of countries. Even some PTTs that were trying to expand overall basic services were accused of insensitivity to needs of business customers for advanced, digital business services.

Even when their management was seen as capable, PTTs often had severe restrictions on their investment budgets due to the public sector financial crises that characterized quite a few countries, particularly in Latin America and Africa, in the 1980s. In some countries, this public sector financial crisis aggravated another problem: the exposure of PTTs within governments to day-to-day politics, where PTTs were often seen as being a "cash cow" for the government treasury or where PTT telephone tariffs were seen as a convenient target for reduction under inflation control programs.

Large users began to try to circumvent PTT prices and restrictions by creating private networks using leased lines. This led PTTs to fear lost revenue from traffic that was shifted away from their public networks, so most placed restrictions on leased circuits, third party traffic and resale and shared use of leased circuits and bypassed local carriers with private networks. To accommodate this concern, the International Telecommunication Union (ITU) specifically permitted restrictions on leased lines designed to push users back into public switched services. This continuing concern was explicity expressed by a number of countries at the ITU World Administrative Telephone and Telegraph Conference (WATT-C) meeting in 1988.

Pressures are being placed on the traditional structures and practices of PTTs by a variety of sources. Perhaps the main point of pressure was from national and multinational users who wanted newer, cheaper services and equipment. These include both business users and residential and rural users, whose needs often differ, even conflict. In Europe, Asia, and Latin America, user pressure was also reinforced by intense regional competition for business and investment, which places pressure on key elements of infrastructure, like telecommunications, to conform to world standards of service and pricing. However, other pressures included regional regulatory trends, like the European Economic Community (EEC) Green Paper, which will force changes in Eastern European PTTs if their governments do in fact wish to join the EEC (Ungerer & Costello, 1988). These competitive pressures force governments to try to make the telecommunications infrastructure more efficient. There are also both ideological and practical pressures from trading partners and commercial and development banks, such as the World Bank, to reduce the size of government operations, debt, and expenditures, especially in sectors such as telecommunications, where would-be private operators and investors can be found.

Pressure is also exerted by the perceived success of liberalization and privatization trends and models in other countries. For instance, in Europe, Asia, and Latin America, the perceived success of the U.S. model has certain effects. In Asia, the particular form of structural change in telecommunications in Japan has had considerable impact, as has the model of the United Kingdom in Europe and the models of Chile and Mexico in Latin America.

The most crucial pressures for change in PTTs in most countries are national economic pressures. In Latin America and Asia, the growth of urban middle and working classes has led to an exponential increase in demand for residential telephone service. The new civilian regimes in Brazil, for example, have instituted government public relations campaigns with an emphasis toward extending at least one telephone to nearly all villages demonstrating a major public commitment. The late 1970s and 1980s brought a series of studies by the World Bank (Saunders et al., 1983), the ITU (Maitland, 1984), and scholars such as Hudson (1984), which affirmed the importance of telecommunications, particularly the extension of residential and rural service, for economic and social development. These studies, both macroeconomic and case studies of various sectors of developing economies, also put pressure on communications and economic planners to rapidly expand the reach of telecommunications. Although this attention was originally focused on helping PTTs improve growth and service quality, institutions such as the World Bank also began to question whether PTTs were efficient enough to achieve the infrastructure required for development goals (Saunders et al., 1983).

Simultaneously, a few countries' PTT structures responded reasonably well to increasing demand. Brazil increased the number of basic lines from 2.38 million in 1972 to 8.4 million in 1981, but telecommunication investment and growth fell off during the economic stagnation and decline of the 1980s. This was true of many countries in Latin America, as the region went through what some now call the "lost decade." In fact, a common irony in countries such as India (Fernandes, 1989), the Philippines (Sussman, 1991), Nigeria (Sonaike, 1989), Brazil (McAnany, 1989), and Mexico (Mody, 1989b; Mody & Borrego, 1991) was that PTT growth in the 1960s and 1970s, particularly in satellite systems and other major capital expenses, was often financed by loans, which added to the overall debt of some developing economies. That debt became crippling in the 1980s in Latin America and Africa and has contributed toward the logic of privatization in several countries, because neither new debt nor internal savings are available to finance expanding basic service in rural or residential areas.

In contrast, a number of East Asian countries continued their economic growth and their investment in telecommunication throughout the 1980s. South Korea invested heavily throughout the 1980s and went from 2.8 million lines in 1980 to 15.9 million in 1992 (Park, 1993). In both Brazil in the 1970s and South Korea in the 1980s, the investment in basic services was made almost entirely from internal funds, cross-subsidies, and bonds held by users. Brazil did increase its debt, however, with major projects like BrasilSat (McAnany, 1989).

In some cases, like Brazil and Mexico, countries have either privatized or are considering privatization to break that stagnation in investment that beset telecommunications during bad economic times in the 1980s. Their goal is to acquire new capital for investment, both by the initial sale of shares and also, as in the cases of Mexico and Venezuela, by requiring large domestic and foreign strategic investors (those who contracted to buy large blocks of shares) to invest a certain proportion of the revenues in expanding the basic telecommunications network or to create a certain number of new basic lines annually (Casasuz, 1990; Tandon, 1992).

In countries like South Korea, which maintained economic growth in the 1980s, varying degrees of liberalization are being considered to encourage independent management abilities of telecommunication entrepreneurs and to reinforce international competitiveness (Park, 1993). In 1993, the Korean government also introduced gradual privatization to "eradicate inefficiency and enhance management efficiency" (Park, 1993).

Large business-user pressures for change on PTTs are also very strong. These users include multinational firms that want similar systems, services, terminal equipment, and prices in all the countries in which they work. However, in quite a few countries, particularly in Europe, Asia, and Latin America, more important are national firms that want newer services, lower tariffs or prices, and better equipment so that they can compete with similarly equipped and served firms in world markets. Particularly in countries with large state sectors, pressure for telecommunications reform can also come from intragovernment users, such as state banks, holding companies, and even industries, that may also want new services and lower costs to improve their own internal efficiencies. In China, for instance, this has led to a decentralization of financing and decision making, which permits local governments and local public companies to find their own financing and make their own decisions about investment (Sun, 1991).

Pressures for change in telecommunications often reflect larger economic pressures and changes. Most acutely, it seems that quite a few countries in both Asia and Latin America have undergone several major shifts in thinking and strategy, essentially shifts in development paradigms, that bear on telecommunications structures and strategies. Several countries have shifted from focusing on internal markets with import substitution strategies to trying to position themselves for greater integration in the world economy with an export orientation. Several of the East Asian countries (Hong Kong, Singapore, South Korea, and Taiwan) are seen as having done this very successfully in the 1980s. Their examples are being watched carefully in countries like Mexico, Brazil, and India, which have been reviewing their own emphasis on internal markets and, as in Mexico, often making dramatic changes in orientation (James & Dietz, 1992).

Several countries have also shifted their overall strategies about foreign investment, moving from suspicion and tight controls to welcoming foreign investment for capital, technology, and job creation. This trend can be observed in countries as diverse as India, Jamaica, Peru, and China. This is so widespread, in fact, that it has increased competition among countries to offer incentives and favorable conditions to attract foreign investment. Severe competition by countries for business has affected telecommunications in several ways: pricing, equipment, digitization of networks, tolerance of private or bypass systems, and so on.

At the most dramatic point, several countries in Europe, the Middle East, Asia, Latin America, and the Caribbean now compete to be regional telecommunication hub locations. (With a regional hub, for example, a company would send all trans-Pacific data and voice traffic to Hong Kong for subsequent routing around Asia, rather than directly sending traffic to Hong Kong, Malaysia, Singapore, and Japan.) This tends to also attract related activities, data processing centers, service centers, and even regional headquarters. Corporations tend to select countries with favorable rules as hubs. Companies tend to examine several factors: tariffs, both intercontinental and intraregion; equipment prices, quality, and rules on interconnection; approval of leased lines and tolerant rules on how they can be used; approval of bypass systems; and provision of a range of digital services. The trade literature contains numerous examples of multinational corporations picking one country hub location over another for various combinations of these factors.

Besides those countries that compete to be telecommunications hubs or, as in the case of several Caribbean nations, data entry and processing centers, many countries find themselves competing for a range of other kinds of foreign investments, particularly in manufacturing industries. Potential investors in these areas also demand up-to-date and low-cost telecommunications services. For example, the perceived need to supply telecommunications services for the increased numbers of foreign investors anticipated under the North American Free Trade Agreement is one reason why Mexico moved relatively early to liberalize and privatize its telecommunications company, Telmex.

Regional standardization pressures are a major pressure for telecommunication restructuring in Europe and may become a similar pressure elsewhere if regional regulatory regimes grow in Asia or Latin America. The EEC pushed member nations for liberalized common standards by 1992: separating regulation from operation, opening terminal equipment markets, opening value-added services to competition, opening satellite services to competition, and bringing prices closer to costs (Ungerer & Costello, 1988). Beyond formal pressures from regional groups like the EEC, however, there is also a strong sense of intraregional competition. There is pressure on telecommunications operators by manufacturing, banking, and other firms operating in Europe, Asia, or Latin America to offer services and prices equivalent to what neighboring countries offer. For example, the introduction of competition in Great Britain reduced international costs between the United Kingdom and the United States, which led some firms to route traffic through hubs in the United Kingdom rather than directly to other European countries. That loss of traffic and revenue then forced other countries, such as France and Belgium, to lower their prices to compete. Similar competition can be seen among Hong Kong, Singapore, South Korea, and Japan, for instance.

NEW PATTERNS FOR TELECOMMUNICATIONS OPERATORS

There has also been a ripple effect of other countries' examples. The U.S. experience of deregulation and competition is perceived to have fostered increased use of new technologies, more services, growth in the telecommunications sector from competition, and lower business tariffs. The EEC clearly reacted to a perceived need to provide an equal infrastructure for its firms to enable them to compete with U.S. firms in global markets. The discussion of reforms or restructuring of telecommunication in Brazil reacts to the example of Mexican privatization, in both its positive and negative aspects.

Range of Liberalization Strategies

There are several strategies and several ranges of options that countries are taking in the liberalization and restructuring of telecommunications. Whereas some countries have privatized as almost a first step in reform, others have seen privatization of the telecommunications operator as a last step, perhaps one that they do not need to take.

In some ways, although we are looking at countries outside Western Europe, it is interesting to examine the sequence of steps of liberalization advocated by the EEC as part of the 1992 reforms, specified by its Green Paper on Telecommunications (1987) and subsequent documents (Ungerer & Costello, 1988). These lay out a logical series of steps in increasing liberalization:

- Separation of regulation and operation.
- More open interconnection of user equipment to lines.
- Liberalized use of leased lines.
- Permission for private value-added networks to compete.
- Privatization or competitive offering of specialized services, such as cellular and mobile.
- PTT offering more competitive services, while *not* changing owner-ship patterns or allowing major competition (France is a prototype of this approach).
- Tolerance of facilities-based private networks.
- Permission for competition in long-distance service.
- Competition in basic local or regional service.

(Ungerer & Costello, pp. 185–226)

A number of Third World and Eastern European countries that have not yet privatized their telecommunications operators, and who may well not do so, are still adapting some of these steps of liberalization to address specific problems. In particular, quite a few countries are opening terminal markets, permitting private competition in value-added and cellular ser-vices, and permitting some facilities-based private systems.

As in Europe, one of the first steps that many countries take is to liberalize markets in terminal equipment. For example, Park (1993) found that in 1982 this was the first liberalization step taken in South Korea. This meets one of the most immediate demands of national and particularly multinational users, that they have a choice of handsets, office systems, modems, faxes, desktop computers, and so on. This is particularly impor-tant to multinationals that want to be able to use similar if not identical equipment worldwide. Still, this represents a major compromise for those countries that have required users to purchase or lease nationally built terminal equipment in order to pursue import substitution of national equipment for imported equipment. In Brazil, for example, debate over this step raged from the early 1980s until substantial liberalization was permit-ted in 1991, which represented a major backing away from Brazil's goal of autonomy in computer and telecommunications equipment (Straubhaar, Zoninsein, & Senger, 1992).

In Western Europe, Eastern Europe, Latin America, and Asia, in quite a few countries, it seems that government telecommunications operators are realizing that they do not have the capital, technology, or even personnel to quickly expand into a range of new services, such as cellular telephony, digital overlay networks, computer-enhanced or value-added services, and very small aperture terminal (VSAT) or other satellite services. Business, both multinational and national, is placing demands on PTTs to offer these services. Pragmatically, quite a few countries seem to be deciding to allow competition into or liberalize services in these sectors. The problem faced by several countries is how to regulate these services, particularly how to try to get these new operators to continue to contribute to the building up of the basic telephone network, as well: in effect, how to obtain a cross-subsidy from private competitors to the basic system.

A number of Eastern European, Latin American, and Asian countries have allowed foreign investors or even foreign operators into cellular telephone system construction and operation. Hungary, Poland, Russia, and Czechoslovakia have created joint ventures with foreign firms in cellular. So have Mexico, Brazil, Malaysia, Indonesia, South Korea, and others (Ambrose et al., 1990). However, some Third World countries, including some of the faster growing ones, such as Taiwan, have not allowed foreign companies into cellular telephony yet. The most avid foreign investors in cellular telephone systems have been the U.S. Regional Bell Operating Companies (RBOCs). Ambrose et al. (1990) showed that Pacific Telesis (Japan, Germany), Bell South (Argentina, France, Mexico, New Zealand, Uruguay), and US West (United Kingdom, Russia, Hungary, Czechoslovakia) are among those with operations in cellular. But other international companies, such as France Telecom and Telefonica of Spain, have been investing in cellular as well.

The question of whether to tolerate private networks or even partially private systems, such as private earth stations for satellite services, has been difficult for several Third World nations. They fear, correctly, that private networks are built up as a means of avoiding higher priced services their PTTs offer. Most Eastern European and Third World PTTs still count on higher priced business services as a source of cross-subsidy for expanding and maintaining basic telephone services, so this is a substantial problem for countries that are still far from the goal of universal service. This fear over potentially extensive and severe loss of revenue surfaced openly at the ITU WATT-C in 1988, where national PTTs were loathe to permit competition with private operators who might divert revenues from them.

However, multinational and other large users claim, also correctly, that they frequently build private systems to obtain more sophisticated and more internally standardized systems than Third World PTTs can offer them. Such needs were also recognized in the 1988 WATT-C agreements. Since the early 1980s, many large international telecommunication users have been creating private networks over leased lines. Both in Europe and in the Third World, this led to years of conflict over what could be done with private leased lines. Prior to the late 1980s, both European and Third World governments tended to place very specific restrictions on use of leased lines. Liberalization of this point was another key aspect of the EEC liberalization schedule for 1992. Some Third World countries have tended to follow this trend.

In practice, it seems that several countries are slowly compromising on private networks. Whereas previously governments such as Brazil's had forced companies to use any equipment or service that the government considered sufficiently similar to what the company desired, governments are now more likely to let companies create private networks to provide services for themselves that are in fact different from what government telecommunication authorities can provide. This is manifested in liberalized rules on use of leased lines, in increased permissions for use of private VSAT or other satellite terminals and networks connecting user facilities, and so on.

In particular, a number of Third World countries had built national satellite systems in the 1980s, including Brazil (McAnany, 1989), China, India, Indonesia, and Mexico (Mody, 1989b). These countries often found themselves with an excess of unused satellite capacity. They were particularly open to private satellite earth station networks that used some of their excess transponder capacity. Before privatization in Mexico, for example, companies could purchase and operate earth stations as long as they gave the title of ownership to Telmex.

Some of the stages in liberalization taken in industrialized countries of the EEC, Japan, New Zealand, Australia, and the United States seem unlikely in Eastern Europe or the Third World. Those include competition in either long-distance service or basic local telephone service. In most developing countries, even those such as Mexico that have privatized the telecommunications operating company, monopoly in long distance is still considered likely in the near future. There are at least two reasons. For nearly all countries, cross-subsidy between long-distance and local service is seen as a key tool for expanding the number of basic lines available. Second, in those countries that do privatize, a guaranteed monopoly on long-distance service for a certain number of years has often been an incentive for strategic investors to purchase large groups of stock with a reassurance that they can recoup their investment from long-distance (and business service) revenues. This trade-off can be observed in privatization contracts in Jamaica, Mexico, and Venezuela, among others (Casasuz, 1990; McCormick, 1993).

Deregulation or Re-regulation?

In quite a few countries, one of the basic first steps in overall change is increasing the distance of the telecommunications operation from the government. In Europe, this usually involved moving the PTT from inside a ministry to independent status as a public (nongovernment) corporation. This has come to be called "corporatizing" government industries—putting them outside direct government control, making them act like independent corporations, and making them subject to regular corporate law and regulation.

Another key step is the creation of a separate regulatory body outside PTT, either inside a ministry or as a separate regulatory authority, such as the U.S. Federal Communications Commission or the British Oftel. This is necessary to achieve equal, arms-length treatment of the PTT and other carriers by the regulators. If competition is to be created in sectors of telecommunication, such as value-added services or cellular telephony, new entrants or competitors are likely to demand that they be regulated by some entity independent of the PTT with which they are competing.

For developing countries that are considering opening up liberalized competition in some sectors or privatizing the main telecommunications operator, creation of an adequate regulatory authority seems to be crucial. If value-added services or cellular telephony are opened to competition, for example, most developing countries are still probably going to want the new entrants to help pay a cross-subsidy to assist with the expansion of the basic telecommunications infrastructure that the new services will in fact probably interconnect with and draw on. To ensure that such levies are fairly set up and actually paid, an effective government regulator seems necessary. Even if extensive competition does begin, a regulator must still set technical standards and rules for competition.

As we show later several cases of privatization have largely failed to achieve their objectives because no regulator existed to set development goals for privatized telecommunications companies or to ensure that stated goals were complied with. Eastern European and Third World countries are often considering privatization as a means of acquiring additional investment for the telecommunications system. However, as we see, that requires carefully structuring the privatization process and then monitoring the outcome. As the experience of the Philippines shows, an unregulated private monopoly cannot be counted on to do much investment toward achieving development goals, beyond that which is profitable (Sussman, 1991).

PRIVATIZATION

Goals

The privatization of telecommunications has been pursued for a series of reasons and with very variable success. One of the clearest elements in the overall pattern about privatization is that much depends on the goals of privatization and how they are structured, pursued, and enforced.

In nearly all countries, one of the principal goals of privatization is to raise capital. However, that capital can be used for very different purposes. In some countries, such as Jamaica and Argentina, the primary goal has been to pay off national debts. In other countries, such as Mexico and Venezuela, the primary purpose for raising capital has been to invest in new infrastructure or to acquire technology to improve infrastructure. Even then, goals must distinguish between raising money to improve infrastructure for business services or for residential and/or rural services.

In some countries, such as South Korea, a primary stated goal of privatization is to improve efficiency (Park, 1993). Saunders et al. (1983) demonstrated that managerial and financial efficiency in PTTs tend to be lower than in commercial organizations. In particular, the World Bank and other students of PTTs think that management will be depoliticized by taking PTTs out from under direct control of either government bureaucrats or politicians. However, this same issue has also been addressed in East Asia and Europe by the corporatization of PTTs, taking them out of ministries and restructuring them as independent public corporations to act and even compete in the marketplace like other corporations.

Forms of Privatization

For many of the reasons already outlined, a number of countries are either committed to or are considering the privatization of ownership in government telecommunications authorities or PTTs. Once the decision to privatize has been made, there are several approaches possible. What clearly needs to be achieved, but which has not always been clear in the past, is a careful fit between goals and forms of privatization.

From the foregoing discussion of reasons for privatization or liberalization, we might restate several separate goals that should be addressed in different ways. First, most countries seem to want to get increased capital for investment for expanding basic residential and rural telephone service. Second, most countries also want increased investment for starting or expanding more advanced business services. Third, many countries are worried about increasing the efficiency of management, resource use, personnel, service offerings, and marketing and customer relations.

Fourth, some countries want to get capital from the sale of telecommunication assets to pay for either telecommunication-specific or general national debt.

Similarly, there are a number of forms of privatization, some of which meet certain goals better than others. Most PTTs privatize by becoming private or partially private stock companies. Stock can be sold in a general public share offering, like British Telecom (BT), for example, or to limited specific categories of people. Restrictions or subsidies might be placed on stock sales, again depending on goals. For instance, maintaining national control is a high priority for the Japanese and Koreans, so the original stock sale of Nipon Telephone and Telegraph (NTT) and the forthcoming sale of stock in Korea Telecom (KT) was limited to national citizens. The KT sale is also structured to give advantage in purchase to lower middle-class and working-class individuals or families to increase social participation in privatization (Park, 1993). A number of PTTs have offered a partial preferred sale of stock to employees, partially to defuse union opposition by spreading benefits and partially, as with KT selling to the lower middle class, to increase popular participation, or what Chile has promoted as "people's capitalism."

Stock sales tend to be to individuals, pension funds, and other traditional investors of various scopes and sizes. Some countries in privatization have, instead, sought out "strategic partners," groups that would acquire relatively large numbers of shares or parts of overall ownership as part of a specific contract, which might also specify further obligations on the part of the strategic investors, even after the sale of stock. For instance, in Mexico the major domestic and foreign strategic investors have signed a specific contract that guarantees them a 6-year monopoly on fixed line services in return for a continuing obligation to invest a certain sum of money (approximately $10 billion U.S.) in the company and install a certain number of new lines in various parts of the country (Crockett, 1990).

These kinds of "strategic investor" arrangements essentially become joint ventures with other companies. In choosing strategic partners, there seem to be several considerations. Large national companies, banks, financial groups, pension funds, and so on have the advantage of maintaining national ownership, control, and financial benefit. However, they may also have more limited capital, access to technology, and management expertise than potential foreign partners. National companies may also have political connections and liabilities. That raises fears of insider deals among friends, political patronage, corruption, and the possibility of national conglomerates that might become too powerful, either economically or politically. (Foreign firms can also further corruption, insider deals, and so on but they do not have quite the same domestic political implications as favorable deals to already powerful national groups.) For these kinds of reasons, for example, the private television quasi-monopoly Televisa in Mexico was told not to try to bid on the Telmex privatization. Some Brazilians objected, for similar reasons, when a consortium including NEC and the owner of an equivalently powerful television oligopoly, TV Globo, won a cellular radio concession for the city of Rio de Janeiro (Rossi, 1989).

Foreign "strategic investors" or joint-venture partners also offer both positive and negative considerations. Foreign companies may well have more capital for purchase of new equipment and investment after the initial purchase of stock. If the foreign partners are already telecommunications firms, such as several U.S. RBOCs, Cable and Wireless (U.K.), France Telecom, or Telefonica (Spain), which have all been involved in several privatizations so far, then the foreign partners may well offer specific management expertise and technology as well as capital. However, the inclusion of foreign partners raises difficult questions of national sovereignty, control, and benefit from profits. Even though some observers consider the privatization of Telmex to have been handled in a manner that did raise considerable investment capital, the fact that profits are now being repatriated to France (via France Telecom) and the United States (via Southwest Bell) is beginning to raise complaints in Mexico that profits from an evidently profitable operation are now leaving the country (Barrera, 1991).

To link forms of privatization with goals, it seems clear in terms of raising capital for investment or getting access to technology, the sale of stock to domestic and foreign strategic investors has certain key advantages. They may be bargained into buying at a higher price. More importantly, as in Mexico and Venezuela, they may be required under contract to continue to invest in the telecommunication operation for a period of years. For instance, the Mexican strategic partners, Grupo Carsa (Mexico), France Telecom, and Southwestern Bell (U.S.) were required to invest approximately $10 billion (Crockett, 1990) in considerable network expansion in order to maintain their monopoly on all fixed link services until 1996.[1] However, the investors still felt that they could recoup their investment in 5 years. To enforce such a continuing investment requirement also requires efficient and ongoing regulation.

To limit concerns about foreign domination of key national infrastructure and assets, some countries, such as Korea and Japan, have excluded foreign purchasers altogether. Excluding foreign investors entirely may not be an option for quite a few countries, such as Mexico, where various studies have noted that domestic capital markets were simply too weak

---

[1]The number of lines in service must expand 12% yearly until 1994, all towns of 500 or more must have service by 1994, public telephones must increase from 0.8 per 1000 to 5 per 1000 by 1998, and waiting time for a new line must be reduced to 6 months by 1995 (Tandon, 1992).

to supply the desired capital plus commitment for continued investment. Most other developing countries will probably find themselves in similar straits—it will be difficult to get capital plus continuing investment needs from domestic capital markets. Some countries have limited the number of shares that foreign investors may hold or have required that they be minority partners in consortia led or controlled by national firms. In Mexico, for instance, the strategic investors hold a total of 20.4% of voting shares and the consortium is led and controlled by the Mexican partner (Casasuz, 1990). The danger, of course, is that foreign firms may find pliant domestic partners who would in fact give them control. To avoid that, if possible, would require careful initial and ongoing regulation. That kind of regulation of foreign partners is something that countries will have to learn, although several countries, such as Brazil, Mexico, and most East Asian nations have had considerable experience with joint ventures or foreign shareholders in other sectors of their economies. Previous experience with government regulation of foreign actors in telecommunications equipment industries shows some record of success, particularly in the larger countries with attractive internal markets, such as Brazil (Fadul & Straubhaar, 1991). The experience of smaller countries, such as those in the English-speaking Caribbean, in effectively regulating foreign investors has been much more problematic (McCormick, 1993).

If countries are primarily interested in reducing national debt, then the primary consideration is simply getting the best sale price. Jamaica (McCormick, 1993) and Argentina (Abdala, 1992) entered into contracts with strategic foreign investors to buy large shares of the telecommunications companies, supposedly because that offered the most immediate return. Domestic sales of individual shares or small blocks of shares cannot usually suffice. The whole problem is that lack of domestic capitalization led to reliance on borrowing, thence to debt problems. Argentina, for instance, privatized most government firms, including telecommunications, quickly between 1989 and 1992 to meet goals of "short term revenue maximization, foreign capital attraction, and external debt repurchase" (Abdala, 1992). In fact, Argentina is also taken as an example of a process that was overly rushed by preoccupation with foreign debt, resulting in an inadequate regulatory structure and disappointing results from the privatization contract (Abdala, 1992).

Privatization and Regulation

A number of studies of various privatization processes of telecommunication companies in developing countries have observed that the role of regulation and the structuring of regulatory authorities has been neglected (see Part III of this volume). Whereas the United States has pursued and promoted the strategy of deregulation, most other countries have actually re-regulated or reformed regulation, often creating formal regulatory authorities to replace political or ministerial discretion over government operations (Abdala, 1992).

In particular, it seems that deregulation has only really been pursued in advanced industrial systems, where reliance on market competition instead of regulation builds on existing infrastructure development. In the United States, United Kingdom, and New Zealand, which have introduced the most widespread liberalization and reliance on market forces, what might be effectively called deregulation, universal service has been long achieved and the initial logic of natural monopoly no longer seems as strong. Such complete liberalization requires potential for cost-effective competition, requires effective competitors, and works best to develop advanced business services.

In contrast, re-regulation, or, in most cases, the new development of effective regulation separate from political decision making seems to have been imperative in most restructuring. Abdala (1992) argued that, at least in Latin America, experience shows that regulatory structures need to be constructed before privatization is begun. The core of the argument for strong regulation revolves around the continuing need for nearly all developing countries, except perhaps those of East Asia, to strongly pursue rapid expansion of service into rural and residential areas. The economic and social needs are such that expansion is very pressing, as discussed earlier, and the near-future profitability of such investment is not great enough to automatically lead a private firm to make such investments.

For systems still trying to achieve universal service, regulation seems to be required to ensure that a private or partially private operator reinvests not only in profitable business and long-distance services but also in extending residential and rural service. Regulation may also be required to get adequate technological upgrading by partners in privatized companies. This is particularly true for private monopolies, and it seems, from most of the current developing country telecommunication privatizations, that monopolies are still being granted in most cases, often as an incentive for further investment, as in Mexico. For systems permitting private monopoly, regulation is needed to set fair price and profit levels (like U.S. rate of return or price cap rate regulation). However, if telecommunications systems plan on permitting competition in some areas, such as cellular telephony, value-added services, or digital overlay networks, then regulation may be required to get private competitors to contribute toward investment in expanding the basic system.

A question still being debated concerns the appropriate form of regulation in developing countries. In part, that seems to depend again on a country's goals. Abdala (1992) and others argued that an independent regulatory authority, patterned on the FCC in the United States or Oftel in the United Kingdom, would probably be most effective. However, analysts in several countries are also worried about the ability or will of many developing governments to effectively permit a strong, independent regulator. Abdala (1992) cited the "capture theory" of regulation, originally applied to industrialized nations, where he saw regulatory agencies as beginning their existence with the public interest in mind but eventually surrendering to the interests of the firms they regulate. Scholars in Latin America have taken that further, observing that frequently the entire state apparatus has been captured by the interests of a few leading economic families or groups (Abdala, 1992; James & Dietz, 1992). In Brazil, for instance, the communications minister between 1985 and 1990 was Antônio Carlos Magalhaes, a powerful politician allied with Roberto Marinho, the owner of TV Globo, the main communications conglomerate. He was accused of biasing several licensing and purchasing decisions in favor of a consortium between Marinho and NEC of Japan (Rossi, 1989). This had led several political parties in Brazil to campaign for "deprivatizing the state," that is, reducing the control that private interests have over it. That may make the prospects for creating an independent government regulator difficult. Taken to its logical extreme, such an approach to telecommunication regulation would have to be combined with large-scale reform of the state mechanism itself in many countries. It certainly argues for taking the regulatory authority out of existing communications ministries, particularly when ministers are often political appointees with power and patronage interests.

Another approach is to regulate telecommunications through licenses or contracts enforced by judicial mechanisms. That might be particularly useful in countries where the judiciary is stronger or more independent from private interests than the executive branch. In Jamaica, a study by Spiller and Sampson (1992) found that executive branch regulation was inconsistent and politicized, which led to privatization of telecommunications via a license and contract that would ultimately be enforced by the courts. A more cynical interpretation might be that Cable and Wireless, the British-owned private operator of the Jamaican system, might prefer to have a contract that ultimately could be decided in British courts, because those are the final courts of appeals in Jamaica, a British Commonwealth nation. In fact, in many developing nations, courts are incredibly overloaded and years behind in their case loads, so that the judicial system might be an unlikely source of effective regulation, particularly in an area where technology and services change so quickly.

Corporatization

There are a variety of reasons why many countries resist privatization, as already noted. Some of them have instead pursued a mixed, limited liberalization, opening only those areas seen as most crucial or pressured.

A 1991 ITU study found that most (77%) of the developing countries responding to a survey about telecommunications reported forms of restructuring other than privatization. For instance, some countries are opening up potential for private investment in new value-added services where demands are high, and the present PTTs are limited in their ability to gear up new services quickly. Quite a few countries have also moved toward liberalization of terminal equipment purchase and interconnection, use of private networks over leased lines, and other areas where competitive pressures from other countries' moves force action. However, Mody, Tsui, and McCormick (1993) and others have noted that the dominant trend seems to be corporatization.

In Europe, Eastern Europe, and the developing nations, many PTTs tend to retain control of the basic network and at least basic fixed services and to define basic services broadly. For example, France would like to define its packet-switched network as a basic service, even though others consider that a value-added service likely for liberalization. Some state telecommunications companies are taking a more business-oriented approach while retaining some cross-subsidy to expand basic services (as in most developing nations) or even developing more advanced services (like Minitel development with subsidy by France Telecom). In structure, this is usually accompanied by "corporatizing" the PTT, putting the PTT outside government and under corporate law, and separating telecommunications operations completely from regulation. This involves "restructuring state telecom departments as independent commercial entities, where the market rewards them or penalizes them for the quality and quantity of service provision, staff training, and relations" (Mody et al., 1993, p. 264). Characteristic of this approach are more aggressive, business-oriented offerings by the PTT itself, an attempt to make PTT services more competitive to resolve business demand, offering closer to cost-based pricing, extensive digitization of the network, and offering of more digital business services. Some of the primary examples of this approach are France Telecom, Hong Kong Telecom (owned by Cable and Wireless), Belgian Telecom, and Singapore Telecom, although the last is debating whether to stay with the "corporatization" approach or actually privatize.

Corporatization frequently represents a political compromise between those demanding reform or restructuring of telecommunications administrations and those who wish to preserve the state sector. In some countries, like Brazil, sectors of both the left and the military still see telecommunications as strategic in the sense of national security, requiring state control. In other countries, many argue against transferring public assets to the private sector and/or losing the potential profit that telecommunications brings the state. Frequently, too, both technocrats and labor unions are loathe to risk loss of jobs and decision-making control potential in the sale or transfer of state assets in privatization.

One of the problems in corporatization is whether government agencies or ministries and large public sector enterprises are susceptible to such thoroughgoing change in labor, operations, technical, and managerial aspects as efficiency and financial viability might require (Mody et al., 1993). Civil servants do not necessarily become market-oriented managers, labor unions do not automatically become more efficient in work practices, and politicians do not stop seeing the public enterprise as a zone of control and patronage.

CONCLUSIONS

There does seem to be a certain convergence between a number of countries on a number of general steps in what is best termed liberalization of telecommunications. To meet the demands of both national and multinational business actors working within a world economy, certain restructuring steps are becoming widespread (see Table 1.1), such as opening rules on terminal equipment, allowing more use of private networks on leased lines, permitting private user ownership of certain kinds of private facilities like VSAT terminals, permiting private companies to enter or compete in cellular telephony, and permitting competition in some value-added services. Even countries with strict national monopolies and import substitution policies, such as Brazil, are opening these areas.

One noticeable trend is that many countries are in fact limiting the degree of liberalization well short of privatization. One of the dominant patterns is corporatizing the PTT into a more market-oriented enterprise, while liberalizing competition in various areas. In some countries, this might be seen as a transition to privatization, and, in others, an alternative response to the same pressures while meeting counter-pressures to keep PTT assets under public control.

Privatization of the primary telecommunications operator does not seem quite as pervasive and, indeed, is being specifically rejected in some countries, such as Costa Rica and Uruguay. Examples of both success and failure are appearing in privatization efforts. That is to say, some countries are now having second thoughts and are unhappy with some of the outcomes, as in Argentina. That is principally due to lack of clarity in the initial goals of privatization, the haste or care with which privatization was pursued, the fit between goals and the form and mechanism of privatization, and the degree to which a regulatory mechanism was

TABLE 1.1

Stages of Telecommunications Restructuring in Various Countries

	Regulator and Operator Separate	Terminal Equipment Open to Compete	Others Compete in Value-Added	Competition in Voice Services	Privatization of Phone Company
EUROPE					
United Kingdom	Yes	Yes	Yes	Yes	Partial gov't ownership
France	Yes	Yes	Yes	No	No
Italy	For 3 of 4 entities	Opening	Opening	Limited	Partial in 3 of 4
Sweden	Yes	Yes	Yes	Cellular	Under debate
Germany	Yes	Yes	Yes	Cellular	No
Hungary	Soon	Likely	Yes	Not yet	Up to 30%
Czechoslovakia	Not yet	Probable	Likely	No	Being considered
Poland	Likely soon	Probable	Likely	Unlikely	Under debate
NORTH/SOUTH AMERICA					
Argentina	Yes	Yes	Yes	No	Telefonica, France Tel.
Mexico	Yes	Opening	Some areas	Cellular	S.W. Bell
Brazil	Yes	No, may open	No, may open	Cellular	Some debate
Chile	Yes	Yes	Likely	Unlikely	Sold to Telefonica
Venezuela	In process	Yes	Yes	Cellular	Likely
Costa Rica	Yes	Yes	Being considered	Unlikely	Being considered
Canada	Yes	Yes	Opening	Will likely open	Regionally varied
ASIA/PACIFIC BASIN					
Japan	Yes	Yes	Yes	Yes	Partial gov't ownership
Singapore	Yes	Yes	Some areas	No	Under debate
Malaysia	Yes	Mostly	Some areas	No	Under debate
Hong Kong	Yes	Yes	Some areas	No	Yes
South Korea	Yes	Increasing	Yes	Pending	Partial gov't ownership
Australia	Yes	Yes	Yes	Satellite leased	Aussat, yes
New Zealand	Yes	Yes	Yes	Yes	Yes

created or refined that could ensure that goals were met. In particular, although initial financial goals seem to have been met in most privatizations, the desire to ensure the continued expansion of basic services, particularly into less profitable residential or rural areas, has not been met in several cases, such as Jamaica and Argentina.

The need for great care in the elaboration and enforcement of regulations, particularly in the structure of the regulatory mechanism, seems the clearest result of this brief survey of trends. Some countries that are liberalizing have not created regulatory mechanisms that can handle and guide independent, private operators, even in specific sectors, such as cellular telephony. For example, the process of giving licenses for private cellular services in Brazil has been delayed, costly, and politically charged. More dramatically, some countries that have moved further to privatization of the main operating company have not clarified their goals ahead of time and created regulatory mechanisms adequate to ensure that those goals are pursued.

REFERENCES

Abdala, M. A. (1992). *The regulation of newly privatized firms: An illustration from Argentina*. Paper presented at the Latin American Studies Association, Los Angeles.

Alencastro e Silva, J. A. (1990). *Telecomunicações—histórias para a História* [Telecommunications—Stories for history]. São José dos Pinhais-PR-Brazil: Editel.

Ambrose, W., Hennemeyer, P. R., & Chapon, J. (1990). *Privatizing telecommunications systems—opportunities in developing countries*. (Discussion Paper No. 10). International Finance Corporation, Washington, DC.

Barrera, E. (1989, May). *Telecommunications in industrial enclaves: The Maquiladora industry in the US–Mexico Border*. Paper presented at the 39th Annual Conference of the International Communication Association. San Francisco.

Barrera, E. (1991, May). *Telecommunications, the global market and the third world: The Mexican strategy*. Paper presented at the 41st Annual Conference of the International Communication Association. Chicago.

Bauer, J., & Straubhaar, J. (1994). Telecommunications in Eastern Europe. In C. Steinfield, J. M. Bauer, & L. Caby (Eds.), *Telecommunications in transition*. Thousand Oaks, CA: Sage.

Casasuz, C. (1990). *Privatization of telecommunications—The case of Mexico*. Paper presented at the World Bank Conference on Structural Reform of Telecommunications, World Bank, Washington, DC.

Commission of the European Communities. (1987, June 30). *Toward a dynamic European economy—Green paper on the development of the common market for telecommunications services and equipment* (EEC No. COM(87) 290). Brussels: Commission of the European Communities.

Crockett, B. (1990, December 24). New TelMex management bodes well for net's users. *Network World*, p. 15.

Dizard, W., & Svensrud, S. B. (1987). *Gorbachev's information revolution—Controlling Glasnost in a new electronic era*. Boulder, CO: Westview-Center for Strategic and International Studies.

Fadul, A., & Straubhaar, J. D. (1991). Communications, culture and informatics in Brazil. In G. Sussman & J. Lent (Eds.), *Transnational communications: Wiring the Third World* (pp. 214–233). Newbury Park, CA: Sage.

Fernandes, C. P. (1989). Communication technologies and economic development in India. *Media Development, 36*(1), 29–31.

Hardy, A. P. (1980). The role of the telephone in economic development. *Telecommunication Policy, 4*, 278–286.

Hills, J. (1990). The telecommunications rich and poor. *Third World Quarterly, 12*(2), 71–90.

Hills, J. (1993, March). [Comments]. Conference on privatization in Eastern Europe and the Third World, East Lansing, MI.

Hudson, H. (1984). *When telephones reach the village*. Norwood, NJ: Ablex.

James, D. D., & Dietz, J. L. (1992). *Latin American lessons from the Far East: Substance or illusion?* Paper presented at the Latin American Studies Association, Los Angeles.

Lerner, N. C. (1988, October 24). Formidable aspirations lead Brazil forward. *Telephony*, pp. 60–63.

Maitland, D. (1984). *The missing link—Report of the independent commission for world-wide telecommunications development*. Geneva: International Telecommunication Union.

McAnany, E. (1989). Brazil, satellites and debt: Who trades what to acquire new technologies. *Media Development, 36*(1), 6–10.

McCormick, P. (1993, March). *Case study of Jamaica*. Paper presented at the Conference on Privatization in Eastern Europe and the Third World, East Lansing, MI.

Mody, A. (1989a). Information industries in the newly industrialized countries. In R. W. Crandall & K. Flamm (Eds.), *Changing the rules: Technological change, international competition, and regulation in communications*. Washington, DC: Brookings Institute.

Mody, B. (1989b). Satellite debt in Mexico's manana. *Media Development, 36*(1), 14–15.

Mody, B., & Borrego, J. (1991). Mexico's Morelos satellite: Reaching for autonomy. In G. Sussman & J. Lent (Eds.), *Transnational communications—Wiring the Third World* (pp. 150–164). Newbury Park, CA: Sage.

Mody, B., Tsui, L., & McCormick, P. (1993). Telecommunication privatization in the periphery: Adjusting the private-public balance. *International Review of Comparative Public Policy, 5*, 257–274.

Park, R. A. (1993). *Telecommunications in South Korea*. Unpublished manuscript, Michigan State University, Department of Telecommunication, East Lansing.

Quandt de Oliveira, E. (1992). *Renascem as Telecomunicações*. São José dos Pinhais-PR-Brazil: Editel.

Rossi, C. (1989). Juiz anula concorrencia que Marinho vencera—Uma história estranha do edital até o vencedor. *Folha de São Paulo*, p. C-1.

Sampson, A. (1973). *The sovereign state of ITT*. New York: Stein & Day.

Saunders, R., Wellenius, B., & Warford, J. (1983). *Telecommunications and economic development*. Baltimore: Johns Hopkins University Press.

Sonaike, S. A. (1989). Telecommunications and debt: The Nigerian experience. *Media Development, 36*(1), 2–5.

Spiller, P., & Sampson, C. (1992, September). *Regulation, institutions and commitment: The Jamaican telecommunications sector*. Paper presented at the Twentieth Annual Telecommunications Policy Research Conference, Solomons, MD.

Straubhaar, J., Zoninsein, J., & Senger, E. (1992, May). *Successes and failures of the Brazilian informatics policy: Import substitution in question*. Paper presented at the International Communication Association, Miami.

Sun, L. (1991). *Diversified financing and telecommunications development in China: Its implications in competition, demand and technology*. Unpublished doctoral dissertation, Michigan State University, East Lansing.

Sussman, G. (1991). The transnationalization of Philippine telecommunications: Postcolonial continuities. In G. Sussman & J. Lent (Eds.), *Transnational communications—Wiring the Third World* (pp. 125–149). Newbury Park, CA: Sage.

Tandon, P. (1992). *Welfare consequences of selling public enterprises—Mexico.* Paper presented at the World Bank Conference on the welfare consequences of selling public enterprises, Washington, DC.

Tunstall, J. (1986). *Communications deregulation.* New York: Basil Blackwell.

Ungerer, H., & Costello, N. P. (1988). *Telecommunications in Europe.* Brussels: Commission of the European Communities.

2

Public–Private Relations in International Telecommunications Before World War II

Daniel R. Headrick
Roosevelt University, Chicago

The starring role now held by the telephone was once occupied by the telegraph. In the United States, the Bell System did not seriously rival Western Union until the early 20th century. France defended its telegraph network against the inroads of telephony until the 1960s, whereas Russia and Eastern Europe are only now switching over. What is true of national networks is even more so of international, especially intercontinental, communications. Although transatlantic telephony became possible by the late 1920s, it remained costly and erratic until the 1960s. In almost every respect, international telecommunications before World War II was synonymous with telegraphy.

This chapter deals with international telegraphy before World War II as a case study in the relations between private and public ownership of an important communications network and for possible comparison with the current wave of the privatization of telephone systems. What I intend to show is that private and government ownership of international tele-communications were never mutually exclusive alternatives but points along a spectrum of increasing government involvement.

International telecommunications before World War II were scarce and costly. Until the late 1920s, intercontinental communications required enormous investments in cables and radio transmitters, yet provided very limited capacity; when this capacity was stretched to its limits due to a political crisis, governments used their power to acquire the channels they needed. Although every decade saw important innovations, heavy fixed costs slowed the pace of diffusion; some cables laid in the 1880s were still in use 50 years later. The rate of technological change affected entrepreneurship, monopolies, and government interventions in the industry.

In the fluctuating symbiosis of private and public interests, several factors influenced the degree of public control and participation: the commercial potential of a given communications system, the political traditions of various countries, the level of international tension, and the rate of technological change.

1850–1868: THE PERIOD OF TRIAL AND ERROR

Soon after the introduction of the electric telegraph in the 1840s, entrepreneurs saw the need to prolong their lines underwater. Submarine telegraph cables, however, were much riskier than land lines, not only because they cost much more per mile but also because they were extremely difficult to repair. The very first cable, laid by John and Jacob Brett across the Straits of Dover in August 1850, revealed the risks involved when a fisherman hauled it up in his net a few hours after it was laid. In contrast, the second cable, laid a year later in the same place by Thomas Crampton, worked well for over 30 years.

At the time, only France and Britain could contemplate investing in this radically new technology, and they took two very different approaches. France had been the pioneer in telecommunications with the Chappe aerial telegraph invented during the French Revolution and operated thereafter as a government service reserved for official messages. In response to an entrepreneur's offer to set up a second network open to the public, the government passed the law of May 3, 1837, forbidding privately owned telecommunications.

This precedent influenced its attitude toward overseas communications. Even as France was replacing its Chappe network with electric telegraphs, thoughts turned to communications with its most important colony, Algeria. Instead of granting the concession for a transmediterranean cable to a private company, the government decided to hire private firms to lay its cables but retain the ownership, the administration, and the risks. Unfortunately, the technology was not yet mature, and the first cables between France and Algeria, laid between 1854 and 1864, failed after a short time. Alarmed by this string of failures, the French government allowed a British firm, the Marseilles, Algiers and Malta Telegraph Company, to lay and operate a cable on that route in 1870. When this one proved reliable, the government reverted to its traditional stance; it purchased a France–Algeria cable for the postal and telegraph administration in 1871 and others in 1879, 1880, 1892, and 1893.

In contrast to France, Great Britain started with a tradition of private ownership, which it gradually modified. The first submarine cables to France and Ireland were private and worked well. When Britain and France went to war with Russia in 1854, however, the British government wanted a cable from Romania (where the European land network ended) to the Crimea; because this was strictly a temporary wartime need that no private firm would invest in, the government purchased the cable on its own account.

Shortly thereafter, the Indian Rebellion of 1857 broke out. The British government, appalled that news of the uprising took 40 days to reach London, desperately sought better communications. Entrepreneurs, who had offered to lay a cable to India for several years but only with subsidies or a guarantee of profits, now saw their opportunity. Instead of purchasing a cable, the government granted a concession to the Red Sea and India Telegraph Company with a 4.5% dividend guarantee. Although the cable was defective from the start, the British government had to pay the shareholders £36,000 a year for the next 50 years. This experience caused a powerful aversion to any further government involvement in cable enterprises.

In the late 1860s two events changed the cable business. After the failure of the first transatlantic cable of 1858, engineers learned how to make and lay reliable cables, such as the successful Atlantic cables of 1865 and 1866. Equally important was the Telegraph Purchase Act of 1868, by which Britain nationalized all domestic telegraph companies, paying their stockholders £8 million, which they invested in overseas cable companies.

In contrast to international telecommunications, domestic telegraph networks were government entities throughout the European continent from the beginning and, after 1868, in Britain as well. Only the Americas had private telegraph companies. As international telegrams had to pass from one domestic network to another, their linkages required international cooperation. In the 1850s, Prussia, Austria, and several smaller German states founded the Austro–German Telegraph Union, while France and its Latin neighbors formed the Western European Telegraph Union. Under French auspices, the members of the two unions founded the International Telegraph Union (ITU) in 1865, the first permanent international organization. Three years later, the ITU created the International Bureau of Telegraph Administrations in Berne to oversee the day-to-day technical and financial issues involved in international cooperation. Having nationalized its domestic telegraphs, Great Britain now joined the ITU. In 1872 private companies were admitted to the ITU as observers without voting rights. The culminating conference, held in St. Petersburg in 1875, set the ground rules for international telegraphic

34

HEADRICK

cooperation for the next 30 years. All the European states were represented, but the United States declined to participate, as it had no state-owned telegraphs.

1866–1898: THE PERIOD OF EXPANSION
AND BRITISH HEGEMONY

From the late 1860s to the end of the century, the British submarine cable industry went through a prolonged boom. Much of this expansion was organized by John Pender, a Manchester cotton manufacturer who invested his fortune in the Atlantic cables of 1865–1866, thereby attracting the confidence of other investors. His Anglo–American Telegraph Company was unable to monopolize the North Atlantic, however, for such a lucrative route soon attracted British, French, and American competitors. By the 1880s, two American firms, Western Union and Commercial Cable, dominated the cartel that regulated prices and competition across the Atlantic.

The North Atlantic route was the closest any market came to a pure capitalist environment in the telegraph business, as both the British and American governments adopted a hands-off policy toward cables. Within Europe, all domestic networks were public monopolies, and on the short sea routes between European countries, cables were jointly owned by the states on whose coasts they landed.

Elsewhere in the world, the situation was never as clear-cut as in Europe or on the North Atlantic. In the late 19th century, the rest of the world had neither the capital to invest in cables nor the ability to operate let alone manufacture and lay them. Yet other parts of the world offered not only lucrative business opportunities but also, in that age of imperialism, potential targets for conquest and annexation. The British—and eventually other European nations as well—saw telecommunications as an essential element of global commerce and empire building.

Within their colonies, the European powers actively built telegraph networks for security and administrative reasons, well ahead of commercial demand. Algeria acquired two telegraph networks, a Chappe semaphore network that stretched 1,498 kilometers in 1854, supplemented by an electric network that began that year and expanded to 3,179 kilometers by 1861. The Indian network, started in the 1850s, quickly became the longest in the colonial world; by 1900 it operated 84,700 kilometers of lines and 4,949 telegraph offices. French Indochina, which began later, had 11,951 kilometers of lines and 224 offices in 1902. Other colonies followed suit, more or less in the order in which they were conquered. Sub-Saharan Africa lagged behind, however, due to its poverty and low population density; some parts of Africa were not tied to the world's network until the late 1920s. All colonial networks were government agencies.

Although colonial telegraphs were mainly used for local and imperial communications, they nonetheless had international repercussions, for the ITU admitted colonies as though they were sovereign states. Thus India joined the ITU in 1868 and the Dutch East Indies in 1872. This enhanced the influence of the colonial powers, as the French director of Posts and Telegraphs explained in 1894: "The adherence of our colonies is of great interest, because every administration that joins has the right to vote in international conferences, and every new adherence of one of our colonies has the effect of giving one more voice to France in the voting which follows the debates" (*Affaires politiques*, 1894).

Between distant parts of the world, communications relied almost entirely on cables, which Britain dominated, not by its voting power, but through its private enterprises. As he had on the North Atlantic, John Pender pioneered submarine cable communications to the rest of the world. Beginning in 1870, he founded a string of cable companies stretching from Britain to India, which he consolidated into the Eastern Telegraph Company. He quickly followed their success with further cables to Australia, East Asia, and South America. By 1892 his Eastern and Associated Companies owned 45.5% of the world's cables, all connected in one building in London.

Eastern was entirely private, in accord with the British government's policy of no intervention and no subsidies. Yet even in the purest laissez-faire atmosphere, cables were a means of communicating information, and information had strategic value. Hence, there arose between the British government and the Eastern group something less than government ownership or regulation, yet more than untrammeled free enterprise. It was a community of interests, a symbiosis.

Wherever they landed, submarine cables needed landing permits. Although Britain granted permits freely to all comers, other countries were choosier. The British government made one exception to its policy of nonintervention; in 1867, it promised to help companies "by using the good offices of the Government with foreign governments upon whose territories it may be necessary to land cables." Within the British Empire that meant the first British company on the scene got the government's blessing and—since there was seldom enough traffic to justify a second cable—a de facto monopoly on that route.

In its dealings with other countries, the British government also favored Eastern. The Treasury informed John Pender during the invasion of Egypt in 1882 as follows:

While her Majesty's government . . . do not deem it expedient that anything in the nature of a guarantee of monopoly should be given to the Companies whom you represent, they are able to assure you that, in their opinion, it would be highly inexpedient to encourage, upon light grounds, competition

against a Company in the position of the Eastern Telegraph Company which
has embarked large capital upon existing lines; and these considerations
would apply with especial force to competition which might be threatened
from foreign sources. (quoted in Headrick, 1991, pp. 36–37)

Once it had laid a reliable and efficient cable on the vital commercial and
strategic route to India, Eastern became part of Britain's foreign relations
establishment. The natural affinities that derived from their common
interests were reinforced by Pender's habit of filling the boards of directors
of his companies with aristocrats connected to the Foreign and Colonial
offices. In exchange, the government obtained several invaluable benefits:
fast and reliable communications and the security of knowing its messages
were handled by British nationals employed by a British firm.

Compared to Eastern, other British firms fared poorly, for they served
markets other than the British Empire. Three small companies served the
Caribbean, but they ran into financial difficulties and were bought up by
Eastern at the turn of the century. A rival, the India Rubber Company,
laid one cable along the West African coast connecting the colonies of
France, Spain, and Portugal and another to South America. It, too, lost
money and had to sell its African cables to Eastern in 1884 and its South
American cable to France in 1902. Thus Eastern benefited not only from
its own successes but also from the failures of others. Given Britain's
global reach in the late 19th century, this meant Eastern enjoyed a monopoly of telecommunications with Australia and South America, and a
near-monopoly with India, China, and Japan.

Before the turn of the century, only three other nations, Denmark, the
United States, and France, had cables. The Great Northern Telegraph Company had obtained the concession for a land line across the Russian Empire
from St. Petersburg to Vladivostok, from where it laid cables to China and
Japan. Although legally Danish, it was partly owned by, and closely tied
to, the Eastern group, which could never have obtained the Russian
concession in its own right because of Anglo–Russian political differences.

The other two cable-owning nations, the United States and France,
represent the extremes on the public-private spectrum of that era. The
United States government had no interest whatsoever in cables and did not
even regulate landing rights. American cables were all owned and operated
by a few large corporations—the Western Union and Commercial Cable
companies across the Atlantic, and the Central and South American and
the Mexican Telegraph companies in Latin America; they operated according to the rules of robber-baron capitalism of that era, switching from
cut-throat competition to cartels and back.

Unlike the American cables, the French ones were consistently unprofitable. The French government was inclined to own and operate its own
cables, just as it did its domestic telegraphs. In those few cases where cables

landed only on French territory, as between France and Algeria, it did just that. Elsewhere, however, such a policy was untenable, for governments that would readily allow a foreign company to land cables and open offices in their territory would never grant such a privilege to the agents of a foreign power, not even telegraph clerks. Furthermore, operating overseas cables was a risky business into which the Chamber of Deputies hesitated to sink public funds. Thus France had to modify its policy and tolerate private enterprise in this one area of telecommunications.

The result was a reluctant and gradually increasing government intervention in the cable business. The Société du Câble Transatlantique Français laid a cable to America in 1869 but lost money and was sold to the Anglo–American Company four years later. The Compagnie Française du Télégraphe de Paris à New York started up in 1879 but survived only with government subsidies. A third, the Société Française des Télégraphes Sous-Marins, was founded in 1888 to link the French colonies in the Caribbean and the Pacific to the Eastern trunk lines. It too floundered and was merged with the Paris–New York Company in 1895 to form the Compagnie Française des Câbles Télégraphiques, a subsidized and government-controlled enterprise.

The world's 246,871-kilometer-long cable network of 1892 was distributed as shown in Table 2.1. This table clearly shows the British hegemony and, within that, the dominance of the Eastern Group over global telecommunications.

1898–1914: INTERNATIONAL TENSIONS AND GOVERNMENT INTERVENTION

Before 1898, most of the world's telegraph cables were private, unsubsidized, and lightly regulated. Telecommunications was a business, not a weapon. This changed at the turn of the century, as growing international

TABLE 2.1
The World Submarine Cable Network, 1892

	Private	Government	Total
Eastern & Associated	45.5		
Other British companies	17.5		
Total British	63.1	3.2	66.3
American	15.7	—	15.7
French	5.4	3.4	8.8
Danish	5.2	.1	5.3
Other	.2	3.7	3.9
TOTAL	89.6	10.4	100.0

Note. From U.S. Department of the Navy (1892). Figures are given in percentages.

tensions led to increasing government intervention in international communications. Intervention took several forms, from subsidies for particular cables to regulation and control of the whole industry to outright ownership. The trend had two causes. As the global network of commercially profitable lines was filled in, demand arose from areas that were commercially unremunerative but politically or strategically important. In turn, the connections between strategy and communications grew stronger as international rivalries intensified.

Here we will consider Britain because it owned two thirds of the world's cables and pioneered in radio-telegraphy as well; France and Germany because their rivalry with Britain drove them to create rival networks; and the United States, the last bastion of free enterprise in telecommunications.

For 30 years after the Red Sea cable fiasco of 1859–1860, the British government studiously stayed out of the cable business. After all, private capital was laying reliable and efficient cables to every significant colony, dominion, and trading partner in the world; furthermore, it was a time of peace and no foreign power threatened Britain's command of the sea. Yet the government's attitude began to shift in 1889 when the Colonial Defence Committee persuaded the Cabinet that the Navy needed a cable from Halifax to Bermuda, the only naval base not linked to Whitehall by cable. It was the first cable subsidy in peacetime and cost the Exchequer £8,100 a year.

In 1891 the Cabinet appointed a committee "to consider the question of telegraphic communication with India in time of war." The committee was basically satisfied with the security of Britain's cable communications, but recommended a new cable from East Africa to Mauritius which cost £28,000, to which Britain contributed £10,000 and its colonies the rest. A few years later, after a war scare with the United States over Venezuela, the government agreed to pay £8,000 a year toward another strategic cable, from Bermuda to Jamaica.

In the year 1898 three events hastened the involvement of European governments in international telecommunications. The first was the Spanish–American War, during which the U.S. Navy, with no cable ships or experience with cables, nonetheless managed to cut communications between Spain and its colonies of Cuba, Puerto Rico, and the Philippines. The American willingness to cut cables belonging to neutral nations sent tremors through the world's colonial and naval establishments. If the United States was able and willing to isolate an enemy's colonies, how much more efficient would the British be in time of war!

The second event was the Fashoda incident, in which French and British troops confronted each other along the upper Nile in the Sudan. As the British had the only telegraph line to Fashoda, the French commander had to travel to Cairo to communicate with his government. France thereby suffered a humiliating diplomatic setback.

Finally, in the war between Britain and the Afrikaners, the British imposed a heavy-handed censorship on all messages between South Africa and the rest of the world, and also between France, Germany, and Portugal and their respective colonies in southern Africa. These three events made it plain that the cable business was not a neutral public utility but a tool of strategy.

As international relations became more strained in that fateful year, the British cabinet appointed a committee "to consider the control of communications by submarine telegraph in time of war." This committee did not limit itself to defensive measures, but advocated an aggressive policy of cutting enemy cables and censoring neutrals and proposed several new cables of purely strategic value. Under pressure to duplicate the vulnerable lines to India that passed through Egypt, the government subsidized Eastern to lay cables from Cape Verde to South Africa and from there to Ceylon, Singapore, and Australia. These cables gave Britain several alternative means of communicating with its eastern empire.

Although British strategists were now satisfied, the Dominions were not. Australians, in particular, felt vulnerable at the far end of a long string of cables and resented having to pay Eastern's exorbitant monopolistic rates. Britain's Colonial Secretary, Joseph Chamberlain, pressed the Cabinet to fund a Pacific Cable from Australia to Canada—the All-Red Route—to round out Britain's global network. Eastern refused to participate in the project, however, rightly fearing it would cut into its profits. Hence the Pacific Cable inaugurated in 1902 was jointly owned and operated by the governments of Britain, Canada, Australia, and New Zealand. The British government had finally entered the cable business.

As World War I approached, Britain alone possessed a secure cable network with duplicate routes to every strategic point on earth, protected by the Royal Navy and serviced by two thirds of the world's cable ships. It alone could sever any potential enemy's communications with the rest of the world. That position of power was the result of a judicious mix of private enterprise, government subsidies, and public ownership.

In contrast to the self-satisfied British, the French felt very insecure in their overseas empire. During the era of imperialist expansion in the 1880s and 1890s, the British had occasionally delayed telegrams between France and its agents overseas for a few hours, arousing suspicions among colonial lobbyists. None of these incidents, however, could overcome the government's reluctance to spend money on cables. Then came the events of 1898, demonstrating how vulnerable France was to its main colonial rival. The result was a sudden surge of interest in telecommunications. In 1900 the French Chamber of Deputies passed a bill to develop new lines of communication to the French empire. In order to avoid the British cables to Africa, the government purchased the cables of the West African Telegraph Company and relaid them to connect the French colonies along the West African coast to Dakar. It also laid cables from French Indochina to Amoy, the southern terminus of the Russo–Danish line, and to Borneo, linking up with Dutch, German, and U.S. lines.

All the neat categories used to describe cables were blurred when the French government purchased the South American Telegraph Company, which owned a cable from West Africa to Pernambuco in Brazil. Was it British or French, public or private, or all of the above? Whatever the category, the French government's foray into the cable business cost it dearly, for private customers preferred the more reliable Eastern lines, and besides, there was never much traffic between Senegal and Dahomey, or Indochina and Borneo. The French government could not even claim a strategic benefit, for its cables were still easy to cut and its East Asian messages had to transit through the lines of other countries. In any event, the security issue was laid to rest with the Franco–British Entente Cordiale of 1904. All that France obtained, at a cost of £240,000 to £320,000 per year, was a measure of national pride.

The events of 1898 inflamed German politicians as much as it did the French, with similar results. By the turn of the century Germany had sufficient traffic with North America, and even with South America, to justify laying new cables. The surge in traffic coincided with a storm of anti-British indignation in the German press and in the Reichstag. In 1899 the Reichstag voted a credit of £67,000 to the Deutsch–Atlantische Tele-graphengesellschaft (or DAT) for an all-German Atlantic cable. This cable was manufactured and laid by a German firm, for unlike France, Germany had a powerful electrical industry fully capable of competing with the British. Laid in 1900 from Emden to New York via the Azores, it was so successful that the DAT duplicated it 2 years later.

If the Atlantic cables were profitable, the same was not true of other German ventures. In the Far East, Germany's desire to have non-British communication channels led it to form an alliance with the Netherlands to lay cables from the Dutch East Indies and from China to the German island of Yap, and from there to Guam, where American cable could carry messages to the United States and on to Europe via the German cable. Like the French ventures in the same region, the Deutsch-Nieder-ländische Telegraphengesellschaft was a political statement rather than a business enterprise, and it cost Germany £76,250 and the Netherlands £18,750 a year in subsidies.

That left Africa and South America, where all the concessions were held by British or French firms. Although the Germans were able to obtain landing rights in the Spanish Canary Islands and in Liberia, they could proceed no further without the cooperation of Britain or France. So strong was anti-British feeling in German commercial, naval, and colonial circles that Germany allied itself in 1909–1911 with France—for telegraphic purposes only—to lay cables from Liberia to German Kamerun and to Brazil. The cable to Brazil would have paid for itself if the war had not intervened, but the African cable was most certainly a political not a business proposition.

More curious still is the story of the American Pacific cable. Until 1898 neither U.S. entrepreneurs nor the government showed much interest in Pacific cables. That all changed with the United States' acquisition of Hawaii and the Philippines. Suddenly it seemed insulting to U.S. pride to have its communications between Washington and Manila transit through London, Bombay, Singapore, and Hong Kong. It also looked as though the Far East would soon offer lucrative opportunities to U.S. businesses.

On several occasions in 1899 and 1900 Presidents McKinley and Roosevelt appealed to Congress to fund a U.S. cable across the Pacific. Many entrepreneurs had offered to lay such a cable with a government subsidy, but Congress rebuffed all such schemes. Unlike the European powers, the United States was not yet ready to pour public funds into political ventures of dubious commercial value.

In 1901, however, John Mackay, president of the Commercial Cable and Postal Telegraph Companies, proposed to lay a Pacific cable without any subsidies whatsoever. He did not even seek a landing permit, on the grounds that Hawaii, Guam, and the Philippines were now U.S. territory, hence the Pacific Ocean was a "navigable water of the United States" open to any U.S. company. Two years later, the cable from Honolulu reached Manila, and three years after that, it was extended to China and Japan.

Whereas politicians and the public, in that era of frenzied nationalism, thought in terms of national companies, capitalism had already outgrown the nation-state. Twenty years later, Mackay's son confessed to a Congressional committee that only 25% of the Commercial Pacific Cable Company actually belonged to his father's Commercial Cable Company; another quarter of the shares were owned by the Danish Great Northern Company and the other 50% by the Eastern Group. In other words, the "American" Pacific cable was really a front for Eastern. Furthermore, the Eastern, Great Northern, Commercial Pacific, German–Dutch, and Indo–European companies, along with the Chinese and Russian telegraph administrations, had formed a secret cartel that imposed higher rates on telegrams between Europe and the Far East via the Pacific than via India or Russia, thereby ensuring the continued predominance of Britain's interests in global communications.

1897–1914: THE EMERGENCE
OF RADIO-TELEGRAPHY

Unlike cables, which appeared in an era of peace, radio-telegraphy or wireless (as it was then called) was born into an age of tensions and jingoism. It had three characteristics with important political ramifications. First, it was a competitor to cables. Second, it could communicate with ships. And third, its waves carried across borders, inviting eavesdropping. Hence it did not go through a long period of privately sponsored development, but was immediately seized upon by governments as a new instrument of political and military rivalry.

Guglielmo Marconi may or may not have "invented" radio, but he certainly was the first to recognize its business potential. As soon as his device showed promise, he approached the Italian government but was rebuffed. He immediately moved to Britain and presented his invention to the British government. William Preece, chief engineer of the Post Office, encouraged Marconi's early experiments, but turned against him when Marconi founded the Wireless Telegraph and Signal Company. For the next 30 years the Post Office remained Marconi's sworn enemy. Marconi, however, had made a valuable friend, namely Captain (later First Sea Lord) Henry Jackson, who helped him become the exclusive supplier to the Royal Navy. When we speak of "the government" versus "private enterprise," we must always bear in mind that these are oversimplifications covering, on both sides, a multitude of conflicting interests.

Prevented by the Post Office from establishing communications within Great Britain, Marconi built his business at sea. In 1900 he founded Marconi's International Marine Communication Company, not to sell equipment but to place Marconi sets and operators on ships and at strategic shore stations. He quickly signed up most large shipping lines and Lloyd's insurance group. Intent on creating a monopoly, he forbade his operators to communicate with radios made by other companies.

This angered the Germans, or rather, it added one more item to their growing list of grievances against Britain. One of the witnesses at Marconi's first demonstrations in 1896–1897 was Professor Adolf Slaby. He rushed home to Berlin, joined forces with Count von Arco and the AEG company, and designed a wireless set that circumvented Marconi's patents. Meanwhile another inventor, Professor Braun, obtained the backing of Siemens und Halske, Germany's other giant electrical equipment manufacturer. Both companies found ready support in the German army and navy. In 1903, in order to compete more effectively with Marconi's Marine, the German government ordered Slaby-Arco-AEG and Braun-Siemens-Halske to merge into a new company, Telefunken. From the very beginning, international competition was the driving force behind the remarkable cooperation between German inventors, industry, and the government.

The German government quickly reacted on the international front as well. In 1903 it called an International Radio-Telegraph Conference to attack Marconi's nonintercommunication policy. In 1906, with the help of the United States, it finally succeeded in overturning it. From then until the outbreak of war, Telefunken became a major manufacturer of mobile sets for army use and Marconi's rival in supplying ship's radios to third countries. Yet Marconi's maintained a comfortable lead by virtue of its close ties to the Royal Navy and to British shipping and maritime insurance interests, then dominant on all the oceans.

In the United States, unlike Germany, the development of radio was frenzied and chaotic. Marconi came to the United States in 1899, but was rebuffed by the navy. Instead, the navy preferred to await further technological advances, postponing a choice as long as possible. In fact, the U.S. Navy pursued a two-faced policy, successfully resisting the introduction of radio into its ships while trying (but failing) to obtain a monopoly of radio communications everywhere else.

Meanwhile, the rest of the U.S. government showed little interest in radio, refusing to regulate the industry, insisting, in international conferences, on minimal regulation and the loosest wording in order to encourage the spontaneous effusion of inventions and competition so dear to turn-of-the-century U.S. society. In this period, American inventors certainly made tremendous contributions to the new technology, even if their inventions fell into the hands of unscrupulous corporations. Among these corporations, the most successful and, arguably, the least scrupulous was American Marconi, which, like its British parent, specialized in maritime communications.

In 1912, the U.S. Navy finally awoke to the military value of radio communications. In avoiding any dealings with American Marconi, it reflected not so much a technological or business decision as a latent hostility to a British-owned firm and perhaps also to its Irish–Italian founder. By then, U.S. continuous-wave sets, especially Reginald Fessenden's high-speed alternator and Cyril Elwell's arc transmitter, had proved themselves superior to Marconi's spark transmitters. The navy, fired with the enthusiasm of the newly converted and backed by a generous Congress, not only equipped all its ships with wireless sets but also built a string of high-powered transmitters stretching from Puerto Rico to the Philippines, creating the first truly global radio network.

Radio not only communicated with ships at sea, it could also reach across oceans and continents, a matter of great interest to all those nations that had long complained of the Eastern cable hegemony. As early as 1901 Marconi claimed he was able to transmit across the Atlantic, and from 1907 on he did in fact operate a commercial service with powerful stations in Ireland and Nova Scotia. He offered to build an Imperial Wireless Chain as a private venture, paralleling the Eastern network. But the British government, torn by conflicting advice from the Dominions, the Royal Navy, the Post Office, and the cable companies, procrastinated until 1913, finally awarding the construction contract to Marconi but reserving the operation for the Post Office. Only two stations were built, in England and in Egypt, before war put a stop to the scheme.

The German government was more motivated, for its communications with its colonies and with America depended on vulnerable cables and British good will. It therefore subsidized Telefunken to build the world's most powerful transmitter at Nauen near Berlin in 1906. By 1914 it had just opened stations in New York, in its African colonies, and in China and the Pacific when war broke out.

By August 1914, maritime radio was a well-established industry, one in which Marconi's faced serious and growing competition from German and U.S. firms. Yet these were not truly free enterprises, for much of their business was with their respective navies and the networks they built (and sometimes operated) were heavily oriented toward strategic and imperial, rather than purely commercial goals. All of them were, in one way or another, extensions of their governments' policies. When war broke out, they became branches of government, immediately in Europe and eventually in the United States as well.

1914–1939: CABLES, RADIO, AND GOVERNMENT INTERVENTIONS

The First World War taught all the belligerents the immense strategic and political value of communications and the importance of close business-government cooperation. It also taught their armed forces that radio messages, no matter how cleverly encoded, were vulnerable to the attacks of code-breakers. Finally, the British control over the flow of war news overseas helped bring the United States into the war, thereby demonstrating the power of propaganda.

After the war, most countries continued the secret surveillance of international cables. Yet, the victors were also eager to return to something like a prewar "normalcy" in business affairs. This desire was thwarted by two major changes that revolutionized the telecommunications industry. One was the growing wealth of the United States compared with the economic exhaustion of its competitors, especially Britain. The other was the rapid rate of technological change. Together, they achieved what France and Germany had failed to do before the war, namely overthrow the hegemony of British firms in telecommunications.

The early 1920s were a period of such booming demand for communications that new competitors could arise without threatening the prosperity of established firms, as the United States showed in 1919. The U.S. Navy, which had taken control of all U.S. radio stations (except the army's) during the war, tried to hold onto its monopoly, but Congress would have none of it. Admiral Bullard and Commander Hooper, alarmed at the reappearance of American Marconi, persuaded the General Electric (GE) Company to refuse to supply alternators to American Marconi and to create the Radio Corporation of America (RCA) in its place. RCA, a private company, was thus founded as a tool of the government in its struggle against British interests. It quickly bought out American Marconi and signed agreements with the other U.S. electronics giants GE, AT&T, and Westinghouse.

Before the war, Marconi's had tried to create a unified network with its own transmitters and offices in every country, just like Eastern's, but this proved impossible. In the postwar atmosphere of heightened nationalism, even the Dominions demanded control over their own communications and considered a British company somehow "foreign." Furthermore, in response to Marconi's repeated offers to create an Imperial Wireless Chain, the British government procrastinated, because of a topheavy bureaucracy and the opposition of the Post Office and the cable companies, and because it found itself with too few resources spread too thinly over too many obligations.

Marconi's was thus obliged to accept a lesser role, similar to that long held by post offices, namely as correspondent with its equivalents abroad. Within a few months, all the international radio-telegraph companies signed traffic agreements to handle one another's radiograms. Furthermore, to maximize the efficiency of their service, they also signed patent-pooling agreements.

These agreements not only gave RCA a major share of the world's radio traffic, they also opened the door to two other giant firms. One was Telefunken, which benefited from a burst in demand after the Armistice and from the disappearance of Germany's Atlantic cables, confiscated by France and Britain. The other was a new French firm, Compagnie Générale de Télégraphie Sans Fil, or CSF, founded by the French government to amalgamate numerous smaller firms and connect France with its colonies.

The four big companies, Marconi's, RCA, Telefunken, and CSF, attempted to divide up the rest of the world. They succeeded in Latin America, long the scene of Anglo-American rivalry over cables. There the AEFG Consortium (so-named after its four members: America, England, France, and Germany) founded a "national" subsidiary in each country, with headquarters in the capital city and a well-placed local politician as president but actually owned by the consortium and managed, discreetly, by RCA.

Thus, while politicians spoke glowingly of their national communications and their liberation from the British hegemony, behind the scenes the companies were creating multinational cartels, often fueled by American capital. Already in the 1920s, there were signs of an emerging postnational postcapitalist global communications system.

Despite all the competition from radio, the cable interests felt no pain until 1927. On the contrary, demand was so strong that they eagerly ordered the latest technological advances: regenerative repeaters, automatic printers, and "loaded" cables capable of carrying eight times more traffic than the prewar cables. Western Union, DAT, the Pacific Cable Board, and Italcable (Mussolini's state-run entry into the cable business) all bought new cables and, for a time, profited handsomely.

Then came a shock that sent the entire industry into a tailspin: shortwave. Technologically, shortwave was a minor invention, a variation on the longwave systems in common use at the time. But from an economic point of view, it was truly revolutionary, for it was so cheap that it threatened to make obsolete not only all existing longwave systems but all cable networks as well. Ironically, it was pioneered by none other than Commendatore Marconi himself, who confessed: "I admit that I am responsible for the adopting of long waves for long-distance communication. Everyone followed me in building stations hundreds of times more powerful than would have been necessary had short waves been used. Now I have realized my mistake" (Aitken, 1985, p. 272).

Soon after the war, amateurs had shown that radio waves in the under-200-meter bands could be heard—albeit with great difficulty—across the ocean. In 1923–1924, Marconi began experimenting with short waves and found that the distance they traveled was not proportional to the energy consumption or cost of the transmitter. In the midst of building the Imperial Wireless Chain for the British Post Office, he offered to start over with shortwave transmitters at a cost one twentieth that of longwave ones.

The impact was almost instantaneous. Within a couple of years, shortwave stations were set up in every country, even destitute colonial backwaters that could never have afforded a longwave station. With shortwave, radio networks could offer rates one sixth that of cables. By 1927, cables had lost half their traffic to radio, and in some places three quarters or more. Needless to say, the Eastern Group was the most directly affected. Faced with imminent bankruptcy, Eastern announced it was planning to distribute its cash reserves to its shareholders, sell its cables to foreign companies, and close down.

This prospect galvanized the British government into action, for cables were one of the mainstays of British security. As British officials knew from their code-breaking successes in World War I, radio is never completely safe from enemy cryptanalysts, and only cables are 100% secure. In 1928, the British government called an Imperial Wireless and Cable Conference to prevent the loss of its cable network. The conference proposed a radical solution: the merger of all British telecommunications systems—the Eastern cables, Marconi's radio, the Post Office's shortwave network, the Pacific Cable Board, and others—into one giant conglomerate called Imperial and International Communications Ltd. (I&IC, later renamed Cable and Wireless). This company started life with 253 cable and radio stations and over half the world's cables.

Other countries reacted half-heartedly to the news of the British merger. In the United States there was talk of a similar merger, but the need was not pressing, for U.S. security hardly depended on its cables. Furthermore, the spirit of competition was so ingrained in U.S. politics that nothing came of the merger proposals. Small mergers occurred in France and Germany, but had little impact, for they, like all other independent countries, gleefully greeted shortwave radio as their liberator from the thralls of the British cable hegemony.

In theory a private company, I&IC was really an arm of the British government, tightly controlled by the Imperial Communications Advisory Committee and forced to subsidize its now obsolete cable network (including cables to places like Zanzibar and Fanning Island) with the profits of its radio operations. To do so just as the Depression was drastically reducing world trade imposed a heavy financial burden. As shortwave stations proliferated around the world, they established direct links, bypassing the British firms entirely. As a result, Britain lost its long-held primacy in international telecommunications. Yet, by shifting from competition among private enterprises to a state-managed conglomerate, it preserved a cable network that contributed substantially to its security in the Second World War. Who is to say that this was not a wise trade-off?

CONCLUSION

The history of international telecommunications before World War II shows that public and private interests were not opposites but formed a symbiosis and that governments got increasingly involved in telecommunications, especially after 1898.

There were several reasons for this trend. One was the heavy fixed costs that left telecommunications companies vulnerable to technological changes and eager for government protection. Second, telecommunications networks formed "natural" monopolies in places that could not support several competing networks, as between Britain and Australia;

such situations invited government intervention to subsidize the initial service, duplicate lines, or regulate prices. Third, telecommunications systems were public utilities, for their social benefits often outweighed their private ones. Thus the public expected service to outlying areas and small customers at the expense of the more profitable lines or of the government; hence the demand of out-of-the-way towns, islands, and colonies for the same service as more profitable areas enjoyed.

But the major justification for government involvement was not technical or social but international, for telecommunications networks carried information, an essential ingredient in national security, not only in wartime, but in periods of international tension as well. Government involvement in international telecommunication was very much a function of the political atmosphere of the time. It is no surprise that the French and German governments invested heavily in telecommunications after 1898 and that the British government took over communications when its security was threatened. In contrast, in the United States, which felt safe behind two oceans, only the Navy showed much interest in government control.

If there were both economic and political pressures for government involvement, what, then, explains the persistence of private enterprise? Some companies, like the South American Telegraph Company or Imperial and International Communications, were really governmental entities hidden behind a corporate mask in order to preserve a diplomatic fiction; others, however, like Western Union and Telefunken, were truly private. What explains the persistence of true private enterprises in telecommunications is technological change. Except in wartime, governments of capitalist nations preferred to leave invention and development (and their risks) to the private sector and hesitated to invest public monies in a new technology until it became important enough to be considered either a public utility or a factor in national security. In the century before World War II, governments had to balance the need to encourage innovation and entrepreneurship, on the one hand, with the desire to control the channels of information, on the other. The three characteristics of telecommunications networks—as technological systems, as business enterprises, and as channels of information—produced a complex and changing interaction between private and public interests.

ACKNOWLEDGMENTS

I wish to express my gratitude to Professors Bella Mody, Joseph Straubhaar, and Johannes Bauer of Michigan State University for organizing the conference at which this paper was presented, and also to Peter Lewis, Bjorn Wellenius, and Rohan Samarajiwa for their perceptive comments.

REFERENCES

Affaires politiques 2554. (1894, March 24). [Letter]. (Dossier no. 4). Available in the Archives Nationales Section Outre-Mer, Paris.

Aitken, G. J. (1985). *Syntony and spark: The origins of radio*. Princeton, NJ: Princeton University Press.

Headrick, Daniel R. (1991). *The invisible weapon: Telecommunications and international politics, 1851–1945*. New York: Oxford University Press.

U.S. Department of the Navy. (1892). *Submarine cables*. Washington, DC.

II

MAJOR POLITICAL FORCES

3

Telecommunications and Global Capitalism

Gwen Urey
California State Polytechnic University, Pomona

Restructuring of telecommunications service has occurred globally for several years, with heterogenous outcomes among both industrialized and developing countries. This chapter examines telecommunications change in the context of global capitalism. Telecommunication lubricates capitalist expansion by facilitating the globalization of both production and marketing for large corporations. But telecommunication also lubricates the processes by which institutions of capitalism in developing countries become integrated into a capitalist world economy. The following section explores intersections between the globalization of the telecommunications industry and the globalization of capital markets. Then the chapter focuses on the political economy of telecommunications, developing a theoretical approach to relations between industrialized and developing countries in the context of financial integration. I argue that integration, as practiced, follows colonialism and imperialism as a mode of international exploitation. In the next part, exploitative dimensions of change in both capital markets and telecommunications are related. The chapter concludes with remarks on strategies for creating alternative orientations for telecommunications development.

TELECOMMUNICATIONS

Much of the impetus for restructuring telecommunications originates within the large and dynamic firms dominating the industry in industrialized countries. Thus restructuring appears more internal to the econo-

53

mies of industrialized nations and more driven by external forces in the case of developing countries. Broader forces related to the globalization of the capitalist mode of production also exert specific influences on the telecommunications industry, but the manifestation of these influences also varies according to a country's integration into globalized sectors of the international economy, as well as the size and sophistication of its telecommunications industry and other factors.

Most governments of developing countries have restructured in some way their national telecommunications industries; these nationally based industries now take a wide variety of public/private and national/foreign ownership and operating forms. Some governments liberalized only telecommunications; others did so as part of a broader national liberalization. An International Telecommunications Union (ITU, 1991) study found that economy-wide reform preceded telecommunications restructuring in the majority of developing countries. They also found it unclear "whether there is a general tendency towards privatization," but their data make it quite clear that there is a general tendency toward liberalization, if liberalization is defined to include "sector . . . reorganization [and] (de)regulation, market-oriented investment and finance, and a wider offering of services in the increasingly competitive drive across and within national borders" (pp. 1–7).

Telecommunications investment in developing countries necessarily involves telecommunications equipment manufacturers from industrialized countries and increasingly involves multinationals providing telecommunications services as well. The scale, capital intensity, and monopolistic history of these corporations argue against trusting them to behave as "supply" should under conditions of perfect competition. Despite competition for international markets, the concentration of telecommunications capital in a handful of giant telecommunications service and equipment firms contradicts the definition of "supply" in a competitive market model. The proliferating joint ventures among these firms further erodes the argument that they might behave according to theories of market equilibrium, as does their record of retaining near-monopolistic status in many of their home markets (Morgan & Sayer, 1988). Thus, liberalization policies that induce developing countries to behave as "demanders" in a telecommunications goods market will not necessarily improve the allocation of global telecommunications resources. As developing countries' governments liberalize telecommunications industries, we need a more complex understanding of "supply" that places the corporations comprising "supply" in the context of the changing logic of global capitalism.

Intersection of Capital Markets and Telecommunications

This chapter also addresses the globalization of capital markets, which together with trends in telecommunications make the story of global financial integration—a story with profound consequences for global inequality. The chapter opens a window on the divide between the included and excluded members of an integrated global economy by examining two fundamental qualifications for membership, that is, the conditions of local capital markets and of telecommunications systems. Integration requires that capital market players learn more geography, but theirs is a very opportunistic geography. As Tynan (1993), a financial analyst, remarked in an article on global banking:

> International business makes an important contribution to the balance sheet of many banks; for some it is their raison d'etre. This is one indication of the continuing integration of the world economy. Another aspect of this trend is the increasing similarity of countries' economic policies, recently resulting in the global slowdown. *Banks have been left with few locations where they can head in search of income to offset loan losses as the capital structure adjusts* [italics added]. (p. 27)

One dimension of financial geography relates to the presence in eight financial centers of the most "global" big banks. The most global 50 have offices in London, New York, and Tokyo; they are likely to also have offices in one or more of the other financial centers—Zürich, Frankfurt, Paris, Hong Kong, and Singapore. Only three of the 50 banks are from developing countries: Bank of China, with 47.1% of its business overseas in 1991; Korea Exchange Bank, with 19%; and Banco do Brasil, with 13.5% (Tynan, 1993). So, although in a few cases "globally integrated" originates in the developing world (or the "newly industrialized world"), it is defined by connectivity to London, New York, and Tokyo. Contradictory images of a global economy that is at once fully integrated but in which developing countries are fully marginalized as insignificant or threatening infect much of the financial trade literature. For example:

> The global economy is most fully realized in financial markets, which every year get closer to the ideal of untrammeled flows of capital across transparent borders to markets governed only by supply and demand. Global capital markets are large: $250 trillion flow through the international bank-payments system yearly, more than ten times the value of all the goods and services produced in the world and fifty times the value of the global trade in goods and services. Capital markets in the developing economies were dominated by the debt crisis which suddenly emerged in 1982. For several years, many analysts feared that massive defaults could severely damage the whole global financial system. But in the later years of the decade, the debt problem came under control, and *capital markets in the developing world were mostly perceived as insignificant* [italics added]. (Orr, 1992, pp. 187–188)

The same contradictions distort images of global telecommunications networks. No measure is taken of places outside the perimeter of the global telecommunications network. For example:

[W]e see the expansion of management information systems into networks extending far beyond the individual organization. The new systems, via networking and personal computers, also extend access to higher-level management. This results in enhanced control over data resources, cash flow, manufacturing operation, supplier or customer transaction, or market information, to name just a few instances. As we witness the globalization of industry, much of this is made possible by advanced telecommunications. For example, whether General Motors is manufacturing auto bumpers in Indiana or Matamoros, Mexico, the plants are integrated into the same highly efficient telecommunications network. If they needed to do so, this integration could be in operation anywhere in the world, simply by linking into the global network. If facilities are not readily available, a satellite earth station can do the job. (Williams, 1991, pp. 29–31)

Williams paints the vision toward which capitalists seeking new markets and production sites drive, one in which access to the global network makes local conditions irrelevant. Setting up a satellite earth station is inexpensive (as cheap as a couple thousand dollars); but access to satellite capacity to relay communications back to a home office is not ubiquitous, especially in particularly poor regions, including much of Africa. *Bypass*, or the technical ability of a firm or competing service provider to circumvent the common carrier, was at the heart of the MCI challenge to AT&T and the subsequent divestiture of AT&T. In developing countries where the public network cannot meet demand, corporations capable of bypassing the local network can often also compete with the local carrier, selling their extra capacity (legally or not).

These images of a financially and technologically "seamless" global economy proliferated during the late 1980s. Also during the 1980s, evidence of the rigid "seams" between rich and poor regions of the world emerged (World Bank, 1990). Indicators of telephone development generally parallel other development indicators, but the indicators themselves may be unusually meaningless in the case of very poor countries.[1] Figure 3.1 plots the telephone density in the early and late 1980s against the 1991 Human Development Index (HDI), a composite measure of income, quality of life, structure of opportunity, and participation. As the figure demonstrates, there is a correlation between these two indices, but it degenerates significantly at the lower ranges of either index.[2] As the

---

[1]Densities, for example, 0.2 telephones per 100 people as in much of sub-Saharan Africa, are particularly silly in sparsely populated areas. None of the indicators takes agglomeration effects into account—that is, the effect of who else participates in the network on telephone utility.

[2]Statistically, 86% of the variance in HDI can be predicted by the log of the telephone density. If the lower ranges are isolated, the figure drops below 30%. Telecommunications experts invariably include a plot of phone density vs. GNP, but without qualifying the skewness.

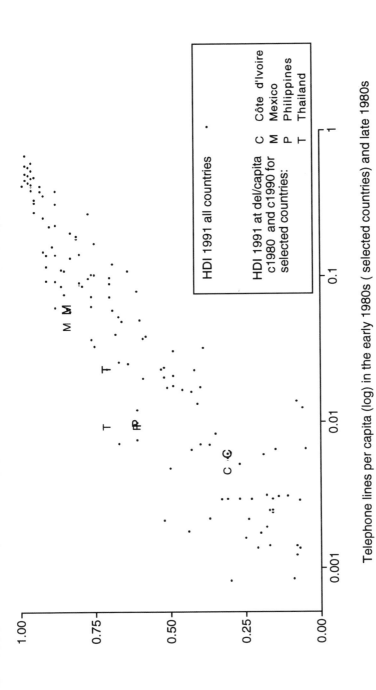

Telephone lines per capita (log) in the early 1980s (selected countries) and late 1980s

(Data sources: International Telecommunications Union, *Stars Database*, 1992; UNDP, 1991)

FIG. 3.1. Telephone lines per capita versus Human Development Index ($r^2 = .864$).

highlighted data points for Côte d'Ivoire, the Philippines, Thailand, and Mexico demonstrate, these four countries represent a wide range of HDI values, but there is little apparent relationship between the HDI value and the change in telephone density between the early and late period. National-level telephone densities provide no information about common distributional biases: especially urban, governmental, and class biases.

"Thin" capital markets in poor countries coincide with "thinly" developed telecommunications infrastructure. More importantly, the "thinness" of both inhibits integration to thickening global systems. Paradoxically, the insertion of single "thick" strands from a global system into local environments may relieve pressure on the "thin" systems to expand. Integration of a limited portion thereby retards development of the whole.

Intersection of Globalizing Telecommunications and Globalizing Capital Markets

Two important relationships exist between capital markets and telecommunications. First, telecommunication is a precondition for the operation of modern capital markets: This connection implies potential convergence of interests among those who control telecommunications development and those who control capital market development. Second, telecommunications systems require large capital outlays: new investment in telecommunications systems requires the organization of finance capital, often on an international scale. The architects of capital markets and of telecommunications systems come together in the larger capitalist project of financial integration.

These "architects" come from telecommunications industries, large corporations that heavily use telecommunications—including big players in the financial services sector, governments, and increasingly, supranational organizations. On the corporate side, they come predominantly from transnational corporations. On the government side, they come from agencies endowed with varying levels of centralized power: in countries where a Post and Telecommunications model remains, for example Germany (at the time of this writing), one centralized agency maintains both ownership and operating responsibilities; in countries where the private sector provides both telecommunications service and equipment, such as the United States, a national level agency (Federal Communications Commission in the United States) is charged with regulating private-sector operators. The tasks of public-sector actors include the development of national regulation and international "agreements," two critical components of the infrastructure of global capital.

Telecommunications service industries remain more dependent on national markets than most other information technology industries, largely because of the extensive tradition of public ownership. This is

3. TELECOMMUNICATIONS AND GLOBAL CAPITALISM 59

changing: Since 1980, pressure from the United States, Britain, multinational users of telecommunications services, the World Bank, and others has "forced the pace of liberalization," and shifted the balance toward private control (Morgan & Sayer, 1988, p. 103). Worldwide liberalization of telecommunications services represents foremost a realignment of governmental interest in telecommunications and secondarily a generalized dismantling of state enterprises, or broad deregulation along the lines of the U.S. airline industry. Telecommunications "liberalization" is accompanied by the construction of a concept of telecommunications as a "strategic industry" and a redefinition of the role of government in shaping the industry's direction. As the generalized liberalization phenomenon adds momentum to a shift toward private ownership, public interest in telecommunications enhances the power of telecommunications corporations. The role played by telecommunications corporations in creating this condition should not be overlooked nor should the contradictions inherent in the condition be underestimated.

Historical Context

The conclusion of World War II opened the door to an era of accumulation in an international economy. Multinational corporations induced rapid technological change in order to transform distinct national and international activities into more highly integrated modes of production and marketing. Postwar technological dynamism fueled transformations in finance and telecommunications at the national levels, but the legacy of national-level organization and regulation prohibited corporations in these sectors from globalizing as rapidly as multinationals in other sectors. Changes in the global composition and distribution of capital during the 1970s led to demand for global financial restructuring in the 1980s. At the same time, international demand for telecommunications services stimulated regulatory changes in developing countries that complemented the technology-leveraged liberalization happening in the United States and other industrialized countries. During the 1980s, telecommunications and capital market interests succeeded in making the integration of capital markets and of international telecommunications liberalization policy goals in many countries.

A political economy of telecommunications

The means of communication make up a form of fixed capital, which has its own laws of realization. (Marx, 1939/1973, p. 523)

Telephone systems can be considered as bundles of commodities: A telephone service commodity is produced by a telephone company that sells it at a price sufficient to cover costs and realize a profit. Theoretically, a competitive market in these commodities will induce the producers to minimize costs through increases in productivity and efficiency. Indeed, the large U.S. providers highlight such efforts (and the benefits for shareholders) in their annual reports.[3] Where markets are not competitive, as in the case of public or private monopolies, prices are regulated. But to idealize telephone services as commodities is to ignore important aspects of telecommunications as a means of production as well as the scale of investment required to produce a telephone system.

In this section, I distinguish between telephone service as a commodity and telephone systems as fixed capital. In the abstract, commodities and fixed capital occupy discrete roles in the overall circulation of capital. As it circulates, capital increases through transformations from finance capital to productive capital to commodities and back to finance capital. Increases occur at each moment of transformation: for example, when a bank loans money to US West to finance the construction of a cellular phone system in Hungary, the bank's capital increases with the return of interest; as the principal is transformed by US West into productive capital—the cellular system—lease of access to the system can be viewed as a transformation of productive capital into commodities, and US West's capital increases as the leases return money in excess of the original principal. This cycle can also be portrayed differently by categorizing the telephone network as a form of fixed capital and viewing leased access, especially to other productive capitalists, as a second transformation of financial to productive capital, with the lease including returned interest. Finally, usage—charges for each call and minute—is sold as a commodity, and some of the profits are transformed back to financial capital. Figure 3.2 illustrates these cycles.

A telephone system is a joint form of capital, representing both fixed capital, because it comprises "instruments of labour actually used to facilitate the production of surplus value," and also consumption fund because it is used in final consumption (Harvey, 1982, p. 205). The stock of telecommunications fixed capital perpetually fluctuates: the allocation of telephone networks to either fixed capital or to the consumption fund rises and declines with each change in one implying adjustment in the other (Marx, 1939/1973). Telephone equipment, like much information technology, clearly illustrates the vulnerability of the consumption fund to spot appropriation.

Corporations have far greater power to control the fluctuation of telecommunications assets between the consumption fund and fixed capital.

---

[3] See chapter 4 on RBOCs. See also the 1992 annual reports for the Regional Bell Operating Companies: Ameritech, BellSouth, Bell Atlantic, Southwestern Bell, Pacific Telesis, US West, and NYNEX.

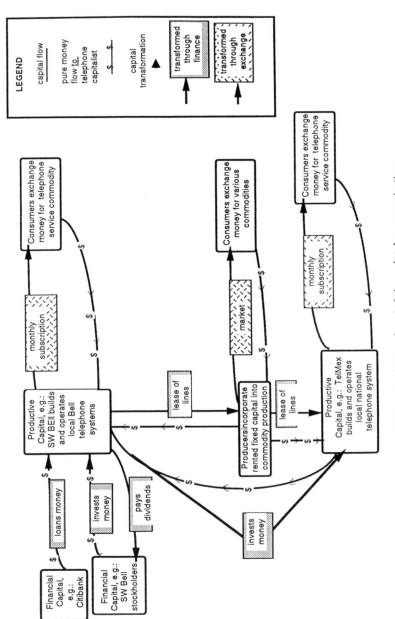

FIG. 3.2.   The circulation of capital through telecommunications.

LEGEND

capital flow

pure money
flow to
telephone
capitalist

capital
transformation

transformed
through
finance

transformed
through
exchange

Consumers exchange
money for telephone
service commodity

monthly
subscription

Productive
Capital, e.g.:
SW BEll builds
and operates
local Bell
telephone
systems

lease of
lines

Consumers exchange
money for various
commodities

market

Producersincorporate
rented fixed capital into
commodity production

lease of
lines

Consumers exchange
money for telephone
service commodity

monthly
subscription

Productive
Capital, e.g.: TelMex
builds and operates
local national
telephone system

loans money

invests
money

pays
dividends

invests
money

Financial
Capital,
e.g.:
Citibank

Financial
Capital, e.g.:
SW Bell
stockholders

61

One outcome of the project of telecommunications companies in the United States today to combine telephone service and entertainment services is to increase the consumption fund portion of the U.S. telephone network during a period of investment, when costs can be passed through to households. Once investment has occurred, we may see pressure to increase the fixed capital portion of the network, as this financial analyst suggests:

> Clearly, the ultimate infrastructure will take years to build—regulatory barriers must be broken and technological hurdles overcome, but multimedia's arrival is not far off.... [T]he true market value of the information superhighway and its services come into question. Near term, *entertainment applications can be counted on to generate revenues and will therefore fund and facilitate the societal goal of equal access to all information for everyone* [italics added]. (Swenson, 1993)

Creating Fixed Capital

Fixed capital formation requires:

> a certain level of productivity and of relative overabundance, and, more specifically, a level directly related to the transformation of circulating into fixed capital.... *Surplus population*, as well as *surplus production*, is a condition for this. (Marx, 1939/1973, p. 707)

Surpluses dedicated to fixed capital formation are *sunk costs*; when the surpluses are social surpluses, sinking them into fixed capital is inherently political. Surplus population and surplus capital may be marshalled through "direct appropriation and primitive accumulation," as was most important during the colonial era of telephone and telegraph building. David Harvey cited the case of the Irish becoming the "railroad navies and construction workers of the world after the potato famine"; it is also true that colonial telegraph telephone networks were built through direct appropriation of indigenous people's labor power (Bright & Bright, 1899). Another avenue of direct appropriation that may grow in importance is the creation of fixed capital by "changing the uses of existing things" (Harvey, 1982, p. 217). Marx's (1939/1973) examples included the "cottages of weavers," converted to fixed capital out of the consumption fund. A conceptual foundation of "telecommuting" is the availability of household telecommunications, use of which can be appropriated by corporations to deliver value and surplus value from the "cottages" of service and professional workers.

Fixed capital in telephone networks today is being formed primarily from the process of overaccumulation and its attendant surpluses of money capital as well as labor power, commodities, and productive capacity (Castells, 1989; Harvey, 1982). Overaccumulation of telecommunications capital itself has occurred in the United States, thus driving producers to seek markets elsewhere, especially in developing countries. Overaccumulation of telecommunications labor power has resulted in hundreds of thousands of laid-off workers in the sector, giving companies leverage to "resist attempts to make them accountable for their performance in adhering to internationally recognized workers' rights" (Postal Telegraph and Telephone International, 1993, p. 12).

Table 3.1 shows the sheer quantity of capital accumulated by the 12 telecommunications companies in the "global 100." Nine of the 12 largest telecommunications service providers are U.S. firms, whose developing-country activities are summarized in Table 3.2. Corporate strategies for continued growth include partnerships and acquisition of cable television companies (such as Bell Atlantic's bid for Tele-Communications, Inc.) and cellular companies (such as AT&T's acquisition of McCaw for several billion dollars) and also expanding into international markets for both traditional and new services. All the regional Bell holding companies still rely primarily on regionally defined monopolies in basic telephony for most of their revenue, but all now stress in their annual reports that they are *global* companies. The most pronounced assertion is Southwestern Bell's proclamation on the cover of its 1992 annual report that Telmex is one of the company's "engines of growth."

Resolving crises in the industrialized world by investing in telecommunications and other fixed capital in developing countries may deeply influ-

TABLE 3.1
Telecommunications Service Firms in the Global 100
(all values in U.S. $ billions)

Company	Market Value	Gross Sales	Assets	Return on Equity
Nippon Telephone & Telegraph	140.5	59.7	107.0	4.5%
American Telephone & Telegraph	82.4	64.9	57.2	21.0%
British Telecom	40.7	20.7	33.6	10.4%
GTE	33.0	20.0	42.14	18.9%
BellSouth	25.6	15.2	31.46	11.7%
Bell Atlantic	23.3	12.6	27.9	16.5%
Southwestern Bell	23.0	10.0	23.8	14.4%
Ameritech Corp. (New)	19.5	11.2	22.8	21.9%
Pacific Telesis Group	19.3	9.9	22.5	10.5%
US West	17.8	10.3	28.0	14.3%
NYNEX	17.2	13.2	27.7	13.7%
Hong Kong Telecommunications	16.5	2.8	2.5	52.8%

Note. Compiled from data from Mead (1993).

TABLE 3.2
Diversification among Telecommunications Service Providers

Company	Number of Subsidiaries	Selected International Operations
AT&T	19	Equipment: Owns 60% of Taiwan Telecommunication, and owns manufacturing and assembly facilities in Singapore, Indonesia, Thailand, and Mexico. In joint venture with Goldstar computers (Korea). Service: 5% ownership of Venworld with GTE and 3 Venezuelan companies. Venworld owns 40% of the Venezuelan PTT; 19.5% of UTEL, a joint venture with Ukrainian State Committee of Communications.
GTE	355	Venezuela with others, operates national carrier (wire-based and cellular systems). Argentina—won license to build extensive national network.
BellSouth	347	No information
Bell Atlantic	110	Mexico—1USACELL cellular (duopoly with Telmex).
Southwestern Bell	11	Telmex
Ameritech	18	Polska Telefonica Komorkowa provides cellular service. Joint venture with France Telecom and domestic carrier. NetCom, wireless telephone system with Singapore Telecom
Pacific Telesis	No information	No information
US West	49	Russia, Hungary, Czech Republic and Slovakia—building and operating digital cellular systems. Russia and Lithuania—international long distance switches. Poland—telephone directories. Hungary (and western European countries)—investing in cable television franchises.
NYNEX	17	Thailand—in a consortium to build and operate 1–2 million lines.

Note. Compiled from data from Mead (1993); annual reports.

ence the course of capitalism in those countries. This is so because these massive investments also entail local resources that may not be in surplus at all. When fixed capital is thus introduced into a developing country's

> industrial cycle, . . . it engages the production of subsequent years. . . . The anticipation of further fruits of labour . . . has its roots in the specific mode of realization, mode of turnover, mode of reproduction of fixed capital. (Marx, 1939/1973, pp. 731–732)

Thus, Harvey's (1982) generalization that "the more capital circulates in fixed form, the more the system of production and consumption is locked into specific activities geared to the realization of fixed capital" spans the borders between industrialized and developing economies (p. 220). This has implications that may limit the possible paths of industrialization in both kinds of countries. One implication is to increase the physical bonds of globalization; for example, if U.S. telephone companies are investing in distant countries, one sure consequence is that the investment will include an increase in the international connections between the foreign country and the United States.

Independent Versus Enclosed Forms of Fixed Capital

Realizing value from fixed capital becomes more difficult when it is of an "independent" rather than an "enclosed" form. Telephone systems are enclosed fixed capital only when a producer owns its own complete system. More generally, telephone systems are fixed capital of an independent form: The circulating capital embodied in the system is owned by someone other than the capitalist who makes use of it in production. Owners of independent fixed capital loan it to producers; owners received the equivalent of interest on the fixed-capital loans, but cannot claim all the surplus value that the fixed capital helps produce. It is like a banker loaning money; in fact, Harvey (1982) interpreted Marx (1939/1973) to mean that independent fixed capital "functions as a material equivalent of money capital" (pp. 227–228). A portion of the surplus value produced with the fixed capital remains "to be competitively divided among the remaining capitalists as they struggle to equalize the rate of profit" (Harvey, 1982, pp. 227–228).

Capitalists will press for the formation of independent fixed capital because individual rates of profit may be increased by substituting independent for enclosed forms of fixed capital. But potential owners of independent fixed capital only will emerge (a) given the certainty of extracting interest in "loans" of fixed capital over a long turnover period and (b) given a state of sufficient accumulation.

Realization problems and obstacles to formation are more serious inhibitions to the development of independent form of fixed capital, including telephone systems, in developing countries than elsewhere. Accumulated local capital may not be sufficient or available for making initial investments and inflation, and other financial tendencies may introduce great uncertainty for the realization of interest. Furthermore, if the consumption fund portion of existent telephone systems dominates the fixed capital portion, then political pressure for investment may be less effective than in the cases in which the interests of productive capital are at stake.

FINANCIAL INTEGRATION AND GLOBALIZED INEQUALITY

Defining and Measuring Financial Integration

Financial researchers usually apply concepts of integration to a single country vis-à-vis one or more other countries. Oxelheim (1990) distinguishes between direct and indirect financial integration, and arrays levels of integration on a continuum ranging from perfect integration to perfect disintegration (or segmentation). Direct financial integration means capital market integration and is measured against the "law of one price," for financial securities. If capital market integration is "perfect," then "an investor can expect the same return on investments on different markets (and borrowers the same loan costs), after the requisite adjustment has been made for risk."[4] The promotion of financial integration is one component of an anti-interventionist policy framework. For example, Oxelheim's definition of *direct* financial integration includes the stipulation that integration must not be effected through state policy:

> Market A is perfectly (directly) financially integrated with market B if the interest rate on market A—after exchange rate expectations and risk premiums have been allowed for—is the same as the interest rate on market B at every moment in time, *and if the politicians on market A have not themselves decided that this should be so* [italics added]. (p. 69)

Of course, such stipulations parallel the assumptions of a "perfect" market and bear little similarity to reality. For example, notes such as this one on the relationship between politics and monetary policy are common: "Many believe that the high interest rates [in Mexico] that have restricted credit ... will be lowered later in the year in an attempt to bolster the economy just as the ruling party's presidential candidate ... is announced" (DePalma, 1993, p. D1).

Indirect financial integration occurs when goods or foreign exchange markets indirectly link the return on investment in one country to the return on investment in another. Use of a common currency (monetary integration) eliminates a barrier to financial integration but bears no implication of integration between two economies; for example, use of the dollar as a common currency facilitates capital movement between Hawaii and Maine but provides no evidence that the economies of the two are integrated. Conversely, the concurrent stimulation of the dollar as currency in Cuba *and* of telecommunications trade between the United

---

[4]The adjustment for risk is a problem for many developing countries.

States and Cuba represent two parts of "one comprehensive strategy" designed to politically and financially integrate certain segments of U.S. and Cuban societies in order to bolster opposition to Fidel Castro (Toricelli, 1993).

Perfect total financial integration means both direct and indirect integration prevail; that is, the existence of

> perfectly integrated goods and foreign exchange markets, and such a highly coordinated economic policy that the gap between political risk premiums is zero. If perfect total integration is global, then the world will consist of *one* financial market composed of perfectly linked national capital markets under strict purchasing power parity. (Oxelheim, 1990, p. 5)

Analysis of interest rate differentials underlies most measures of direct financial integration (Goldstein, Mathieson, & Lane, 1991). Many choose the rate on treasury bills as the data source with the greatest comparability from one nation to another (Oxelheim, 1990). A "global interest level" is used as a reference point against which to measure national interest rates. The global interest level is a weighted average of rates in the largest Organization of Economic Cooperation and Development (OECD) countries (Oxelheim, 1990).

The Integration of Developing Countries

Two kinds of conclusions about financial markets in developing countries can be drawn from the findings of studies of financial integration: (a) For small open economies that are "integrated," interest rates are exogenously determined; (b) Measures of converging interest rates are made after eliminating three factors that are all or partially endogenously determined and that are of particular relevance to developing countries: risk premiums, transaction costs, and exchange rate expectations (Oxelheim, 1990).

Risk premiums calculated for political risk in a country and exchange rate and inflation risks pertain to integration. Risk premiums also include premiums for commercial and financial risk that are not of interest to a study of integration. Political risk premiums can be high: Oxelheim cited studies showing that corporate decision makers "take political risks very seriously and the international investor has a high aversion to this kind of risk" (Oxelheim, 1990, p. 59). In a chapter titled "Governments as Perpetrators of Political Risk," Agmon derived an algorithm for estimating political risk to an exchange rate between a large and a small country. The source of the risk in the model is the government of the small country (Agmon, 1985).

68

UREY

Premiums for political risk are systematically biased against developing countries:

In estimating the political risk premium we are certainly justified in basing our view on a link between rising indebtedness and a higher propensity to political intervention. A country's indebtedness, as we have seen during the 1980s, can result in payment difficulties which seriously reduce that country's economic-political room for manoeuvre. Consequently there is a greater likelihood that restrictions and controls will be imposed, implying greater political risk. The premium can be calculated on a basis of the country's gross international debt, its net foreign assets, and some other expression for possible financial difficulties. (Oxelheim, 1990, p. 59)

Financial literature legitimizes explicit systematic discrimination against developing countries. The IMF researchers link "stage of development" to a concept of credit-worthiness with the euphemistic "contagion effect":

One key issue is whether perceptions of credit-worthiness are subject to 'contagion effects' in the sense that an otherwise credit-worthy country's access to international credits is curtailed because other countries at a similar stage of development or with a similar external debt position are experiencing external payments difficulties. (Goldstein et al., 1991, p. 38)

On the flip side, analysts also note that the wealthy inhabitants of many developing countries have greater information about and access to international financial markets than before and are becoming "increasingly sensitive to differences between financial conditions in domestic and external markets" (Goldstein et al., 1991; Varman-Schneider, 1991).

Nonetheless, trends in developing countries are toward openness, in part due to pressure from the International Monetary Fund (IMF; Quirk, Gilman, Huh, Leeahtam, & Landell-Mills, 1989). Openness empowers corporations over people. Citizens sacrifice power directly to corporations and also indirectly, as governments sacrifice authority to multilateral trade organizations which serve the interests of transnational corporations (Stallings, 1992). Corporations paved the road to the international "NGO" frontier: in the areas of finance and telecommunications, for example, transnationals explicitly or implicitly share influence with national governments in organizations such as the World Bank, the Global Agreement on Tariff and Trade (GATT), and the ITU to manage transborder issues, accelerate development of a large international capital market, and explore the possibilities of space without sovereignty (off-shore production, orbiting satellites).

Corporations need such organizations in order to engage each other politically, to focus their pressure on governments, and to supersede nation-states on some matters. As a vice president for the European Bank for Reconstruction and Development said in reference to development in Eastern and Central Europe:

[H]owever paradoxical it may seem, the transition to market economies cannot be effected by the private sector alone. The market by itself cannot create the market. A market can operate effectively only if a legal and institutional framework exists to organise, monitor and channel market forces. (Sarcinelli, 1991, p. 33)

Even in their rhetoric, some of these organizations reveal antilabor, antipopular biases. For example, Sarcinelli began his speech with reference to changes in Eastern Europe: "A social, economic and geopolitical order is dissolving as the domain of freedom is being enlarged." He went on to blame recessions in those countries on factors including labor's agenda: "the democratisation of local political scenes led to a growing deterioration in labor discipline resulting in strikes, lower productivity, etc." (Sarcinelli, 1991, p. 31).

Capital Markets

Capital markets are about money, timing, and location. Private companies and public bodies need money to meet day-to-day expenses and also for long-term plans: "to undertake fixed capital formation, to finance expansion and to develop new processes or new products" (Foley, 1991, p. 7; Grou, 1985). Companies must generally have working capital— money on hand to carry on business (especially, to meet payroll). Companies in advanced economies often have internal capital to finance expansion. Foley (1991) reports evidence from the United Kingdom and the United States demonstrating that "sources of external finance play a relatively minor role in the provision of investment funds" (Foley, p. 9). Internal capital is an important but distinct component of "global capital," influencing capital markets but not necessarily available as "supply" to the market. Harvey (1991) noted that "the last decades have witnessed an increase in the concentration of multinational capital; the difference is that this power is now increasingly organized through networks of seemingly autonomous firms and activities" (p. 73). A lack of internal capital inhibits autonomous investment decision making by firms and creates the demand side of capital markets. Demand includes money-poor companies everywhere as well as companies who have only poor money (nonhard currency), which circulates only within a limited range.

Of course, it also takes money to live. Individuals and households use money to meet day-to-day consumption expenses and for longer term plans of growth and change. Consumption requires a constant supply of money to households, which motivates the continuous exchange of labor power for money. (Labor thus perpetually underwrites the quality of money as a vessel for capital.) Household demand for money integrates consumption into capital markets without empowering households as capitalists. (Households use money as capital only in select cases, as in the purchase of education.) Interest on loans for housing, durables, and other consumer debt accrues to some capitalist and relates back to the production cycles (including the labor component) for housing, and so on. Households supply money to capital markets via savings institutions, but the common accounting identity that sets savings equal to investment is misleading (Davidson, 1992). The concept of *financial surplus* implies savings. Both households and businesses save by deciding not to spend current income on the products of industry, or also, in the case of businesses, on resources of industry.

Money's circulation is a special form of capital circulation (but recall that money is only one of several forms of capital). Access to money must come at the right time. In this vein, capital markets

> facilitate the transfer of investible funds from economic agents in financial surplus to those in financial deficit. . . . As a result, companies, governments, local authorities, supranational organizations and so on have access to a larger pool of capital than would be available if they had to rely exclusively on generating their own resources. (Foley, 1991, p. 6)

Figure 3.3 illustrates how the flow of capital from the supply side to the demand side is typically depicted. Investors provide capital to companies or public authorities in the primary market in exchange for shares (e.g., equity in a company) or bonds (debt—e.g., financial claims against a public utility). After initial purchase in the primary market, investors buy and sell securities (shares or bonds) in the secondary market. Because it provides liquidity, the secondary market "makes the primary market operate more effectively [because it] provides a facility for the continuous reallocation of financial assets among various investors" (Foley, 1991, p. 6).

On the demand side, the capital market is biased toward large, well-established companies. As Foley (1991) stated:

> Small private companies will generally have to rely on a limited group of existing shareholders, bank loans and/or leasing arrangements. Medium sized public companies with a market listing are able to sell new shares or issue debt to existing shareholders or to new subscribers. Large well established publicly quoted companies are able to tap a much greater variety of sources including in some cases foreign and international capital markets. (p. 21)

On the supply side, the capital market is biased toward institutional investors, which fall into four categories: insurance companies, pension funds, unit trusts, and investment trusts. Foley (1991) further stated that:

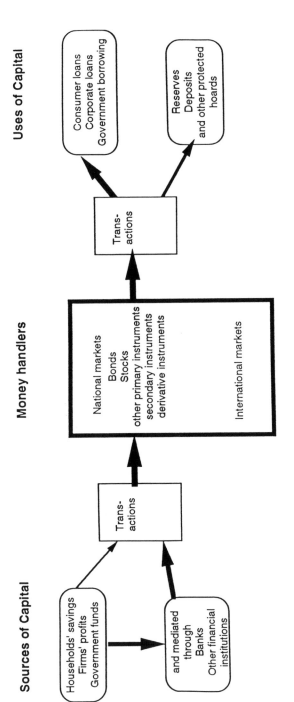

FIG. 3.3. Capital flows through capital markets.

One of the most significant features of capital markets over the last thirty years has been the rise of institutional investors. This phenomenon has been a characteristic of nearly all capitalist economies and is particularly marked in those economies with well developed stock markets. The growth of institutional ownership has led [commentator J. A. Kregel] to describe the system as "money manager" capitalism. (pp. 174–175)

The influence of institutional investment reaches some middle-income developing countries. For example, the recent boom in Mexico's stock market, where foreigners accounted for 20% of the capitalization in April 1993, includes capital from large investment fund companies in the United States and pension funds: Kemper Financial Companies; Scudder, Stevens, and Clark; Fidelity Investments; and Capital Research Managers "have become major Mexican players" (DePalma, 1993). Fidelity, Scudder, Merrill Lynch, and Bear, Stearns have "amassed [most of] $9 billion worth of Mexican Government securities," tripling U.S. holdings of these securities from 15 months earlier (Uchitelle, 1993, p. A1). Between 1970 and 1991, developing countries issued nearly $80 billion in bonds and $18 billion in commercial bank notes. Despite growth in developing country markets, however, between 1988 and 1992, less than one fifth of the world's developing countries raised capital in the bond market, and most that did fall into the "middle" income range.

Although Fig. 3.3 implicitly suggests that "capital use" must equilibrate with "sources of capital," there is in fact *no* mechanism to ensure equilibrium. The figure has no time dimension, and, at any moment, "sources of capital" may in fact be greater or less than "capital use." The space dimension is also problematic: both the "sources of capital" and the "uses of capital" have international dimensions: A global equilibrium could be comprised of many local disequilibria.

Foreign Direct Investment

Capital markets in poorer developing countries exist in the shadow of the multinational investment market and are not a significant portion of foreign investment in most countries. Foreign direct investment (FDI) is more about the globalization of production rather than of finance, and may in fact inhibit the development of local capital markets. Nonetheless, FDI integrates commodity markets and thus indirectly contributes to financial integration specifically skewed toward multinationals. As employers, producers, and raw material consumers, multinationals can act on many levels to reduce political risk premiums attached to investment. Harvey (1991) remarked that, since 1973,

[T]he rapid dispersal of production to underdeveloped regions [of] the developing world enabled companies to employ geographic mobility as a threat . . . in bargaining sessions with unions that had long confined their political vision to the nation-state. Furthermore, national governments found it harder and harder to control international capital flows and, irrespective of their political inclinations, were increasingly forced to discipline labor in order to attract multinational investment. Class struggle within nation-states sharpened at the same time that both governments and working-class movements found their maneuvering room reduced. (p. 68)

Foreign direct investment grew rapidly in the 1980s, and the relative preference for countries in the Americas declined with rising investment in Asia. As the total invested climbed, the proportion invested in a handful of American economies fell from over half between 1978 and 1982 to less than a third after 1988. The Asian share absorbed the difference: Asian economies have accounted for more than 40% of FDI in developing countries since 1988. The shift reflects, in particular, less FDI in Brazil, although Brazil continues to account for a large share of FDI profits, and the rise in FDI in Malaysia and China. Indonesia rarely accounts for more than 3% of FDI, but between 1975 and 1991 consistently accounted for a larger share of FDI profits than any country except Brazil from 1985 to 1989 (World Bank, 1992).

Clever capitalists adjust the organization of multinational firms to ensure profits return from subsidiaries to headquarters. For treasury officers of large multinationals, harvesting profits from subsidiaries is a game of shifting information asymmetries and power relations with their subsidiary staff as well as a political game with government bureaucrats and leaders. Textbooks and primers describe techniques for winning the game: pooling the finances of more than one subsidiary, automation, evaluations, and so on (Business International Corporation, 1987, 1988). The general trend of decentralizing authority to subsidiaries is least pronounced in financial operations (OECD, 1987).

Distributional Effects of Capital Flows

In most developing countries, domestic (non-MNC) capital demand exceeds domestic supply. Under conditions of unmeetable demand, allocation of scarce money capital, especially hard-currency money, is excruciatingly political, and real interest rates are high. The history of money capital circulation in developing countries during the second half of the 20th century reflects evolution in local social structure and shifting relations with other countries. After World War II, the (Marshall Plan) development model gave agency to the state in dealing with multilateral and aid sources of capital. The focus on large industrial and infrastructural projects channeled large blocks of capital through the state. Arrighi (1991) argues that through the 1950s and until the mid-1970s, capital flowed from rich areas to poor areas and stimulated changes that:

cut across the great West–East and North–South divides and [were] primarily the result of purposive actions aimed at narrowing the gaps that circa 1950 separated the wealth of the peoples situated on the privileged side of the two divides (the West/North) from the relative or absolute deprivation of the peoples situated on the underprivileged sides (the East and the South). The most important of these purposive actions was the pursuit of economic development by governments. By internalizing within their domains one or another of the features of the wealthier countries, such as industrialization and urbanization, governments hoped to capture the secret of their success and thus catch up with their wealth and power. (pp. 39–40)

Arrighi's (1991) data demonstrated that inequalities between rich countries and poor countries declined steadily during much of this period: Inequality between Latin America and "core" economies declined steadily from 1948 to 1960 and from 1970 to 1980; in southern and central Africa from 1960 to 1970, and every other region except South Asia from 1970 to 1980 (Arrighi, 1991). He also showed that intraregion income inequalities during the last 50 years generally declined in rich regions but rose in poorer regions (Arrighi, 1991). The agency gained by the state during the "development era" was effectively retained by local elites in control of internationally recognized political structures. Thus, when development capital contracted in the 1980s, social structures that had evolved to distribute inflowing capital more inequitably distributed deficits (poverty) equally inequitably. Since 1980, famine- and disaster-relief have grown as modes of commodity circulation, while capital circulation in developing countries has become more constrained. As Arrighi (1991) put it, "Solvency rather than development has become the password" (p. 51).

The supply of concessional capital is small. Capital flight reduces the domestic savings portion of domestic capital and the depth of local capital markets. Discrimination constrains access to international capital. The technological aspect of integration with richer countries is expensive. But, the primary agents of the technological layer of integration also happen to be very wealthy telecommunications corporations engaged in fierce competition with each other. Furthermore, even in the most bloated bureaucracies, traditional state telecommunications enterprises have still been profitable (if not dynamic). As an example, the ITU found that the 60% rate of internal financing for telecommunications projects exceeded the rate for most infrastructural investments of similar scale (ITU, 1991). Thus, telecommunications corporations have competed to supply capital for recent projects in selected developing countries (in exchange for a share in controlling interest).

TELECOMMUNICATIONS AND INTEGRATION

The Meaning of Recent Investment

Recent investment in telephone systems in developing countries allies holders of surplus capital with capitalists who will benefit and whose modes of production allow for paying interest on the use of fixed capital of an independent kind. The boom in telephone investment suggests that something has changed in the relationship between productive and financial capitalists. After lagging in telephone investment relative to investment in other fixed capital, what caused the reversal of this tendency? Who is restructuring telecommunications in developing countries?

A partial answer to this question lies in the parallel thrusts toward globalized finance and the privatization of the telephone service industry. The holders of surplus capital and the recipients of loans for fixed capital investment are governed by the same global economy but by different rules in each domestic economy. Thus, new telephone lines in Mexico are financed in part by Mexican rate payers, in part by purchasers of Telmex stock (either on the Mexican Bolsa or the New York Stock Exchange), in part by holders of Southwestern Bell stock or France Telecom stock. But to whom does the lien on future labor fall? First, on the current and future ratepayers in Mexico, whose incomes must rise if they are to demonstrate the demand corporate strategists have envisioned. Second, if demand falls short and future revenues do not match expectations, domestic and international Telmex, Southwestern Bell, and France Telecom stockholders will find their capital devalued. Third, at least in the case of Southwestern Bell, ratepayers might be at risk if Southwestern Bell can find a way to transfer the loss through creative accounting.[5]

The precondition for the emergence of alliances among telecommunications firms is a new level of global economic integration including the use of several preferred hard currencies. The allying parties include holders of surplus capital on one side and productive capitalists on the other. Either side may be either transnational or domestic, but the store

---

[5]As an example of creative accounting aimed at protecting shareholders at the expense of ratepayers, US West in 1993 created a $5.1 billion loss by decreasing their depreciation schedule. Thus, in 1 year they "lost" 12 years' worth of use-value from their copper wire by reducing its depreciation schedule from 27 to 15 years. They reduced depreciation of digital switches from 18 to 10 years and of optical fiber from 30 to 20 years. Financial analysts note, "This means US West will not have to reduce rates or 'share' profits with ratepayers under incentive plans, since from a regulatory perspective US West's depreciation costs will be going up to the same degree its operating costs will be declining" (Grubman, 1993).

of surplus capital must be largely in hard currency and the expected revenues from production must be composed of enough hard currency for interest payments. Thus, it is likely that (a) the surplus capital will be held by transnationals and (b) the productive capitalists will be important to the export sector. Augmenting the precondition of integration is the fact that telephone service transnationals have a particularly acute capital surplus. Whereas demand for telephone services has been expressed for a long time by productive capitalists in developing countries, only recently have the preconditions for supply existed. Although it was possible previously for capitalists and the rich to pressure governments to develop telephony a little bit, especially in terms of the consumption fund, governments lacked the cash surplus required to extensively modernize national telephone infrastructure. Local productive capitalists were not powerful enough to attract partners from among the transnational service providers, and those providers were themselves less cash-rich.

During the 1980s, local productive capitalists were empowered by the global preference for market-oriented economies (Stone, 1993). At the same time, transnational telecommunications service providers accumulated greater capital surpluses. Given the precondition of integration, these conditions are conducive to the formation of alliances between productive capitalists and telecommunications service providers. Parties to potential alliances can powerfully challenge government monopolies and force governments to yield authority for the creation of new corporate entities to control telephone development.

The local politics of such challenges and responses may vary widely according to bureaucratic stakes in the status quo, consumer interest and power, the adroitness of the parastatal post and telephone (PTT) authorities, as well as the power and alacrity of other players. The powerful players pursue their individual interests—telecommunications service transnationals want to control a market and productive capitalists want reliable, extensive, international, and domestic services. Each will couch their interests in the context of broader social goals, appealing to government and PTT players, appropriating language as convenient in order to rewrite telecommunications regulation. Government and PTT players will also be pursuing their individual and institutional interests which may, in fact, contradict the broad social interests they purport to advance (Cowhey, Aronson, & Szekely, 1989).

There are many possible outcomes of negotiations among these parties, and the complexity of their interaction and the size of the stakes present opportunities for greater mobilization of marginalized parties around issues of telecommunications restructuring. The high stakes, like the large development loans of the 1970s, tie everybody to the outcomes, whatever they are (Schiller & Fregoso, 1993).

The history of information technologies, especially since the invention of the telegraph, supports Harvey's (1984) insistence that

> a constant source of preoccupation under capitalism is the creation of social and physical infrastructures that support the circulation of capital. . . . The multinational corporation, with its capacity to move capital and technology rapidly from place to place, to tap different resources, labour markets, consumer markets and profit opportunities, while organising its own territorial division of labour, derives much of its power from its capacity to command space and use geographical differentials in a way that the family firm could not. (pp. 129, 142)

Physical infrastructure includes telecommunications, but telecommunications have also become integral to the social infrastructures supporting the circulation of capital, not least capital in its money form:

> Each form of the geographical mobility of capital [money, commodities, production] requires fixed and secure spatial infrastructures if it is to function effectively. The incredible power to move money around the world . . . demands not only a well-organised telecommunications system but—as a minimum—secure backing of the credit system by state, financial and legal institutions. (Harvey, 1984, p. 148)

TELECOMMUNICATIONS AS A "STRATEGIC" INDUSTRY

The construction of telecommunications as a "strategic" industry complements the financial integration strategy and operationalizes some of the strategy's contradictions. First, telecommunications industries are everywhere defined as strategic; telecommunications are "engines of growth" for developing economies (ITU, 1991) and "the most critical area for influencing the 'nervous system' of modern society" in industrialized countries (Wallenstein, 1990). But the "strategic" argument comes down to arguing for liberalization of telecommunications in developing countries and for just the opposite in industrialized countries. Scholars and business leaders alike point to the need for new forms of national and international regulation of the industry in the United States, Europe, and Japan.

Constructions of Telecommunications as "Strategic" in Developing Countries

The Maitland Commission's *Missing Link* report focused attention in 1985 on the overall underdevelopment of telecommunications infrastructure in many parts of the developing world. The report argued that telecommunications "should be regarded as an integral part of economic and social development" and established a goal of global universal access by the 21st century. In summarizing his disappointment with the progress toward this goal, Maitland (1992) noted first the political failure to respond to the technological opportunities provided through "the ingenuity of our engineers and scientists and the enterprise of industry and commerce" (p. 17). But it was the political ingenuity of others employed in some of the same industrial and commercial enterprises that transformed the "Missing Link" mandate into the strategic industry mandate, in which rare mention is made of *creating* universal access, and "social" rarely joins "economic" in modifying "development."

Addressing the political obstacles to extending telecommunications services to the least privileged, Maitland identifies the most relevant question as "how can the benefits to the community or nation as a whole of providing these facilities to the least privileged be more readily understood and accepted by those who wield power?" Indeed, even in industrialized countries, marginalized groups, by definition, are irrelevant to arguments about "strategic industries."

"Strategic" Constructions in Industrialized Countries

Debate surrounding "strategic" industries in industrialized countries is shaped by the current predicaments of militarily "strategic" industries and by competition among the high-technology corporations of Europe, Japan, and the United States. Those who argue for government support for "strategic" industries assert that, without such proactive industrial policies, national industries will succumb to external competitors who receive government support. "Strategic" firms can multiply the benefits of strategic industry policies by becoming "multidomestic: presenting themselves as 'good domestic citizens' in as many countries as possible."[6]

This debate remains distinct from the construction of "strategic" in developing countries. Ostry (1991) identified the participants as multinationals and governments of OECD countries:

> [T]he third phase of international linkage is centred on capital and technology flows. To a considerable degree, it has tended to exclude the non-OECD countries. . . . The battle for market share in leading-edge sectors

---

[6]This becomes bizarre for the police of good citizenry. For example, Walt Mangers related his dilemma as a promoter of U.S. firms in the case of handling negotiations with Tunisia. He had to disqualify Northern Telecom, a Canadian company with a U.S. staff, in favor of advocating for AT&T, which, even though it is a U.S. company, had a French staff for the project (Mangers, U.S. State Department, Bureau of Communication and Information Policy, 1993, personal interview).

involves not only competition among multinational enterprises but also rivalry among the different market systems which influence the enterprise's ability to compete. (pp. 82–83)

The Common Denominator: Strategic Corporations

Strong support from their home markets and governments positions multinationals for expansion into developing country markets. Not only does the established home market provide a financial anchor, national policies to support strategic industries provides multinationals in these sectors with diplomatic agency beyond the usual trade advocacy services. In the United States, for example, the State Department's Bureau of Communication and Information Policy duplicates some of the Commerce Department's export promotional work on behalf of telecommunications multinationals and also engages in political work to shape telecommunications regulatory environments in developing countries.

In pursuit of developing-country markets, these firms often work together, forming strategic alliances to compete for large projects. In sectors such as telecommunications, dominated by a handful of multinational firms now in alliance with each other, Soete (1991) warns that increasing oligopoly on a world scale creates serious barriers to technological access:

> [A] geographically concentrated network of strategic alliances raises major issues about access for those countries/companies not belonging to the already existing networks. In the absence of an international regulatory framework, it is likely that such technology networking will increase inequality of access to technology and investment. (p. 61)

These concerns are echoed by Michalski (1991), who discusses implications of national strategic industry policies for broadly defined international welfare, including negative impacts on the international system of trade, investment, technology, finance, and international competition. He notes the concern of developing countries and smaller newly industrialized countries, "about the overall effect of restricted access to government-sponsored science and technology projects on worldwide technological progress and economic development" (pp. 11–12).

CONCLUSION

This chapter has painted an ominous picture of telecommunications technologies in the hands of its current owners. I have argued that telecommunication is key to financial integration and that, in many ways, integration is bad for many people in developing countries. Integration brings together—via telecommunications—capital, capitalists, and investment opportunities from all parts of the world. Integration also thereby disempowers—via exclusive access to telecommunications—the majority of people in developing countries.

But there are reasons to be more sanguine about the potential for people concerned about more equitable social change to also exploit globalized telecommunications. The state of flux and the willingness of the big telecommunications companies to negotiate in the face of "inevitable" change present opportunities for influencing the debate. Activists and researchers have begun to coalesce to promote the need for a right to telecommunicate in a world "integrated" by telecommunications.

A progressive theory of telecommunications development has not been fully articulated, but a progressive practice is under way in some places, as evidenced by:

1. The "indigenous" adaptation of radio and walkie-talkie technologies ("Amateur Radio Link," 1993); such demonstrations of homegrown telecommunications expertise reveal diffused capacity to control technological development.

2. The use of e-mail and fax for communication among human-rights and environmental activists (Annis, 1991) reveals a layer of information infrastructure being used to globalize grassroots political and social movements.

3. Symbolic acts, such as the mayor of Caracas confronting with raised fists the new transnational owners of Venezuela's telephone company (Brooke, 1993), contribute to a long history of opposition to transnational telephone corporations that began with resistance to the Bell Company by European communities in the late 19th century.

4. Advocacy of rights to telephone service by civil rights organizations and researchers demonstrates a growing awareness that access to the information infrastructure should be a right. Documentation of racism institutionalized in the telecommunications industry in South Africa and the United States (Morris, Michael, & Stavrou, 1993; Skrzycki, 1993); litigious acts in the United States, such as the petition filed by a consortium of urban advocacy groups, charging several Bell companies with "redlining" in their planning for "video dial tone" (Campbell & Schwartzman, 1994).

5. The beginning of a globally based labor strategy–making by labor activists (G. Kohl & E. Diaz, personal communication, May 23, 1994) will begin to align telephone workers with others working to "globalize" labor strategies to match the power enjoyed by "globalized" capitalists.

Telecommunications investment will continue apace: today's opportunities to redistribute the stakes, at least in the context of national economies, should be exploited. Longer term strategies for reclaiming social ownership of independent fixed capital should be explored. New forms of social ownership need to be devised to accommodate the many boundaries—geographical, cultural, and otherwise—transcended by independent fixed capital in the form of telecommunications networks. The work has begun in order to mobilize claims to telephone infrastructure among marginalized groups and the poor: There is room for optimism, but not complacency.

REFERENCES

Agmon, T. (1985). *Political economy and risk in world financial markets*. Lexington, MA: Lexington Books.

Amateur radio link aids India's quake relief. (1993, October 7). *The New York Times*, p. A5.

Annis, S. (1991). Giving voice to the poor. *Foreign Policy, 84*, 93–106.

Arrighi, G. (1991). World income inequalities and the future of socialism. *New Left Review, 189*, 39–65.

Bright, E. B., & Bright, C. (1899). *The life story of the late Sir Charles Tilston Bright*. Westminster, England: Archibald Constable & Co.

Brooke, J. (1993, March 26). New in the seat of power: A two-fisted radical. *New York Times*, p. A4.

Business International Corporation (1987). *Coping with global financial turmoil: An action guide for solving today's critical financial problems*. New York: Author.

Business International Corporation (1988). *101 checklists for successful global treasury management*. New York: Author.

Campbell, A. J., & Schwartzman, A. J. (1994). *Petition for relief from unjust and unreasonable discrimination in the deployment of video dialtone facilities: Petition for relief of Center for Media Education, Consumer Federation of America, Office of Communication of the United Church of Christ, National Association for the Advancement of Colored People and National Council of La Raza*. Washington, DC: Federal Communications Commission.

Castells, M. (1989). *The informational city: Information technology, economic restructuring, and the urban-regional process*. Cambridge: Basil Blackwell.

Cowhey, P. F., Aronson, J. D., & Szekely, G. (1989). Changing networks: Mexico's telecommunications options. *University of California, San Diego, Center for U.S.–Mexican Studies Monograph Series, 32*.

Davidson, P. (1992). *International money and the real world*. New York: St. Martin's Press.

DePalma, A. (1993, April 12). Fortunes are cast in Mexican stocks. *New York Times*, p. D1.

Foley, B. J. (1991). *Capital markets*. New York: St. Martin's Press.

Goldstein, M., Mathieson, D. J., & Lane, T. (1991). Determinants and systemic consequences of international capital flows. In *Determinants and systemic consequences of international capital flows* (pp. 19–21). Washington, DC: International Monetary Fund.

Grou, P. (1985). *The financial structure of multinational capitalism*. Dover, NH: Berg.

Grubman, J. (1993). *US West—Company Report*. Paine Webber, Inc.

Harvey, D. (1982). *The limits to capital*. Chicago: University of Chicago Press.

Harvey, D. (1985). The geopolitics of capitalism. In D. Gregory & J. Urry (Eds.), *Social relations and spatial structures*. London: McMillan.

Harvey, D. (1991). Flexibility: Threat or opportunity? *Socialist Review, 21*(1), 65–77.

International Telecommunications Union. (1991). *Restructuring of telecommunications in developing countries: An empirical investigation with ITU's role in perspective*. Geneva: Author.

International Telecommunications Union. (1992). *Telephone indicators database*. Geneva: Author.

Maitland, D. (1992). The missing link revisited. *Transnational Data and Communications Report*, pp. 15–18.

Marx, K. (1973). *Grundrisse: Foundations of the critique of political economy* (M. Nicolaus, Trans.). London: Penguin. (Original work published in 1939)

Mead. (1993, July 12). *US public corporations and Business Week* [CD-ROM].

Michalski, W. (1991). Support Policies for strategic industries: An introduction to the main issues. In *Strategic industries in a global economy: Policy issues for the 1990s* (pp. 7–14). Paris: OECD International Futures Programme.

Morgan, K., & Sayer, A. (1988). *Microcircuits of capital: "Sunrise" industry and uneven development*. Boulder, CO: Westview Press.

Morris, M., & Stravou, S. E. (1993). Telecommunication needs and provision to underdevelop black areas in South Africa. *Telecommunications Policy*, (17), 529–539.

Organisation for Economic Cooperation and Development (1987). *International investment and multinational enterprises: Structure and organisation of multinational enterprises*. Paris: Author.

Orr, B. (1992). *The global economy in the 90s: A user's guide*. New York: New York University Press.

Ostry, C. (1991). Beyond the border: The new international policy arena. In *Strategic industries in a global economy: Policy issues for the 1990s* (pp. 81–96). Paris: OECD International Futures Programme.

Oxelheim, L. (1990). *International financial integration*. New York: Springer-Verlag.

Postal Telegraph and Telephone International (1993). *Structural Changes and Multinational Companies* (27th PTTI World Congress). Cairo: Author.

Quirk, P. J., Gilman, M. G., Huh, K. M., Leeahtam, P., & Landell-Mills, J. M. (1989). *Developments in international exchange and trade systems*. Washington, DC: International Monetary Fund.

Sarcinelli, M. (1991, September). *East European reforms: A challenge for the west and a role for the IBRD*. Paper presented at *Cooperation versus Competition Revisited*, 14th annual SWIFT International Banking Operations Seminar, Hong Kong.

Schiller, D., & Fregoso, R. (1993). A private view of the digital world. In K. Nordenstreng & H. I. Schiller (Eds.), *Beyond national sovereignty: International communication in the 1990s* (pp. 210–234). Norwood, NJ: Ablex.

Skrzycki, C. (1993, September 17). AT&T apologizes for "racist illustration": Company publication depicting caller in Africa as gorilla stirs furor. *The Washington Post*, p. A1+.

Soete, L. (1991). National support policies for strategic industries: The international implications. In *Strategic industries in a global economy: Policy issues for the 1990s* (pp. 51–80). Paris: OECD International Futures Programme.

Stallings, B. (1992). International influence on economic policy: Debt, stabilization, and structural reform. In S. Haggard & R. R. Kaufman (Eds.), *The politics of economic adjustment: International constraints, distributive conflicts and the state* (pp. 41–88). Princeton, NJ: Princeton University Press.

Stone, P. B. (1993). Public–private alliances for telecommunications development: Intracorporate baby bells in the developing countries. *Telecommunications Policy, 6*, 459–469.

Swenson, C. F. (1993, November 8). *Regional bell operating companies—Industry report*. Hancock Institutional Equity Services [electronic format].

Torricelli, R. G. (1993). [Opening Statement]. *Recent developments in Cuba policy: Telecommunications and dollarization.* (Hearing before the subcommittee on Western Hemisphere affairs, 103rd Congress, 1st session). Washington, DC: GPO.

Tynan, N. (1993). The elusive goal: There are fewer places to head for in the search for profits, but some banks are bucking the trend and developing overseas expertise. *The Banker, 143,* 27.

Uchitelle, L. (1993, April 22). High Mexican interest rates are luring Wall Street cash. *The New York Times,* p. A1.

United Nations Development Programme. (1991). *Human Development Report.* Geneva: Author.

Varman-Schneider, B. (1991). *Capital flight from developing countries.* Boulder, CO: Westview.

Wallenstein, G. (1990). *Setting global telecommunications standards: The stakes, the players, and the process.* Norwood, MA: Artech.

Williams, F. (1991). *The new telecommunications: Infrastructure for the information age.* New York: Free Press.

World Bank (1990). *Poverty* (World Development Report and World Development Indicators). Washington, DC: Author.

World Bank (1992). *World development indicators (STARS database).* Washington, DC: Author.

4

Foreign Direct Investment by the U.S. Bells

Sharmishta Bagchi-Sen
State University of New York at Buffalo

Parthavi Das
Michigan State University

The globalization of the telecommunications market has led to the internationalization of the regional Bell operating companies (RBOCs). *Internationalization* is defined as the process by which a firm expands its economic activities beyond its national boundaries (Dicken, 1992). With the limits the RBOCs are facing in the United States as a result of government regulations and the saturation of their domestic markets, RBOCs are investing abroad to expand their markets. As a result of the current phenomenon of privatization in the telecommunications sector, the RBOCs have been able to enter new markets.

In 1984 the Modified Final Judgement (MFJ) was passed, which led to changes in the service provision of the American Telephone and Telegraph Company (AT&T) and the creation of 22 Bell operating companies. The 22 companies were placed under seven regional holding companies that are responsible for providing local service in the United States. These firms are Ameritech, Bell Atlantic, BellSouth, NYNEX, Pacific Telesis, Southwestern Bell, and US West (Rosenberg, Borrows, Hunt, Samarjiva, & Pollard, 1993). This chapter focuses only on the RBOCs and not on other U.S. independent telephone companies. Based on their commonalities in terms of their creation and the current regulatory environment, the RBOCs as a group create a dominant force in the global economy. The MFJ restricted the RBOCs from manufacturing products and customer services equipment, providing information and long-distance services. The RBOCs also have control of regulated operating subsidiaries, unregulated communication subsidiaries, and other unregulated subsidiaries (Rosenberg et al., 1993). The firms hoped the experience they would gain from providing these services in foreign markets will enable them to have an advantage to compete in the United States once restrictions are lifted (Hyde & Martin, 1990). Even after some of the restrictions placed on the RBOCs have been lifted, firms continued to invest in foreign markets because of globalization.

In order to understand the internationalization process, this chapter looks at the firm-specific and location-specific factors that have resulted in the RBOCs' investments. By first looking at firm-specific factors, the chapter examines what advantages the RBOCs have (e.g., capital, knowledge, and technology) that lead to investment. The chapter then looks at the particular characteristics (e.g., market size and political stability) of the host country that attracts the foreign investors. Finally, the chapter makes an attempt to understand the impact of these investments on the host countries. The hope is that by learning more about the strategies of firms, we can have an understanding of how a country can best prepare itself for the entry of a foreign firm.

Elements of Dunning's eclectic theory (1981) are used to explain the reasons for internationalization. Unlike other theories that deal with foreign investment, eclectic theory is the only theory that looks at both the factors that concern both the firm and the location. There are other theories concerning firm-specific factors, such as product cycle theory and the theory of industrial organization, but eclectic theory is the only one that looks at location-specific factors. Although the firm-specific factors are important, in order to have a more sound understanding of foreign investment there needs to be an evaluation of location factors.

In studies of foreign investment in other sectors, the research focus has also been on the developmental role these firms played in the host countries. For example, Franko (1991) found that foreign firms usually entered developing countries in the form of joint ventures. The joint ventures were either in the form of the companies having 50–50 partnerships or minority partnerships. In some cases, licensing arrangements were made between the foreign company and the local firm. Franko noted that the degree to which a company participated in a venture was dependent on the host country as well as each industry-specific strategy.

Because foreign investment in telecommunications is a recent trend, there has not been much research that looks at the role of foreign investment in telecommunications in host countries. A study on foreign investment in telecommunications may or may not have the same findings as studies on foreign investment in other sectors. The purpose of this chapter is then to examine how foreign investment in telecommunications occurs and its possible effects on the host country. The research questions are (a) What are the RBOCs' firm-specific advantages that enable them to invest internationally? (b) What are the location-specific factors that attract international investment?

The data were gathered from trade journals. In addition, a questionnaire was sent to the seven RBOCs to acquire more in-depth information about their process of internationalization. The questionnaire asked the RBOCs to rank reasons for choosing specific countries. It also addressed their preference for deciding to provide basic or cellular service. It asked the reasons why a firm may enter into a joint venture. Finally, the questionnaire asked the RBOCs how their investment strategies have evolved with the passing of the MFJ. In some cases in which firms did not fill out the questionnaire, telephone interviews were conducted with the RBOCs' representatives. Findings based on the questionnaire survey, in which the firm-specific and location-specific factors are more closely examined, are discussed later.

BACKGROUND: THEORETICAL APPROACHES

Dunning's eclectic approach is used as the theoretical framework in which to explain the internationalization of the RBOCs. Dunning's approach evolved out of his own research involving international production. It is eclectic because it draws upon components of several other theories that have been used to explain internationalization. The role of foreign investment in host countries has always been a controversial issue (Jenkins, 1991).

The neoclassical approach sees capital flow as being based on differential rates of return (Helleiner, 1989). In this approach the host country appears to benefit from the capital inflow by experiencing an increase of capital within the country. Other benefits for the host country include an increase in technological knowledge and an increase in training for labor (Jenkins, 1991). The product cycle theory explained trading patterns between developed and developing countries. The theory states that the developing country takes over the production of a product that has already been developed in an advanced country. The assumption is that as the product becomes more standardized and efficient in its production in the host country, the transnational corporation's (TNC) competitive and comparative advantages may decline. The developing country can introduce an efficient production process, and the product remains competitive in the international market. The theory of industrial organization examines international investment operating in imperfect markets. Hymer discussed that if a firm is operating in a foreign market, it is necessary for the firm to have some advantage (Hymer, 1960). Dicken (1992) mentioned that the industrial organization theory was the first approach to indicate a foreign firm's need to have some type of advantage over local firms in the host country in order to operate in that market. Although indigenous firms know more about the local environment, the foreign firm uses its advantages (e.g., technological know-how and marketing know-how) to gain entry into the market.

Although the theory of industrial organization emphasized the importance of ownership factors, the question as to why firms choose to produce in one country than in another remained unanswered. In explaining foreign direct investment, the scope to internalize its advantages became a deciding factor in location choice. Internalization means that the firm does not sell the "advantage" it has to other firms in foreign markets but enters into foreign investment, which is seen as a way of overcoming an imperfect market (Buckley & Casson, 1976). Dunning intertwined the ideas of firm characteristics, location factors, and internalization into his eclectic approach. Before going in depth about the eclectic approach, it is important to mention the other viewpoints that discuss foreign investment.

As mentioned before, there are varying viewpoints concerning foreign investment. The aforementioned theories have often been viewed as proponents for foreign investment. Other theories that discuss foreign investment tend to have a less than positive view on foreign investment. The Marxist approach or neo-Marxist approach views internationalization as being a deterrent to development in host nations (Jenkins, 1991). The Marxist viewpoint states that firms enter other countries in search of new markets or new ways in which to obtain other resources or inexpensive labor. This approach emphasizes that the TNCs "drain" the surplus resources that exist in the host country through foreign investment (Helleiner, 1989). By using the host country's resources, the firm inhibits the host country from using its natural resources to create its own economic progress. As a result, the host country becomes more dependent on the TNCs for the products they provide (Jenkins, 1991).

Another group of Marxists, the neofundamentalists, have a somewhat more positive view concerning TNCs as being "progressive." The main proponent of this viewpoint sees the entry of the TNC into a country as helping the country develop its resources as well as developing its local capital (Warren, 1973). This view suggests that because many TNCs are competing with one another for markets, the host country has the upper hand in deciding which TNC enters and at what price. Finally, he sees internationalization as a way in which host countries, especially the developing nations, can export goods (Warren, 1973). Jenkins (1991) pointed out that Warren's approach is an extreme one among Marxists.

Eclectic Theory

Eclectic theory specifies three conditions that need to be satisfied in order for foreign investment to occur. These conditions are firm-specific advantages, internalization-specific advantages, and location-specific advantages (Dunning, 1981). The firm-specific advantages are the firm's assets (e.g., knowledge, technology, management skills, capital, and access to resources). Once a firm has these advantages, it is then important for the firm to internalize them. Internalizing the advantages means the firm keeps its assets within itself. The firm keeps its advantages to itself by directly providing its services abroad rather than leasing or selling them to a foreign firm (Dunning, 1981). The internalization factors also give the firm a coordinating advantage that enables it to participate in a joint venture or consortia. If there are large costs in the investment, there is the possibility of different firms coming together and forming a consortium in order to share costs (Dicken, 1986).

The third condition deals with the location-specific factors that influence where firms invest. Location-specific factors (e.g., market size, political stability, and local resources) are the characteristics that pertain to the host country where the investment will be made. Location advantages are important to the firm because they indicate how long a firm can continue to grow in a specific market (Dunning, 1981). Dunning's eclectic approach can be used as a "tool kit" in which to explain foreign investment. In using the approach, Dunning points out that it is not always necessary for all the characteristics to be present. Also, there is no exact combination in which the factors have to exist. Finally, the eclectic approach is not static (Dunning, 1981).

The Costs and Benefits of Foreign Investment

Foreign investment has occurred in many sectors prior to the new investment patterns in telecommunications. This section discusses the potential role of foreign investment in development and its benefits and costs. The exchange of capital internationally has existed for an extensive period of time. The earliest form was merchant capital that involved long-distance trade. Merchant capital was the precursor to the rise in capitalism that occurred in Europe. By the 19th century, internationalization of capital became widespread in Europe (Jenkins, 1991).

During this time, European countries were spreading their investments into developing countries. Helleiner (1989) pointed out that it was the colonial history that existed between the European firms entering the developing countries that has created the controversial view concerning TNCs. Initially, the European countries were involved in trading with developing countries. Eventually, their involvement in the developing countries increased. The firms were involved in production, specifically in mining and agriculture (Helleiner, 1989).

At the beginning of World War I, more than 60% of foreign investment was in developing countries. Fifteen percent of the investment was in manufacturing whereas 55% was in the primary sector. During the interwar period, foreign investment grew in the manufacturing sector in developing countries, especially in Latin America (Jenkins, 1991). In the 25 years that followed the end of World War II, there was a significant increase in international production. The advances that took place in the technology of transportation and communications enabled the rapid increase in foreign investment by enhancing the ability to coordinate long-distance operations.

The 1950s and 1960s saw a continued growth of TNCs in developing countries. In the 1970s, foreign investment continued to increase in developing countries as well as did loans from commercial banks. In the 1980s, the developing countries started to incur balance-of-payment problems. As a result, they began to accept more foreign investment to help them with their rising debt. By allowing foreign investment, developing countries have the ability to continue to receive funding from commercial banks as well as from the World Bank (Helleiner, 1989).

Different types of TNCs have participated in foreign investment. Dicken (1992) discussed four different TNCs: the textile industry, automobile industry, electronics industry, and financial services industry. He points out that the impact of these industries in developing countries is often not the same because of the particularities of each industry. The method of entry, the role the government of the host country plays in terms of regulation, and the actual product being produced by the industry influence the pattern of TNC impact in a host nation. In the clothing and textile industry, Dicken stated that each of the individual clothing producers have varying strategies on how they deal with a country. For the most part, the textile industry tends to arrange subcontracts or leasing arrangements in the countries in which goods are manufactured. This industry also operates where there is low-skilled labor. The automobile industry has been dominated by a few international corporations. In this case, the location of where the auto industry decides to invest depends greatly on the government regulation within each country. Government policy determines the amount of access foreign firms have in the country. In France and Japan, the government prefers importing few foreign cars in order for their domestic producers to have a better chance of selling in the market. However, in the United States and the United Kingdom the policy is complete open access to foreign automakers. As a result, there is a good amount of competition for the domestic producers of cars in these countries (Dicken, 1992).

In the electronics industry (e.g., semiconductors, computers, and telecommunications), Dicken (1992) stated that there is more government intervention because these industries are of strategic importance to the host. Since the electronics industry is competitive and research and development is expensive, firms enter the host country in the form of joint ventures to share costs of operation. Finally, Dicken suggested that due to the strong competition in the electronics industry, firms try to diversify their products as well as relocate themselves in order to maintain profitability.

The service sector (e.g., financial, transportation, communication, and health related) establishes itself in foreign markets either through foreign investment, joint ventures, subcontracting, or licensing. In some cases, the creation of a joint venture enables companies that offer complementary services to become a "transnational service conglomerate" (Dicken, 1992, p. 388). As a conglomerate, they are able to provide many services for their client. Dicken elaborated on the financial service sector (e.g., banking and credit services). He sees that these services are able to enter foreign markets more easily because of less government regulations. He also finds that the deregulation occurring in the financial service sector is related to the deregulation that is happening in telecommunications.

The purpose of discussing other sectors that have participated in foreign investment is to use their experience as a basis for comparison to the telecommunications sector. Also in order to have a better understanding of the possible impact of foreign investment in telecommunications, it is necessary to know what are seen as being the most common benefits and costs in general foreign investment.

Generally, TNCs are seen to benefit the countries they enter because they help create economic development. The TNCs bring access to items (e.g., technology, capital, marketing skills, management skills, and training for labor) that are scarce in the host country (Helleiner, 1989). They often increase employment. Helleiner stated that some of the drawbacks of a TNC is that particular factories can cause pollution, the host country's natural resources are reduced, and the host country's national income may also decrease.

Dicken (1992) discussed the pros and cons of allowing TNCs to operate in a host economy. For example, a country could spend extensive capital to set up basic infrastructure, however, the technology brought in by the TNC may turn out to be inappropriate for the host market. As both Helleiner and Dicken stated, more detailed empirical research is needed on the costs and benefits of foreign investment in order to have a better picture of its actual impact.

FIRM-SPECIFIC DETERMINANTS OF FOREIGN INVESTMENT IN TELECOMMUNICATION SERVICE PROVISION

The RBOCs are Ameritech, Bell Atlantic, BellSouth, NYNEX, Pacific Telesis, Southwestern Bell, and US West. These firms have a long history of providing basic service in the United States. Through their experience, the RBOCs have acquired the knowledge and technology required to operate telephone service to customers in their local areas. The RBOCs' operations have generated a considerable amount of revenue. In 1991, RBOC revenues were as follows: Ameritech ($10.82 billion), Bell Atlantic ($12.28 billion), BellSouth ($14.45 billion), NYNEX ($13.23 billion), Pacific Telesis ($9.90 billion), Southwestern Bell ($9.33 billion), and US West ($10.58 billion) (Coy, Hof, & Ellis, 1992).

The RBOCs have capital and technology, but they are in need of new markets. They are unable to expand in the local areas where they already provide service. Also, they are not allowed to implement any other types of service (e.g., cable television) in their local markets because of the regulations in the MFJ. As of July 16, 1992, the U.S. companies were given the permission to offer video services over their lines. Although the companies are able to provide video dial-tone, they can not provide programming because of the Cable Act of 1984. However, the Federal Communications Commission has allowed the telephone companies to have 5% interest in program services. The actual time and cost to create this network is still not known (Farhi, 1992). In addition all seven RBOCs are currently in the process or are already investing in cable television companies within the United States (Rozansky, 1993). The RBOCs have also tried to diversify domestically in the areas of real estate, computer software, and nonwireline services. The RBOCs have had little success in real estate and in computer software. However, in nonwireline services, which is more of a similar market to basic service, they have had measurable success (Rosenburg et al., 1993).

In order to grow domestically, the RBOCs face many challenges. As a result, they have started to invest internationally. The questionnaire addressed specific firm characteristics that led to internationalization. The main characteristics examined pertaining to the company are management and marketing skills, technology, capital, and expectations for return on investment. The questionnaire looked at the reasons for the RBOCs forming joint ventures in their investment with both other RBOCs, local companies, or local PTTs. Questions were asked to see if the RBOCs preferred offering basic or cellular service. Finally, the questionnaire asked the RBOCs if their investment strategies had changed with the passing of the MFJ.

Each RBOC is discussed on an individual basis. The following section takes a closer look at the major international activities of the RBOCs.

Ameritech

Ameritech is primarily responsible for providing service in the midwestern states. Although Ameritech is seen as being a conservative company, it is the second most profitable among the RBOCs (Ameritech, 1992).

Within Ameritech's organizational structure, Ameritech International handles foreign investments involving cellular service, pay television, and joint account management ventures with PTTs and large telecommunications suppliers (Ameritech, 1992).

Ameritech International's general international investment strategies are that it chooses a telephone company in which it can have an active operating role and where the capabilities to manage it are available (Hammer, 1991). When looking at a foreign venture, the company takes a risk analysis approach that examines potential profits as well as shareholder interests. Ameritech also looks at its competitors' bids to see if a particular bid will have an impact on its own international expansion (McClenahen, 1990). One primary concern in choosing a place to invest is the nation's political and economic stability. In particular, Ameritech examines the existing patterns of investment in the nation, potential economic growth in the nation, and the development of the nation's infrastructure. The company also looks at the length of time for a return on the investment (Bande, Foreword, this volume).

In 1990, New Zealand privatized the Telecom Corporation of New Zealand. The company was bought for $2.4 billion U.S. by a consortium that included Ameritech, Bell Atlantic, and two New Zealand firms, Fay Richwhite Holdings Ltd. and Freightways Holdings Ltd. Bell Atlantic and Ameritech each have shares in the company. When the two RBOCs sold off 31% of the company in July 1991, they made a profit of $147 million (Kupfer, 1991). In 1993, both the companies had to sell off some of their shares so that their shares totaled only 24.95% (Ameritech, 1992).

Initially, when New Zealand's plans for privatization were made, there was considerable opposition from parts of the government and the Council of Trade Unions. They were opposed to having U.S. interests in their country. Despite the opposition, the company was sold. However, no significant changes in terms of U.S. domination in managing the company have been noticed. Furthermore, the U.S. companies are expected to provide intelligent networks, caller ID, and call forwarding to customers sooner because of their expertise with these services (Hyde & Martin, 1990).

With their purchase, both Ameritech and Bell Atlantic have a "physical presence" in the Pacific. The Asia-Pacific region has 50% of the world's population but only 17% of the world's telephones. Overall, in New Zealand, the two RBOCs have a "hands-off attitude" when it comes to managing the company. They play more of an advisory role in deciding what types of service would be beneficial for the company (Hyde & Martin, 1990).

In June 1991, Poland issued the first license for a country-wide cellular system. The license went to Telekommunikacja Polska SA, France Telecom, and Ameritech (Polska Telefonica Komorkowa) to build and operate the system. The Polish PO owns 51%, while France Telecom and Ameritech have a joint venture minority interest of 49%. The cellular service is called Centertel. The building of the infrastructure began in June 1992 in Warsaw and was expected to be completed in 1994 (Ameritech, 1992). According to *Datapro*, Ameritech's investment in Poland is seen as use of the "neighborhood syndrome" strategy. This strategy suggests that companies invest in other countries that have the same ethnic population as is served in the home base of the company (Ameritech, 1992).

Bell Atlantic

Bell Atlantic is the second-largest RBOC and is primarily responsible for providing service to the mid-Atlantic region of the United States. The RBOC also serves the U.S. federal government (Striplin, 1992a). Bell Atlantic targets countries that meet certain criteria for privatization and provides investment opportunities in wireless and pay television in order to ensure long-term growth. The company also uses its consulting service and network expertise to generate short-term growth, as well as to identify longer-term opportunities (Striplin, 1992a). The company wants to invest in areas where they can provide the services they already provide in the United States. The company's goal is for international business to account for 10% of total revenue (Hammer, 1990).

When Bell Atlantic decides to invest internationally, the company looks to enter the venture with a strong partner. Bell Atlantic looks for countries with economic and political stability. The company also seeks countries in which there can be deregulation (Striplin, 1992a). One of the factors that is important to the company is access to the parent company's technological and financial resources. The company is interested in entering host countries where state-of-the-art technology can be introduced. Another factor of great importance is receiving an increased return on assets by investing in the host country (Bell Atlantic, 1993).

Since the passing of the MFJ, Bell Atlantic has become more focused in trying to target its opportunities. The most important factor to Bell Atlantic for entering a joint venture is to lower its risk in the investment. The company prefers entering a partnership with the local PTT as opposed to another telecommunications firm that provides similar services. In terms of providing basic or cellular service, it seems that the factors (e.g., cost, existing infrastructure, technology, and government regulation) are the same for implementing either service. Finally, Bell Atlantic has actively sought out international investments. Bell Atlantic is in partnership with US West in the former Czechoslovakia. The two firms along with the Czechoslovakian Ministry of Posts and Telecommunications have agreed to create a cellular data network, named Eurotel, across both the Slovak and Czech republics (Striplin, 1992a). The company plans to invest $105 million in the former Czechoslovakia over a period of 10 years (1991–2001). As a result of Bell Atlantic's investment, it will have 24.5% ownership in both businesses. A company spokesman states that this investment does not affect the firm's capacity to invest (Hammer, 1990). The initial phase for the cellular service began in September 1991 and the data network began in October 1991. The company hopes to serve 3.45 million customers by 1995, which would account for 22% telephone density (Striplin, 1992a).

In the former Soviet Union, Bell Atlantic is working with US West, the Soviet Ministry of Posts and Telecommunications, Millicom, Inc., and four Soviet partners to operate the Moscow cellular system. The cellular system will function at the Nordic standard of 450 MHz. Bell Atlantic also plans to develop a long-distance gateway with the government of St. Petersburg (Striplin, 1992a).

In Argentina, Bell Atlantic International had made a bid to operate the northern half of Argentina's state-owned company; however, the deal failed because of a lack of funding. In Brazil, the company is in the bidding process to obtain a second cellular license being offered by the government (Striplin, 1992a).

Another of Bell Atlantic's major purchases was its investment in New Zealand with Ameritech (refer to Ameritech section). The company decided to invest in New Zealand in order to experience operating in a deregulated environment. It seems that the company chooses to invest in countries that are either already doing well (e.g., New Zealand) or are developing countries (e.g., Eastern European countries). The company plans to continue investing in ventures that enable it to have partnerships with the local entities (Striplin, 1992a).

BellSouth Corporation

BellSouth Corporation is the fifth-largest telecommunications operating firm in the world. This RBOC is responsible for providing services for the southeastern United States (Striplin, 1992b). BellSouth International makes the decisions pertaining to international investments for the company. The company has a three-pronged approach to choosing its foreign investments. The first part is that the company searches for the areas of highest growth in the world. The company has found that the high-growth areas are in Europe, Latin America, and Asia-Pacific region. The company seeks countries where markets are opening up to foreign investment. After locating these areas, BellSouth determines the key markets for opportunity. Each market is examined on a country by country basis. The risks and the opportunities are weighed in each investment.

The company considers such factors as stability of the political and economic environment, the regulatory environment, and long-term potential for growth in the country. The company also examines other factors: the country's need for communications, the technology and technological know-how within the country, and the culture of the country (Schnabel, 1993). Once the location is chosen, the next part involves finding the right partner with whom to work. The selection of a partner is dependent on the location. BellSouth wants to work with well-positioned companies who share Bell Atlantic's philosophy of entering markets of long-term growth. They want to be involved with a company that complements the services it provides (Schnabel, 1993).

The final part is the type of service the company prefers to provide. Schnabel (1993) states that BellSouth's focus is implementing wireless service because it is the direction of the future, especially for long-term growth. BellSouth prefers wireless because it is easier than dealing with wired service. With wired service, the company must involve itself with the existing network infrastructure, which is usually crowded and congested. Even though BellSouth prefers providing wireless service, that is not the only area in which the company focuses. In order to increase the company's assets as well as those for the shareholders, the company invests internationally (Schnabel, 1993).

In Latin America, the company has brought cellular service to five countries. BellSouth International entered these markets by forming joint ventures with other private companies and in some cases with the local government. The consortium Compania de Radiocomunicaciones Moville SA (CRM) is located in Buenos Aires, Argentina. The consortium is comprised of BellSouth, which is a managing partner; Motorola; Citibank; and two Argentine companies, SOCMA and BGH. All of these companies together bid to build and operate the first private cellular business in South America. They have invested approximately $220 million U.S. to create a network to serve 320,000 customers (Striplin, 1992b).

In 1991, Chile awarded a license to the consortium, Cidcom, of which BellSouth is a part, to provide cellular service. In Guadalajara, Mexico, cellular service was functioning by August 1990. BellSouth belongs to the consortium Communicaciones Celulares de Occidente SA de CV, also known as Comcel. Comcel is responsible for providing service to the western part of Mexico. The consortium has been given a 20-year license and is expected to serve 135,000 customers by 2000. In December 1990 in Uruguay, Abiatar, the consortium that BellSouth leads, received the approval to create and operate a cellular service. In November 1991 cellular service began in Montevideo, and Maldonado/Punta del Este (a resort area) received its services in December 1991. In the same year, the cellular consortium in Venezuela began service. TelCel Communications SA was led by BellSouth to operate and develop the network. In order to serve 20 million people, the consortium plans to invest $100 million U.S. (Striplin, 1992b).

Southwestern Bell

Southwestern Bell is the fifth-largest RBOC. The company provides service to Texas, Oklahoma, Kansas, Arkansas, and Missouri (Southwestern Bell, 1992). Southwestern Bell has always had a steady source of revenue from its advertising and publishing activities. According to *Datapro*, the company has adopted a multitier approach in its marketing. This way the company can achieve gradual growth returns through royalty payments on relatively low capital investments (Southwestern Bell, 1992). According to the *Datapro* report on Southwestern Bell, it seems the company is trying to strengthen its regional hold, which explains the company's investment in Mexico. According to Southwestern Bell's Annual Report, the reasons for choosing to invest in Mexico were the favorable business climate, the opportunity to work with a good local partner in Carlos Slim of Grupo Carso, the potential for growth, and the economic ties between the United States and Mexico. The company has close ties to Mexico since Texas, one of the states Southwestern Bell serves, shares the border with Mexico. Also, approximately 20% of Southwestern Bell's employees are Hispanic (Southwestern Bell Annual Report, 1991). *Datapro* sees this as another example of the neighbor syndrome whereby a country in physical proximity may be the recipient of large amounts of foreign investment from its neighbors provided there is political stability and scope for development (Southwestern Bell, 1992).

In December 1990, the Mexican government privatized the national telephone company, Telefonos de Mexico (Telmex). A consortium consisting of Southwestern Bell International Holdings (SBIH), France Telecom, and Grupo Carso purchased 51% of the voting shares of the company. By the end of 1991, the initial investment has doubled and is now worth $2.5 billion U.S. In order to have an internationally competitive telecommunications company, there is a need for continual funding. Southwestern Bell plans to invest $9 billion U.S. for improvements in the network through 1995 (Southwestern Bell, 1992). In 1991, Telmex created more than 670,000 access lines. This represents an overall increase of 12.5% in the number of lines being provided, which is 32% more than in the previous year. Telephone service has increased to 2,000 Mexican villages. The number of cellular customers has doubled to 70,000. The company wanted the current 5.4 million telephone lines to increase to 7.5 million lines. Essentially, Southwestern Bell is assisting Telmex by using technologies and business procedures with which they are already familiar (Southwestern Bell, 1992).

NYNEX

NYNEX serves the states of New York, Connecticut, Rhode Island, Massachusetts, New Hampshire, Vermont, and Maine. Based on the number of customers it serves (approximately 15 million), it is the second-largest RBOC (NYNEX Corporation, 1991). For the most part, NYNEX's overseas strategy is conservative (NYNEX, 1992). According to a company representative NYNEX is capable of providing either basic or cellular service, although another article states that the company is not interested in paying $2,000 a line (NYNEX Corporation, 1992). The company is not interested in the privatization of wirelines in Latin America because landline would require much larger investments than wireless; the company would prefer making investments where it actually has the opportunity to build the system. It foresees such opportunities in Southeast Asia.

NYNEX prefers entering joint ventures with the local PTT (NYNEX Corporation, 1992). The company enters a joint venture because of the foreign ownership restrictions in the license requirement and to reduce the risk of investment. NYNEX has found foreign investment attractive because since the MFJ, there is considerable domestic competition. Along with the competition, the company has experienced an economic downturn. NYNEX tends to provide the service internationally that the company already provides domestically. Currently, the company is building a system in Bangkok, Thailand, serving as a minority partner. In Eastern Europe the company is trying to provide database businesses. Also in Prague, Czech Republic, NYNEX is providing Yellow Pages (NYNEX Corporation, 1992). In Indonesia a grant from the World Bank has enabled Indonesia's Perumtel, the country's telephone company, to receive training from NYNEX on network expansion (NYNEX Corporation, 1991).

Pacific Telesis

Pacific Telesis is the RBOC responsible for providing service to 14.3 million people in California and Nevada (Striplin, 1992c). In comparison to the other RBOCs, Pacific Telesis has not been as aggressive in pursuing international investments. The company has not been as active as other RBOCs internationally because it is trying to first increase its domestic revenue. The company hopes to gain revenues with such ventures as home entertainment through a cable network and information services (Striplin, 1992c).

The factors that are most important to Pacific Telesis for international investment are the firm's financial management, marketing know-how, and ability to offer state-of-the-art technology. These same factors enabled the firm to invest in South Korea as well as in Thailand. Additionally, Pacific Telesis invested internationally in order to be first in its target markets as well as to increase its assets (Kirk, 1993).

Pacific Telesis does not provide basic service outside the United States. The company would rather provide cellular service because it is the newer technology and other telecommunications firms also offer cellular services (Kirk, 1993). In 1984, the year the MFJ was passed, Pacific International created Pacific Telesis International. The company pursued telecommunication opportunities from 1984 to 1986, but they had little success. In 1987, the company restructured its strategy and began focusing on providing wireless service in specific countries. In 1987, Pacific Telesis placed a value-added network in South Korea. The company is currently in the process of trying to obtain the license to provide cellular service. In Thailand, Pacific Telesis is part of the consortium PerCom Services Limited that provides services in national paging. The company is also involved in another paging service called Pacific Telesis Engineering that provides service to the city of Bangkok (Striplin, 1992c).

US West

US West is the largest RBOC in geographic terms. It provides service for about 35% of the United States (US West, 1992). In its international ventures, US West prefers being the minority partner and likes a strong local partner because the local entity would have a better understanding of economic possibilities as well as an understanding of the political environment (McClenahen, 1990). US West also enters a joint venture to lessen the risk of investment. Since the company wants to expand internationally, it plans to use $600 million in international investments by 1995. The goal of the company is to increase investment in international communication activities, even if some of those investments may limit short-term earnings (US West, 1992). *Datapro* sees US West's strength is its ability to deal with large countries based on the fact that US West has successfully dealt with providing service for a large area. The company provides service for fourteen U.S. states. The company's plan is to maintain minority ownership interests in international ventures (US West, 1992). Therefore, a factor of most importance to the company in international investments has been the host country's regulatory environment. The company is also interested in ventures that affect its own long-term growth, such as an increase in its assets.

In 1987, the company began its international activities. US West has two international subsidiaries. US West International looks for investment opportunities for the company, while the subsidiary, Global Alliance, was created to deal with its business partners such as France Telecom (US West, 1992). In Lithuania, US West International is involved in a joint venture with Kaunsa Enterprise of Lithuanian PTT Communications Ministry (KPPT) to develop and manage an international data-switched network. The company has been given an exclusive 15-year license. Initially, the company invested $2 million for 49% interest that will fund 120 voice circuits. This joint venture is seen as a way to increase economic development because contact between individuals in Lithuania and the rest of the world can increase business opportunities. This particular venture is expected to create a model business structure for Lithuania (US West, 1992).

In October 1990, Hungary had one of the first national cellular ventures in Eastern and Central Europe. The company WESTEL Radiotelfon is a joint venture between US West and Magyar Post (Hungarian Postal Telegraph and Telecommunications). It serves 15,000 subscribers in Budapest, and US West invested $5 million for 49% ownership (US West, 1992).

In the former Czechoslovakia, US West is in partnership with Bell Atlantic and the Czech and Slovak Posts and Telecommunications administration. The participants have been working together to build a public packet data network since 1991. US West invested $5.8 million for 25% ownership. Also in the former Czechoslovakia there is cellular service in Prague, Bratislava, and Brno. The national cellular network was built by a joint venture between US West and Bell Atlantic. The name of the company is Eurotel Cellular Service, which had planned to serve a population of 15.7 million by 1991. US West invested $20.1 million for a share of 24.5% (US West, 1992).

US West is aiding in cellular service in St. Petersburg and Moscow. In St. Petersburg, the St. Petersburg Cellular Network, is a joint venture between US West, St. Petersburg City Telephone Network Production Association, and St. Petersburg Station Technical Radio Control. US West has 40% ownership. The system began operating in September 1991, providing service to 750 subscribers. The Moscow cellular system was expected to start construction by 1992. US West is responsible for managing the development of the cellular networks. The members of the consortium are the Russian Ministry of Posts and Telecommunications; Millicom, Inc.; and four Russian partners (US West, 1992).

The country is planning to develop three international gateway telephone switching systems in each of the cities of Moscow, St. Petersburg, and Kiev. The agreement is being discussed between US West International and the Russian Ministry of Posts and Telecommunications. The two organizations are also working on plans for the Trans Russian Fiber Optic project. The project will be the longest fiber-optic line with an estimated cost of $500 million (US West, 1992).

Overall Attributes of the RBOCs. In Dunning's eclectic approach, the firm-specific characteristics that are most important are the size of the firm, capital, technology, and knowledge. The RBOCs are definitely large firms that possess an extensive amount of capital. Since the MFJ was passed, the RBOCs have had control over U.S. local service. The ownership of local lines gives the RBOCs a domestic monopoly ("High court lets," 1993). The RBOCs' monopolistic advantages have given the companies the opportunity to make investments through their subsidiaries. Brown and Crockett (1990) state that since the RBOCs make investments through their different subsidiaries, it is often difficult to know how costs are being channeled domestically or internationally. The RBOCs' corporate structure consisting of regulated and unregulated subsidiaries has enabled them to generate high revenue (Rosenberg et al., 1993).

In general, the RBOCs provide technological services internationally that they already provide well in the United States. They are providing basic telephone service because that is the area in which they have the most experience (McClenahen, 1990). However, it is important to note that the RBOCs are also providing services that they are unable to provide in the United States in order to gain experience. They hope that when regulations are lifted in the United States they will already have the experience needed in order to enter the market (e.g., cable television) (Hyde & Martin, 1990).

Firm-Specific Advantages. In examining the firms, it is apparent that, for the most part, they all are seeking opportunities for international investment in order to increase their company's assets. The MFJ's restrictions have limited their chances to continue to grow domestically. Since they have revenue, they are using their capital abroad.

There were certain attributes that were identified by the firms as their advantages for investment. One of the characteristics that was cited as being important was the financial and technological services the firms possessed. The most important factor appears to be the state-of-the-art technology the RBOC is able to introduce into the host country. Table 4.1 indicates the firm-specific factors that enable the internationalization process of the RBOCs. The firms listed in Table 4.1 are those from which data were collected using a questionnaire survey. The motivational factors are the firm-specific advantages the RBOCs feel they possess to make international investments.

Services Offered. The significance of providing basic or cellular service is evident in that these services are requested in the host country's licenses. For the most part, the RBOCs provide both basic and cellular

TABLE 4.1
Motivational Factors for International Investment

RBOCs Factors	Bell Atlantic	BellSouth	NYNEX	Pacific Telesis	US West
Managerial know-how			X		
Marketing know-how	X		X	X	X
Technology	X	X	X	X	
Price of service	X		X	X	
Parent company resources	X		X	X	
First in target market	X	X	X	X	
Increase in assets	X		X	X	X
Increased rate of growth from investment			X		
Service quality	X		X		

service according to the license. However, more of the RBOCs prefer providing cellular service because for them it is easier to implement.

Modes of International Participation. It is evident that in many of these investments the RBOCs enter the host country in the mode of a joint venture. According to Dunning's eclectic approach, the internalization advantages the firm has enables it to have a coordinating advantage. This advantage allows the firm to work with those opportunities that can benefit it the most. The RBOCs enter the joint venture in order to reduce their risk in investment. The choice with whom to be in partnership (e.g., another telecom firm, a local firm in the host country, or a local PTT in the host country) depends on the RBOC. In most cases, the RBOCs prefer being in partnership with some local entity because the local firm has a sound understanding of the political, economic, and cultural environment. Also foreign ownership restrictions make the firms participate in joint ventures. Table 4.2 illustrates the reasons why the RBOCs enter joint ventures.

All seven of the RBOCs are involved in international investments in different parts of the world. The firm-specific characteristics they possess are capital, technology, and knowledge that allow them to internationalize. The companies are providing both basic and cellular services, but some RBOCs (e.g., BellSouth and Pacific Telesis) prefer providing cellular service since it is easier for them to implement. As a result of foreign ownership restrictions in country licenses, many of the RBOCs are entering international investments in the mode of joint ventures. They also participate in joint ventures to lessen the risk of their own investment. The RBOCs, for the most part, prefer entering joint ventures with local firms or the local PTT because the local entity has a better understanding of the local environment. In order to understand why the RBOCs invest in certain parts of the world, it is necessary to examine location-specific

TABLE 4.2
Motivational Factors for Entering Joint Ventures

RBOCs Factors	Bell Atlantic	BellSouth	NYNEX	Pacific Telesis	US West
Lower risk of investment	X		X	X	X
Partnership with other telecommunication firm		X			
Partnership with firm elsewhere					
Partnership with local entity		X	X	X	X
Partnership with local PTT	X	X	X	X	X
Share overhead costs	X		X		

advantages. The next section includes a discussion of the role of location-specific factors in encouraging foreign investment and the potential impact these investments may have on the location.

LOCATION-SPECIFIC DETERMINANTS OF FOREIGN INVESTMENT IN TELECOMMUNICATION SERVICE PROVISION

In the previous section, the discussion focused on the firm-specific characteristics that enabled the RBOCs to invest abroad. In looking at their major investments, it is interesting to note that the RBOCs seemed to have sectioned off certain portions of the world among themselves. This section discusses the location-specific factors that attract firms for investment. After discussing the attributes of the location, this section uses specific examples from Asia, Eastern Europe, and Latin America to illustrate the internationalization process.

In Dunning's eclectic approach, the location-specific advantages are the market size, government regulation, natural resources, the infrastructure, political stability, and economic stability. In order to have a clear picture of what each location offers, the next section discusses cases in specific countries in order to illustrate the issues. In examining the RBOCs, it is evident that both political and economic stability are important criteria for investment. Obviously, the role the host government plays has an influence on international investment. The current trend of privatization has enabled companies to enter markets that were previously closed to foreign investors (Hammer, 1990).

The trend in privatization, especially in developing countries, has occurred for various reasons. The state has been providing telecommunication services in developing countries. Because telecommunications has functioned as a state monopoly, there has been little regulation concerning telecommunications in these countries (Mody, Tsui, & McCormick, 1993). In developing countries, there has been a lack of investment in the telecommunications infrastructure because higher priority was given to other economic sectors (e.g., electricity and transportation) (Saunders, Warford, & Wellenius, 1983). As a result, the telecommunications infrastructure that was established had outdated equipment and provided poor service. The growth of telecommunications had occurred more in urban areas than in rural areas because the demand for communication was greater in urban areas where there are businesses and elite classes. In the urban areas, the high demand for basic service has resulted in the overuse of the existing equipment. Also, the government, in most cases, has been unable to satisfy the demands for basic service in rural areas as well as provide enhanced services in urban areas (Wellenius & Stern, 1989).

The possibility of investment leading to economic growth was discussed in the Maitland Commission's report in 1984. The commission examined the link between telecommunications and economic growth. It seems that countries with strong economies also had higher telephone penetration for their respective population (Bruce, Cunard, & Director, 1988). Because telecommunications is a sound investment for the enhancement of the economy, most developing countries started looking for funding for both basic and enhanced services. With globalization, there is a perceived need to be able to compete, which also calls for improved telecommunications. However, in order to invest in telecommunications, these countries needed funds that most of them lacked due to a rising debt. As a result, many of these countries turned elsewhere to obtain the financial resources they needed to improve the telecommunications infrastructure. Organizations like the World Bank are willing to provide loans; however, they want the countries to restructure their telecommunications sector. The World Bank feels restructuring would make the telecommunications infrastructure operate more efficiently. Essentially, the Bank recommends that the government should reduce its control on the telecommunications sector and allow it to be more autonomous and commercial. The Bank has also encouraged private investors to enter these markets (Wellenius, 1989).

As mentioned, most of the developing countries have a rising debt problem; this is one of the main reasons that many developing countries decide to liberalize certain sectors that were under government supervision. In Mexico, the state-owned telecommunications (Telmex) was privatized in 1989. A consortium consisting of Grupo Carso, Southwestern Bell, and France Telecom purchased 20.4% of Telmex (Barrera, chapter 6, this volume). In February 1992, Telmex made a net profit of $2.26 billion U.S. for fiscal year 1991 (McCarthy, 1993). In the process of privatization, it is necessary to note that the government does decide the extent to which a foreign firm participates in the telecommunications sector.

Country-Specific Examples

Countries in the Asian-Pacific Region. NYNEX, Pacific Telesis, Ameritech, and Bell Atlantic are the RBOCs with the most significant investments in the Asian-Pacific Region. The firms have invested in the countries that are primarily opening up their markets. Thailand is a country where the opportunities for economic growth are hindered by the poor telecommunications infrastructure. The country's two telephone operators, Telephone Organization of Thailand (TOT) and Communications Authority of Thailand (CAT), lack sufficient funds to improve their networks. Consequently, TOT and CAT turned to private investors to improve their infrastructure. The country has chosen to participate in some build, operate, and transfer deals. In this case, foreign firms build and operate the infrastructure and later transfer the network to the host country ("Thailand: The commercial regulatory environment," 1993). Currently, NYNEX is building a landline network system in Thailand ("NYNEX embarks on a new road," 1992). In Thailand, the country opened its markets and allowed firms the opportunity to build and operate the infrastructure.

Latin American Countries. In Latin America, BellSouth is the leader among the RBOCs, with the majority of international investment. The other primary investor is Southwestern Bell in Mexico. Mexico is another country in which the telecommunications sector was privatized in order to deal with the country's rising debt of $105 billion. The country only had 4.9 telephone lines per 100 population ("Telecommunications in South and Central America," 1990). The consortium's (including Southwestern Bell, France Telecom, and Grupo Carso) initial investment has doubled and is worth $2.5 billion U.S. In 1991 the number of telephone lines in Mexico increased by 12.5% (Southwestern Bell, 1992).

BellSouth International has invested in cellular service in five Latin American countries. Edwards (1993) states that cellular service is an exception because the PTT does not have the resources to provide the service for which there is definite demand. As a result, BellSouth is able to invest in Argentina, Chile, Mexico, Uruguay, and Venezuela because their governments have deregulated cellular operations (Liscio, 1990).

Eastern European Countries. In Eastern Europe, US West is the leader among the RBOCs in providing service. However, Ameritech and Bell Atlantic are providing services in Poland and in the former Soviet Union and the former Czechoslovakia, respectively. With the collapse of communism in this region, there has been a gradual development of a market economy, and firms wish to invest in Eastern Europe in order to attain a market advantage in this region (Williamson, Titch, & Purton, 1992; Lees, 1993).

Lees (1993) states that because the telecommunications infrastructure is in such need of improvement, it is important to ensure there is not uneven development. In Eastern Europe, countries' PTTs are entering joint ventures with private investors especially in providing cellular service. The Eastern European countries seem to be opting for cellular because it is cheaper and easier to implement (Williamson et al., 1992). Consequently, US West has also invested in cellular service in Hungary, the former Soviet Union, and the former Czechoslovakia.

In looking at where the RBOCs are investing, it is also interesting to note where they are not investing. The RBOCs are not in Africa because countries in Africa lack the location-specific characteristics (e.g., economic stability, political stability, and potential market size) that attract telecommunications firms. Although organizations state that privatization of telecommunications will allow foreign investors to come in, private investment will not occur in these countries because the profit margins will be limited.

Overall, the different investments reflect that one of the most important characteristics is government regulation. The host government influences where the firms can invest and how much they can invest. Political as well as economic stability play key roles in internationalization. Firms definitely seek joint ventures where there is potential to grow. Finally, the RBOCs seem to be attracted to locations where they have the opportunity to build and operate the service for some period of time.

Possible Impacts of Foreign Investment in Telecommunication Service Provision

In discussing these international investments, it is difficult to actually determine their impact in host countries because in some cases the contracts have recently been drawn or the services have just started operating. One country where the telephone lines have increased and profits for Southwestern Bell have gone up is Mexico. As it has been stated before, there was an increase of 12.5% in the number of lines being provided (Southwestern Bell, 1992). US West's participation in Westel, the Hungarian cellular provider, has 11,000 subscribers (Williamson et al., 1992). This suggests that increase in services available for individuals will definitely promote development through increased business contacts.

Within the residential and business sectors and the rural and urban sectors, there is going to be more growth in the business sector and in urban areas as opposed to the residential sector and rural areas. In Bell Atlantic's and NYNEX's responses to the questionnaires, these companies saw growth in these areas because telecommunication services know there is definite demand for services in the business sector and in urban areas. Edwards (personal communication, October 29, 1993) reiterated that development will initially occur in urban areas and in business sectors because of high demand. Eventually, he sees telecommunications services diffusing to residential use and to rural areas.

In terms of the development of basic service and cellular service, the RBOCs are willing to provide either basic or cellular service depending on the license they are given. Although basic service is provided, it is more likely that cellular service will be provided because it is cheaper and easier to implement. Finally, the extensive role the RBOC can play in any country is dependent on the regulation in the host country. It is important for the host country to make sure it stays in control to prevent domination from a TNC. In most of the international investment cases discussed, the RBOCs are most often partners with the local PTT. Furthermore, they are in many markets for only a specific amount of time in order to provide the service that should benefit the host country.

As for the RBOCs, their international investments will certainly increase their company's corporate profits. Ameritech received a profit of $147 million when it sold 31% of New Zealand Telecom (Kupfer, 1991) and Southwestern Bell's investment in Mexico doubled to a value of $2.5 billion (Southwestern Bell, 1992). Although all companies admit that they invest internationally to increase their assets, no RBOC states how much their corporate profits should increase. Table 4.3 illustrates the characteristics of the locations that attract the RBOCs. It is based on the data

TABLE 4.3
Locational Characteristics That Attract International Investment

RBOCs Factors	Bell Atlantic	BellSouth	NYNEX	Pacific Telesis	US West
Potential for growth		X	X	X	X
Potential customer base			X		X
Build infrastructure	X		X	X	
Operate infrastructure	X		X	X	
Industrial infrastructure	X				X
Macroeconomic infrastructure					
Political risk		X	X		X
Stability of currency		X	X		X
Length of time on ROI*	X		X		
Existing international development			X		

*ROI = return on investment.

collected through the research. In looking at location-specific factors, the opening of the telecommunications sector to private investment is the most important one to the RBOCs.

CONCLUSIONS

The purpose of this chapter was to understand the internationalization process of the RBOCs, that is, their specific location patterns. In order to explain the internalization process, Dunning's eclectic approach was used as a framework. Dunning's approach examined the firm-specific, internalization, and location-specific advantages. The chapter has provided a comparative picture of the investment strategies of the RBOCs.

According to Dunning's eclectic approach, the RBOCs possess the firm-specific characteristics that are necessary for foreign investment. The RBOCs are large firms possessing extensive reserves of capital. The RBOCs are knowledgeable in the technology that is necessary for providing basic and cellular networks (Dunning, 1981). According to Dunning, when a company internalizes its advantages, it keeps the advantage to itself. The firm does not sell its advantage to foreign firms, instead the firm establishes itself abroad. The internalization advantage also gives the company a coordinating advantage to become part of a consortium. Consequently, in many of the international investments discussed, a considerable amount of joint ventures has been undertaken.

The RBOCs enter joint ventures in order to reduce the risk of investment. The RBOCs also like to be in partnerships with the local entities because they have a firm understanding of the cultural, political, and economic environments. Finally, the RBOCs are often in joint ventures because the host country's licensing agreements have foreign ownership restrictions (Piche, 1993). Dunning's approach to internationalization also considers the location-specific attributes attracting foreign investors. In examining these investments, it is evident that location plays a key role. The RBOCs invest where there is definite potential for growth. They prefer going to places where they have the ability to build and operate the service. In accordance with the eclectic theory, size of market and political and economic stability are also important factors to the seven RBOCs. The most important location-specific factor to all RBOCs is government regulation. Because the countries are opening their markets for private investment, the RBOCs are able to make their international investments.

Because the RBOCs are attracted to locations that provide potential for growth through political and economic stability, it is most likely that investments will continue in the countries of Latin America and Eastern Europe and in certain countries in the Asia-Pacific region. However, foreign investment may not occur in areas like Africa since they lack location-specific characteristics.

In studying the internationalization process, it is necessary to understand why firms choose certain locations for investment. In order to do this information was gathered from trade journals. A questionnaire was also prepared for each of the seven RBOCs to complete. The trade journals provided general information about the RBOCs' current investments. Unfortunately, only three out of seven questionnaires were answered, and two companies allowed brief telephone interviews. The RBOCs with fewer international investments, such as NYNEX, were more willing to participate than companies with high international investment, such as Ameritech. Finally, it was also difficult to obtain any actual dollar figures of investment because the companies considered that proprietary information.

In examining the internationalization process, it is evident that the RBOCs are actively pursuing opportunities to invest abroad. In their investments they are willing to join in partnerships with other telecommunication firms as well as local entities in the host country. The RBOCs are willing to provide both basic and cellular services. Some of the RBOCs, such as BellSouth and NYNEX, prefer providing cellular service because it is easier and cheaper to implement. It is apparent that the RBOCs will continue to invest abroad even if regulations from the MFJ are lifted because these investments are expected to increase their profit margins. The RBOCs' investments are expected to improve the telecommunications infrastructure in the host countries that can then approach foreign investors in other sectors. Restrictions from the MFJ and the competitive global market have led to the RBOCs' search for new markets. Furthermore, the RBOCs' possession of capital and technology enables them to enter deregulating markets in developing countries to provide basic and cellular service. Future research needs to examine how these international investments have impacted the process of development in the host countries.

REFERENCES

Ameritech. (1992, September). *Datapro Reports on International Telecommunications*, pp. 101–108.

Bell Atlantic (1993). *Datapro Reports on International Telecommunications*, pp. 101–106.

Brown, B., & Crockett, B. (1990, June 25). RBHCs' foreign investments rankle U.S. network users. *Network World*, pp. 1, 58.

Bruce, R., Cunard, J. P., & Director, M. D. (1988). *The telecom mosaic: Assembling the new international structure*. London: Butterworth.

Buckley, P., & Casson, M. (1976). *The future of the multinational enterprise*. Bassingstoke & London: Macmillan.

Coy, P., Hof, R. D., & Ellis, J. E. (1992, October 5). The Baby Bells' painful adolescence. *Business Week*, pp. 124–134.

Dicken, P. (1986). *Global shift: Industrial change in a turbulent world.* London: Harper & Row.

Dicken, P. (1992). *Global shift: The internationalization of economic activity.* New York: Guilford Press.

Dunning, J. H. (1981). *International production and the multinational enterprise.* London: George Allen & Unwin Ltd.

Farhi, P. (1992, July 17). FCC clears phone firms to offer video services. *The Washington Post*, p. B1.

Franko, L. G. (1991). New forms of investment in developing countries by U.S. companies: A five industry comparison. In J. N. Sheth & A. Eshgi (Eds.), *Global microeconomic perspectives* (pp. 83–115). Cincinnati, OH: South-Western Publishing.

Griffiths, S. (1992, November). AT&T. *Datapro Reports on International Telecommunications*, pp. 101–111.

Hammer, D. (1990, August 1). Bell Atlantic's global strategy moves ahead with another win. *Telephone Engineer & Management*, pp. 24–31.

Helleiner, G. K. (1989). Transnational corporations and direct foreign investment. In H. Chenery & T. N. Srinivasan (Eds.), *Handbook of development economics* (pp. 1441–1489). Amsterdam: Elsevier Science Publishers.

High court lets 'Baby Bells' branch out despite protests. (1993, November 16). *The Charleston Post and Courier*, p. B5.

Hyde, V., & Martin, J. (1990, October 1). U.S. firms may speed telecom services. *Datamation*, pp. 104(7)–104(12).

Hymer, S. H. (1960). *The international operations of national firms: A study of direct foreign investment.* Cambridge, MA: MIT Press.

Jenkins, R. (1991). *Transnational corporations and uneven development: The internationalization of capital and the third world.* London & New York: Routledge.

Kupfer, A. (1991, November 4). Ma Bell and seven babies go global. *Fortune*, pp. 118–128.

Lees, C. (1993, January). Telecommunications in central and eastern Europe. *Datapro Reports on International Telecommunications*, pp. 121–131.

Liscio, J. (1990, December 10). "Latin America, Si!": Go south, an analyst urges U.S. investors. *Barron's*, pp. 18–19.

Martin, J. (1990, October 1). The telecom bargain hunting continues. *Datamation*, p. 104(8).

McCarthy, C. (1993, January). Mexico: The commercial and regulatory environment. *Datapro Reports in International Telecommunications*, pp. 251–258.

McClenahen, J. S. (1990, September). The bells are ringing: And overseas, they're answering. *Industry Week*, pp. 56–59.

Mody, B., Tsui, L-S., & McCormick, P. (1993). Telecommunication privatization in the periphery: Adjusting the private–public balance. *International Review of Comparative Public Policy, 5*, 257–274.

NYNEX Corporation. (1991, November). *Datapro Reports on International Telecommunications*, pp. 101–107.

NYNEX embarks on a new road. (1992, April 27). *Telephony*, pp. 33–40.

Rosenberg, E. A., Burrows, J. D., Hunt, C. E., Samarajiva, R., & Pollard, W. P. (1993). *Regional telephone holding companies: Structures, affiliate transactions, and regulatory options.* Ohio State University: The National Regulatory Research Institute.

Rozansky, M. L. (1993, October 17). Planned merger to affect millions. *The Post and Courier*, pp. 1E & 3E.

Saunders, R. J., Warford, J. J., & Wellenius, B. (1983). *Telecommunications and economic development.* Baltimore and London: The Johns Hopkins University Press.

Southwestern Bell Annual Report 1991.

Southwestern Bell. (1992, September). *Datapro Reports on International Telecommunications*, pp. 101–108.

Striplin, P. (1992a, March). Bell Atlantic. *Datapro Reports on International Telecommunications*, pp. 101–108.

Striplin, P. (1992b, May). BellSouth Corp. *Datapro Reports on International Telecommunications*, pp. 101–110.

Striplin, P. (1992c, May). Pacific Telesis. *Datapro Reports on International Telecommunications*, pp. 101–105.

Telecommunications in South and Central America. (1990, July). *Datapro Reports on International Telecommunications*, pp. 101–110.

Telecommunications in the South Pacific. (1991, November). *Datapro Reports on International Telecommunications*, pp. 101–106.

Thailand: The commercial and regulatory environment. (1993, January). *Datapro Reports on International Telecommunications*, pp. 101–105.

US West. (1992, January). *Datapro Reports on International Telecommunications*, pp. 101–108.

Warren, B. (1973). Imperialism and capitalist industrialization. *New Left Review*, September-October, 3–44.

Warren, B. (1980). *Imperialism: Pioneer of capitalism*. London: Verso.

Wellenius, B. (1989). Beginnings of sector reform in the developing world. In B. Wellenius, P. A. Stern, T. E. Nulty, & R. D. Stern (Eds.), *Restructuring and managing the telecommunications sector* (pp. 89–98). Washington, DC: The World Bank.

Wellenius, B., & Stern, P. A. (1989). Structural change in telecommunications. In B. Wellenius, P. A. Stern, T. E. Nulty, & R. D. Stern (Eds.), *Restructuring and managing the telecommunications sector* (pp. 1–6). Washington, DC: The World Bank.

Williamson, J., Titch, S., & Purton, P. (1992, July 6). The curtain rises on telecommunications in eastern Europe. *Telephony*, pp. 27–32.

5

Infrastructure for Global Financial Integration: The Role of the World Bank

Gwen Urey
California State Polytechnic University, Pomona

This chapter examines the role of the World Bank in creating the links, outlined theoretically in chapter 3, between new investment in telecommunications and global financial integration in developing countries. The chapter focuses on telecommunications and on the construction of an urgent need for accelerated investment in the telephone infrastructure of developing countries. Development planners, including those at the World Bank, long couched the need for more telecommunications capacity in a familiar language of prerequisites for growth in industry, commerce, services, and other sectors. Since the late 1980s, however, led by U.S. officials and private telephone company interests, the promotion of investment in telecommunications has become increasingly draped in the language of privatization and liberalization. Talk is cheap, and, in fact, there has been much more talk about privatization than there has been wholesale privatization of telephone service supply. Nonetheless, over the last half decade, World Bank time and money have become effective forms of leverage for privatization.

The privatization of telecommunications infrastructure represents an aspect of global financial integration because privatization often involves international joint ventures and foreign direct investment. Furthermore, international service, services for whom business users and the financial community are the primary clientele, and increasing the profitability of the sector have been the emphases of much of the restructuring and growth, and these too are aspects of global financial integration. The privatization of telecommunications in developing countries, often lever-aged by the World Bank, deepens the intersection between developing country telecommunications and global capitalism. Through its telecom-munications work, the Bank engages national governments and telephone administrations in the larger project of global financial integration. This project has little direct connection to the communications and informa-tion-circulation needs of the majority of the world's population. The Bank could address these needs more directly, but it has chosen a different path. This chapter explores that choice.

The World Bank wields enormous influence in the restructuring of telecommunications in developing countries. It has directly influenced structural change in the sector through its $6 billion of telecommunica-tions lending, as well as its power to persuade governments to restructure telecommunications as part of larger economic restructuring or adjust-ment processes. But much of its influence has been indirect, through its research and contributions to developmental telecommunications policy discussions (Najmabadi, Banerji, & Lall, 1992). It participates in these discussions through dissemination of its own publications, written and editorial contributions to the broader literature, and by maintaining a consistent presence at international conferences hosted by other multilat-eral organizations, regional organizations, and trade groups.

I first summarize the history of telecommunications lending by the Bank. I briefly review the Bank's telecommunications policy literature next, with attention both to the ideas circulated and to their ideological underpinnings. The Bank frames a discourse that narrows in scope: at one end is its advocacy of free markets and privatization and at the other end is liberalization with surgically targeted development assistance. The perpetuation of the broad public-sector ownership of telecommunications infrastructure and resources is out of the question, given the interest of the Bank's telecommunication experts in "demolishing" monopolies (Wel-lenius, 1994a).

I conclude with a critique of the Bank's telecommunications activities in the context of broader discussions about the Bank's overall ideas about development policy.

LENDING HISTORY

The World Bank has committed an average of about 2% of its loan monies to telecommunications (Barbu, 1993, p. 11)[1] and has made a total of 125 loans to 56 countries for telecommunications projects. Lending began in

---

[1]According to other Bank sources and methods of calculating assets, the average is consistently 1.5% and has rarely gone above 2% (Communications and Development Team [CDT], 1994, 33).

the 1950s, with a $1.5 million loan to Ethiopia in 1951 and a $300,000 loan to Iceland in 1953. Fourteen early "nonevaluated" loans totaled $174 million and involved 10 countries. Subsequently, between 1967 and 1993, the Bank made 84 loans that have been evaluated and 35 more ongoing loans to a total of 53 countries (Barbu, 1993). This lending history is described in greater detail later.

In addition, since 1988, 12 telecommunications loans have been included as components in larger structural-adjustment and public-enterprise-reform programs of the Bank. Totaling over $1.6 billion, these telecommunications loans for privatization-related activities are on a different scale than the other telecommunications loans and include five of $250 million or more to South American countries. Table 5.1 lists these loans and compares them to the largest and the most recent "standard" telecommunications loans to the same countries. As the table shows, whereas the adjustment-related telecommunications loans to African countries, Jamaica, and Uruguay have not been particularly remarkable in relation to other telecommunications loans, such loans to South American countries bear little similarity to previous telecommunications lending there.

In 1993 dollars, the Bank's portfolio of evaluated or ongoing loans between 1967 and 1993 amounts to $9.1 billion.[2] As Fig. 5.1 illustrates, the largest share has gone to Asian countries, which account for 56% of cumulative lending, followed by African countries with 23%, Latin American countries with 15%, and Eastern European countries with 6%. Figure 5.2 illustrates trends in the relative shares of lending to each region since 1967. Figures 5.1 and 5.2 do not include the adjustment-related loans discussed previously, which heavily favor Latin America, nor do they include telecommunications subcomponents of other types of loans.

It is difficult to discern patterns in the history of the Bank's lending for telecommunications. As Fig. 5.3 shows, lending was somewhat consistent in the early to mid-1970s, but from the late 1970s to the 1990s, telecommunications lending proceeded with enormous year-to-year differences. Nonetheless, in constant dollars, about half (48%) of lending occurred before 1980, and about half thereafter. Figure 5.4 provides a snapshot of the geography of all Bank loan money "sunk" into telecommunications since 1967.

The largest loans went to a diverse set of countries. Table 5.2 lists all countries that have borrowed $200 million or more (in 1993 dollars).

One reason for the apparent lack of continuity in Bank telecommunications lending may be that there has never been an explicit telecommunications policy. Currently, a campaign by telecommunications-related staff to get the board of directors to commit the Bank to formally articu-

---

[2]Current dollars were adjusted using the deflator for producer prices for capital goods.

TABLE 5.1
A Comparison of Standard World Bank Telecommunications Loans and
Telecommunications Portions of Structural Adjustment Loans (SAL)
from the World Bank (current and constant US$)

| | | | Standard Telecom Loans | | | |
| | SAL Loans | | Largest Telecom Loan | | Most Recent Telecom Loan | |
Country	Year	$million	Year	$million	Year	$million
Africa						
Côte d'Ivôire	1992	15.0	1974	25	1986	24.5
in $1993		15.5		66.2		29.9
Ghana	1988	10.5	1975	23.0	1988	19.0
in $1993		12.3		52.8		22.2
Guinea	1992	7.3	1990	11.3		
in $1993		7.6		11.7		
Mali	1988	40.0	1982	13.5		
in $1993		46.8		18.0		
Mauritania	1990	40	none			
in $1993		43.5				
South America and the Caribbean						
Argentina	1991	300	none			
in $1993		316.6				
Colombia	1991	304	1977	60	1980	44
in $1993		320.8		121.4		68.6
Jamaica	1987	20	1967	11.2 (11.1 cancelled)		
in $1993		23.9		41.8		
Peru	1992	300	none			
in $1993		310.7				
	1993	250	none			
in $1993		250				
Uruguay	1993	11	1986	40		
in $1993		11		48.7		
Venezuela	1990	350	1965	37	1971	35
in $1993		380.8		140.4		112.2

Note. Compiled from data in Barbu (1993, p. 56).

lating a telecommunications policy is under way. This campaign has led
to the construction of some in-house history from the record of previous
de facto policy. Bank telecommunications lending is broken into two eras:
the pre-1986 era of the "Old Agenda," and the current era of the "New
Agenda." Under the Old Agenda, the Bank focused on:

building the efficiency of the main public operating entity and funding
major new investments to meet demand. . . . The main assistance instruments utilized . . . were IBRD/IDA [International Bank for Reconstruction

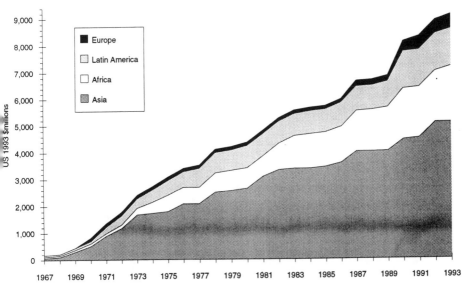

FIG. 5.1. Cumulative World Bank telecommunications loans by region, 1967–1993. Compiled from data in Barbu (1993).

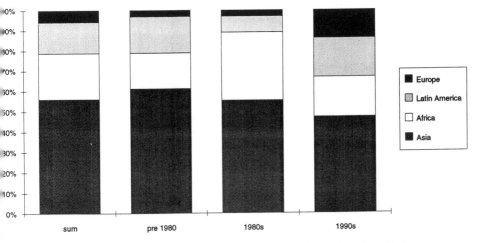

FIG. 5.2. Distribution of World Bank telecom loans by period. Compiled from data in Barbu (1993).

117

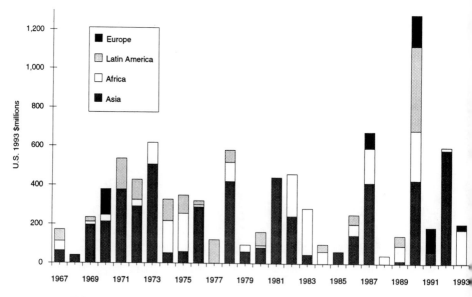

FIG. 5.3. World Bank telecommunications loans by region, 1967–1993. Compiled from data in Barbu (1993).

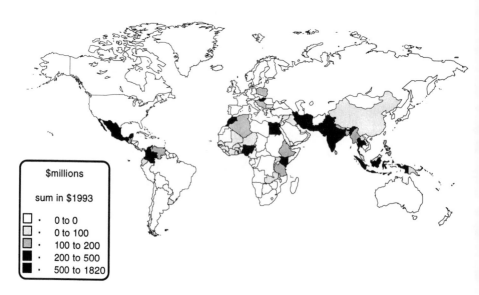

FIG. 5.4. World Bank telecommunications loans, 1967–1993 sum (aggregated $millions U.S.). Compiled from data in Barbu (1993).

118

TABLE 5.2
Countries Receiving Over $200 Million (1993) in Telecommunications
Loans From the World Bank (excluding telecom portions of
SAL and other large comprehensive loans)

Country	Cumulative Sum	Received 1967–80	Received in the 1980s	Received in the 1990s
India	1,819.5	962.8	856.7	0.0
Indonesia	829.7	43.0	17.7	769.1
Thailand	530.1	340.3	189.8	0.0
Mexico	437.3	0.0	0.0	437.3
Pakistan	393.7	218.3	175.3	0.0
Iran	368.0	368.0	0.0	0.0
Colombia	332.2	263.6	68.6	0.0
Kenya	316.8	214.7	102.2	0.0
Egypt	253.9	168.3	85.6	0.0
Morocco	249.6	0.0	149.6	100.0
Hungary	246.9	0.0	83.8	163.2
Nigeria	244.8	0.0	0.0	244.8

Note. Compiled from data in Barbu (1993).

and Development/International Development Association] investment loans and co-financing. (Communications & Development Team, 1994, p. 35)

Under the New Agenda, the focus is first of all "away from the traditional public telecommunications monopoly," and instead on:

overall sector policies, regulatory environment, and institutional structures and . . . in the formulation and implementation of sector strategies designed to meet demand. This approach has included advocating policies designed to promote new entry, competition, and private participation. (Communications & Development Team, 1994, p. 37)

Under the New Agenda, the Bank collaborates primarily with the International Finance Corporation (IFC) and the Multilateral Investment Guarantee Agency (MIGA). Under the Old Agenda, the Bank's multilateral partner was more likely to be the ITU, especially in the preparation of master plans and engineering studies (Barbu, 1993). According to Wellenius, the coziness with the IFC is "never an easy thing," due to the different nature of the relationship of each to local actors in developing countries. The Bank traditionally works with governments under time frames that average a couple years, the IFC works with investors, which requires both shorter time frames (3 to 6 months) and less openness in the process.

In the context of its activities, the Bank has only partially completed its shift to the New Agenda. As Barbu (1993) pointed out, "some of the most recent Bank loans . . . e.g., to Tanzania, Poland and the Philippines . . . still appear to follow the 'old' agenda" (p. 15). For ideological purposes and for the purposes of internal political strategy within the Bank Group, the New Agenda rules. But to maintain its presence "in about one-half of the almost 100 developing countries that are active customers of the World Bank Group" (Wellenius, 1994b, p. 14.3), the Bank must sometimes engage in less political forms of pragmatism.

One form of pragmatism they have exercised is a certain amount of patience. Although firmly maintaining that they will enforce a rule of "no intention to change—no investment financing" (Wellenius, 1994b, p. 14.3; CDT, 1994, p. 38), Bank economists also concede that structural changes must be tailored to circumstances in each country and that some countries face more challenging circumstances than others, including high levels of political risk or the needs of urban and rural poor that may "result in government having to assume a residual financing role." In such cases, Wellenius (1994b) predicted that "The World Bank Group would probably be prepared to give a hand" (p. 14.4).

The Bank exercises patience while inducing institutional transformation by requiring operating entities to adopt new business practices under the tutelage of consultants from advanced industrialized countries. Barbu likened such strategies to " 'sowing the seeds' of structural changes" (1993, p. 15). In Thailand, for example, Bank conditionalities required the Telephone Organization of Thailand (TOT) to set up an Office of Economic Study and a Fundamental Planning Unit. To meet the Bank's conditions, TOT had to contract with Mid-Continental Telephone Consulting Service of the United States for a $1.2 million management information system in 1980 (Telephone Organization of Thailand [TOT], 1981). To meet conditionalities on a later loan, in 1984 they had to contract with AT&T International for over a half million dollars to "achieve the objectives of the telecommunications projects and the telecommunications economic study" (Telephone Organization of Thailand [TOT], 1985, p. 44). Even as a major recipient of Bank loans, Thailand resists conditions and articulates the antagonistic nature of its relationship to the Bank in its official publications. For example, evaluators noted that the Bank exploited its position vis-à-vis Thailand during structural adjustment loan (SAL) negotiations. Thailand's "bargaining position was weak" for SAL I and SAL II and into 1985 when a third SAL was proposed:

For SAL III, the World Bank . . . insisted on the continuation of the conditions that failed in SAL II such as a launch in new initiatives in state enterprise finances. Some of these proposed conditions in SAL III were still political[ly] unpopular, and, therefore, the negotiation could not reach an agreement. (Sahasakul, Thongpakde, & Kraisoraphong, 1989, p. 39)

As another example, in announcing the advent of private-sector participation in the construction and operation of basic telephone service in parts of Thailand, the 1992 Annual Report articulates two forms of skepticism—on page 1, it states, "The goal [of private participation], *if achieved*, will be a historic record" (emphasis added; TOT, 1993, p. 1). The subsequent pages contain a quote from the King criticizing one of the tenets of market orthodoxy:

> To accomplish the task, one must aim at the real objective of the task to successfully achieve [the] utmost benefit for the task itself and the accomplished party. If the task is performed for other objective[s], *like personal interest*, regardless of the numerous outcomes, the task has ineffectively been executed and ... a useless effort of the person to the task. (italics added; TOT, 1993, pp. 2–3)

The Bank is patient in dealing with Thailand and other countries. Among the strategies advocated under the New Agenda are "bottom up" strategies (Wellenius, 1994a; Smith & Staple, 1994). These often include what might be more aptly described as "sideways through" when they mesh with the globalization strategies of transnational corporations. As joint ventures, multibillion dollar international investments in telephone networks in developing countries qualify as "bottom-up." Such deals have included Southwestern Bell's stake in Telmex, GTE's stakes in Venezuela and Argentina, Telefonica's stakes in the networks of Argentina and Chile, and AT&T's Memorandum of Understanding with China. "Bottom-up" should not be confused with "grass roots" strategies designed to empower would-be new telephone subscribers, for Smith and Staple omitted this constituency from their enumeration of "the broad coalition of users and would-be suppliers" who must "champion reform" (Smith & Staple, 1994, p. xvii). Like would-be subscribers, telecommunications workers, especially if unionized, represent not constituents, but rather "issues" to be "adequately addressed ... dealt with ... [and] made aware of ... the tradeoffs and why the government believes they are worthwhile" (Smith & Staple, 1994, p. 80).

Overall, loans buy less and less equipment as Bank lending for telecommunications moves toward the New Agenda. Restructuring of the information infrastructure is itself an information-intensive process and requires spending money on information handlers, including many not locally available. A textbook privatization, for example, requires the services of an international auditor, local and international law firms, an international business-technical consulting firm and an international investment banking firm (Booz, Allen & Hamilton, 1992, adapted by Smith & Staple, 1994).

Finally, another reason for the apparent lack of continuity in Bank telecommunications lending may be that the trajectory of telecommunications from a "luxury" good to a "strategic industry" ran against the prevailing winds within the Bank. When the Bank under McNamara was concerned with poverty, especially during the second half of the 1970s, the government was seen to need telephones, but private telephones were seen as luxury goods. Even now, Wellenius (1994b) noted, "it is hard to argue for more telephones instead of more hospital beds or school classrooms" (p. 14.1). When, under Clausen and during the debt crisis, developing countries faced "a sharp decline in [their] real leverage as their demand for finance now outstripped available supply" countries also encountered Bank reluctance to lend for physical capital formation (Gibbon, 1992, pp. 198–199). If Bank telecommunications experts remain tied to their Old Agenda today, they would now find themselves unpopular in an institution bent on abolishing state enterprises and promoting markets. Telecommunications staff have been subject to many reorganizations; by demonstrating their centrality to the current board of directors' vision, telecommunications staff at the Bank may hope to achieve greater institutional stability themselves. Thus, in order to achieve greater continuity in the future, it is strategic for telecommunications experts to align themselves with the prevailing winds of market liberalization.

INTELLECTUAL HEGEMONY

The purpose of this section is to place Bank ideas in the context of current debate. Bank telecommunications economists loan money to developing countries; they also attend many development conferences and write prolifically. Their ideas appear in Bank series, they contribute to scholarly and trade journals, and they sit on editorial boards of important journals. The Bank is the only organization currently represented by more than one member on the editorial board of *Telecommunications Policy*, for example. (Wellenius & Nulty, Bank staff, are listed on the journal's editorial board.)

Much writing on investment in telephony in developing countries is grounded in a theory of modernization, taking as given that national goals will aim to increase levels of telephone density in parallel with a broader strategy of industrialization, and with western industrialized countries as models:

> The Bank's approach to industrial policy may be described as "moderate neoclassical" which accepts that factor and product markets are not fully

efficient in developing countries and that there is a role for government interventions. However, it strongly prefers functional to selective interventions: governments should make markets more efficient in a neutral way. (Najmabadi et al., 1992, p. 5)

The dominant telecommunications literature in the 1970s and 1980s focused on heightening awareness of the importance of telecommunications to development (Saunders, Warford, & Wellenius, 1983). Since 1990, telecommunications consciousness has been sufficiently raised in the development establishment; the literature currently stresses shaping telecommunications as a strategic sector and overcoming obstacles to investment (Antonelli, 1991).

Two points of view coexist within the dominant school of thought. The first view, which guides the Bank's telecommunications activities, seeks to get the prices right: if telecommunications were turned over from state enterprises and protected monopolies, the theory maintains that supply would be free to meet the enormous pent-up demand existing in most developing countries, economies of scale would be achieved to control costs, and everyone would be better off. The second point of view, which has provided impetus to the International Telecommunications Union's (ITU) development activities, takes as abhorrent the telecommunications "gap" between rich countries and poor and seeks to foster an international commitment to "bridging the gap." Enlightened, informed policymakers should be able to stimulate investment within the context of the existing institutional and industrial structure. If donors and development banks would support telecommunications projects as richly as they have other infrastructural projects and if the revenues from telephone services could be used to guarantee commercial loans, then developing countries could achieve more appropriate levels of investment.

There is great potential for compromise between these two points of view, and the Bank has been an instrumental force of convergence toward a "unified" theory of using "the right prices" to bridge the gap. In fact, repeated use of the phrase "the gap between supply and demand" (e.g., see Smith & Staple, 1994, pp. xii, xiii, xv, xxi), seems a symbolic attempt to co-opt the Maitland Commission's language (Maitland Commission, 1984) in order to silence concern for what Blackman suggests as the possibility of "creating a global information-rich elite while condemning the rest of the planet to the information slums" (Blackman, 1994, p. 4). Blackman edits *Telecommunications Policy*.

Obviously, both the shifting distribution of global telecommunications and information resources and the transfer of control over those resources from one class of manager to another have political and ideological meanings. Nonetheless, Bank authors often explicitly label their ideas: "nonideological." The Bank defines its emphasis on restructuring as a response to nonpolitical technological change and rising demand and to factors that "amplify" technological and market factors in developing countries. Among the "amplifiers" are two distinctly political positions. One specifically embeds Bank-supported telecommunication lending in the context of broad structural reform, including structural adjustment programs. The other explicitly promotes political alliances with powerful liberalizing countries (Wellenius et al., 1993). What these two factors really amplify is the Bank's broader goals of global economic integration on a capitalist model. These two factors link developing-country telecommunication development explicitly to the needs of exogenous private interests—telecommunications companies and financial institutions—in industrialized countries.

Getting the Prices Right

Orthodox economics has changed (a little) since the publication of Stigler's "Economics of Information" in 1961, in which he observed that:

> One should hardly have to tell academicians that information is a valuable resource: knowledge *is* power. And yet it occupies a slum dwelling in the town of economics. Mostly it is ignored: the best technology is assumed to be known; the relationship of commodities to consumer preferences is a datum. (p. 213)

Mainstream economists now assert that "information is today a fundamental factor of production, alongside capital and labor," and that telephony must consequently be considered a necessary precondition for economic development (Wellenius et al., 1993, p. 1). In a capitalist system, where resources are allocated through markets and growth in GDP is the preferred measure of development, telephone service, as a commodity, is an important production input. Telephone service can reduce costs and increase the efficiency of production of basic needs, such as housing, medical care, food, and education. In the words of an International Telecommunications Union (ITU) economist: "You can grow produce without fertilizer, just as you can run an economy without telecommunications, but each is considerably less efficient without these basic inputs" (Richter, 1990, p. 3). The World Bank speaks loudest for "getting the prices right" through the institution of a free market. When telecommunications service is supplied through a government entity, as was traditionally the case in most countries, the move to a market requires the achievement of two broad goals: privatization of as much of the state entity as possible, with as few resulting monopolies as possible; and creation of regulatory institutions to enforce free market requirements.

Bank authorities present these initiatives as technical solutions to problems created by global forces that developing country officials may not understand. As Wellenius (1989) cautioned:

Few governments are well equipped to address the questions [of technological and economic forces that are driving change in the telecommunications sector worldwide]. Many are not even fully aware of the issues, the alternatives, and the risks of inaction. In telecommunications, unlike other sectors (for example energy and agriculture), there is generally no explicit policy.... responsibility for policy and regulation is often vested in monopolistic operating enterprises. In this situation structural change is discouraged, the expertise needed for broader analysis is lacking and decisions for change are shifted to the political arena instead of being handled within the sector framework. (p. 97)

For a developing country professional seeking a Bank loan, Wellenius is the contact person (ITU, 1991). The World Bank claims itself the largest multilateral source of telecommunications funding, and arrogates to itself a monopoly on "legitimate" expertise. Hills (1993) argued that

in the telecommunications sector the World Bank is increasingly occupying a gatekeeper function in the access of developing countries to loans from all sources, and through its control of capital is prescribing privatization as the solution to LDC problems. (p. 53)

Bank economists advise developing countries that social and political considerations no longer have a place in the formulation of telecommunications policy and add that liberalization is an inevitable, nonideologically driven force (Nulty, 1989). Their 1989 Symposium proceedings concludes with a chapter outlining "in an orderly, pragmatic, and nonideological way the basic regulatory tasks that need to be performed and possible schemes for executing them." They use the United States as their model

because the U.S. and Canada are the only countries with a long track record of operating the telecommunications sector as a mixed private–public system of government-regulated commercially oriented companies. Inasmuch as most other countries are moving in this direction, common sense dictates that we should examine U.S. experience closely for useful lessons. (Nulty & Schneidewind, 1989, p. 28)[3]

---

[3]Bank economists further point to the early 20th-century experience in "North America" (presumably they include Mexico), citing the role of "bottom-up" network expansion programs led by independent local telephone companies [and others], but they neglect to review the concurrent history of illegal anticompetitive activity by AT&T and the Bell companies or to recall that the U.S. government in fact took possession of the nation's telephone infrastructure briefly during World War I (Danielian, 1939, p. 245).

But as both an examination of U.S. experience and other Bank telecommunications experts highlight, many of the "basic regulatory tasks" are themselves politically charged. At a high level in the United States, the series of issues dealt with by the Federal Communications Commission, by Judge Greene, and legislatively since the divestiture of AT&T reveal the contention among and between users and suppliers of telecommunications services. For developing countries just setting up regulation:

> Regulation must be broadly understood to include a country's political and judicial competence as well as its commercial customs. Whether or not an effective regulatory regime can be designed for telecommunications will depend on how well the new regime fits with existing governmental institutions. Regulatory designs cannot easily be transplanted unchanged from one country to another. (Smith & Staple, 1994, p. xv)

The Need to Eliminate the Telecommunications Gap

In 1983, the ITU established an independent commission, chaired by Sir Donald Maitland, to investigate worldwide telecommunications development. Their 1984 report, *The Missing Link*, noted global inequality of access to telecommunications and, while urging a greater private sector role in financing telecommunications growth, also stressed the role of governments in taking action:

> We look to governments of industrialised and developing countries alike to give fuller recognition to this common interest and to join their efforts to redress the present imbalance in the distribution of telecommunications, which the entire international community should deplore. (Maitland Commission, 1984, p. 65)

The Maitland Commission's 1992 report was grounded in principles of equity, arguing for aid and mechanisms of redistribution:

> Our shared conviction that the denial of access to information implicit in the gross imbalance in the distribution of telecommunications throughout the world was intolerable led us to set a clear objective: all mankind should be brought within easy reach of a telephone by the early part of the next century. The Commission believes that, while this called for a variety of actions ... the most important single step towards achieving that goal would be to improve arrangements for providing developing countries with the assistance they needed. (pp. 15–16)

Crucial to the Commission's recommendations were some financial innovations that have not come to pass; they advocated for more grants, innovative loans, and the creation of a development tax on international calls to build a development fund (Maitland Commission, 1984). The commission's other suggestions did not rely on the generosity or alternative visions of the powers that be in richer countries, and facilitated the migration of the development efforts of the ITU and its member nations toward the World Bank's vision. Whereas the Maitland Commission stressed improving self-reliance in operations, ITU thinking by the 1990s focused more narrowly on selected operational objectives, especially "human resources development." A former director of the development arm of the ITU, Arnold Ph. Djiwatampu, won his position in a highly contested election on a platform of attracting and gaining trust from the private sector and inducing them to participate in development projects in developing countries.

Coinciding with this development-oriented invitation to the private sector is a broader agenda to increase the private sector voice in ITU governance. Private firms have long been nonvoting, dues-paying members; initiatives to grant them voting status are gaining momentum as evidenced by the May 1993 vote to record private-sector participants' approval and disapproval of resolutions. Resistance comes primarily from the large number of developing countries, who argue that large corporations should be represented by the votes of their home countries. Within the ITU bureaucracy, long committed to reserving voting privileges to governments, some now see a need to bring in the private sector as a matter of institutional survival (D. MacLean, personal communication, May 25, 1993). A 1991 BDT study on restructuring stresses opportunities to increase the private-sector share of new investment in developing countries through strategic restructuring, especially the use of commercial and corporate forms of organization to create a more favorable investment climate (ITU, 1991).

The ITU established in 1992 the World Telecommunications Advisory Council (WTAC), similar in prestige and autonomy to the Maitland Commission, charged with "providing the ITU with strategic advice from the public and private sectors on the telecommunications environment and how, in the light of its dynamic nature, the Union's principal activities could be carried out more effectively" (*Telecommunications Journal*, 1992). In *Telecommunications Visions for the Future*, WTAC asserts the need for a global-scale "social contract" for the globalization of telephone service and they promote the ITU as the institution to shepherd such a contract. In calling for attention to two areas needing global governance—one of technical harmonization and one of distribution—they also warn that the ITU's political structure must accommodate new voices and adapt in other ways before it can accomplish these governance tasks.

The ITU has not maintained a loud voice for redistribution and greater equity in global telephony. In order to continue speaking simply for more telephone lines in developing countries, it has adapted to the pressures of the increasingly powerful and globalized private sector. In doing so, the ITU has brought its development agenda more in line with that of the World Bank. This is a strategic victory for both the large transnational telecommunications companies and for western governments pushing for more "open" trade. The ITU shift reflects also their reaction to threats posed to ITU regulatory agency by regional organizations and trade agreements, such as the General Agreement on Tariffs and Trades (GATT).

This section elaborated the positions of a dominant part of the literature on telecommunications development and challenged the claim that the principles put forth by the Bank are "apolitical." The literature's theoretical strength lies in its consistency with principles of global economic integration, which is presumed to be a universal good. On the practical side, the literature's strength comes from the power of its authors, by virtue of their institutional affiliations, to shape telecommunications policies in many developing countries.

TELECOMMUNICATIONS AND CRITIQUES OF THE WORLD BANK

The following quotes refer to broad critiques of the World Bank:

> The commitment to establish an efficient and well-functioning telecommunications system and the timetable for doing so have often become a litmus test of the government's overall commitment to economic reform. (Smith & Staple, 1994, p. 79)

> Looked at from [a] rational actor model telecommunications privatization produces benefits for World Bank officials in that it ensures that they have seats at the highest policy table of every country wanting a loan. (Hills, 1994, p. 11)

> The World Bank's verdict on the causes of the East Asian miracle was bound to be controversial because the topic is controversial and so is the Bank. The Bank is inherently a political organization yet purports to produce objective economic analyses. Its middle management is comprised of many highly respected economists while its top management is comprised of political appointees who serve at the discretion of the industrialized countries, especially the United States. (Amsden, 1994, p. 2)

Telecommunications experts at the Bank candidly call their approach to telecommunications development an "agenda" and are not a bit shy about pressing to enhance their prestige and autonomy by convincing the Bank's directors to promulgate an explicit telecommunications policy. If the Bank does so, research and lending could focus on the needs of the sector in diverse developing countries rather than on the publication of glossy brochures that demonstrate how snugly they can wrap the telecommunications agenda in the cloak of Bank dogma. Bank telecommunications analysts could strive for greater continuity in their programs and spend less time paralleling the efforts of the ideologues who wrote the conclusions to *The East Asian Miracle* (World Bank, 1993) and the 1991 *World Development Report* (World Bank, 1991).

Telecommunications is not like any other business and public-sector operators are generally not like other publicly owned enterprises. Despite inefficiency, corruption, and political unrest, state-owned telephone operators are more likely than other large state-owned enterprises to operate in the black. More than many other state-owned enterprises, telephone operators are specific targets for transnational corporations in the same line of business. But greater diffusion of access to telephones is an inherently political issue concerning access to information and possibilities for social interaction. The New Agenda should embrace these issues, especially in the context of genuine "bottom up" strategies.

Miracles in Asia

Scholars inside as well as outside the Bank criticize the conclusions drawn by the authors of *The East Asian Miracle*. As Amsden (1994) explained,

> The *Report* is rich in empirical data, but they do not support the Bank's dismissal of industrial policy as 'ineffective.' . . . The greatest disappointment of the *Report*'s market fundamentalism is a failure to study seriously how the East Asian model can be adapted to suit conditions in other countries. (p. 1)

The same could be said of Smith and Staple's (1994) *Telecommunications Sector Reform in Asia: Toward a New Pragmatism*. Like *The East Asian Miracle*, their conclusions stress the need for increasing the role of competition and the private sector. But their examples—New Zealand, Singapore, Philippines, Thailand, and Sri Lanka—provide primarily counter-evidence for their conclusions: New Zealand's regulatory capacity was inadequate, and many issues ended up in litigation after privatization—if we seek lessons for developing countries, what should we learn from New Zealand? Singapore's state telecommunications monopoly increased density rapidly during the 1980s, it has money to burn, its successes "may not be easily transferable," and its actions would likely cook up "a recipe for economic stagnation and corruption" in other countries—what are the lessons here for privatizing or introducing competition in developing countries? The Philippines example provides lessons to avoid; it has been included as an example of how, even under private ownership, "sector performance [can be] suboptimal along several dimensions" (Smith & Staple, 1994, p. xx). Thailand's case seems to have more lessons, but as the results of private-sector participation are not yet in, most of the lessons pertain to situations of government ownership. Recall also that Thailand ranks fourth in total Bank lending for telecommunications.

Further, Smith and Staple (1994) argue that the pace of privatization should be accelerated in the telecommunications sectors of both very poor countries and in countries in which large unmet demand exists. For the former, reform should include a shock-therapy approach: "step-by-step liberalization of the market may be less effective than a more radical open-door policy that offers multiple concessions to unserved or underserved areas" (p. xvi). Among their examples, no countries have taken and proven this path. The poorest country in their study, Sri Lanka, has taken a step-by-step approach and has indeed not yet satisfied demand—but this is no proof that demand would be met by opening the door widely to concessionaires. As the chairman of the Telecommunications Board of Sri Lanka explains, Sri Lanka deliberately resisted the shock-therapy path. He points out that officials in his country and other countries in similar need of telecommunications capital confront

a discernable bias in what is called the management contract, under which the collaborator [major telecommunications organizations in industrialized countries] seems to expect full control for a period [which is] hardly appropriate for a society such as Sri Lanka, which meets relatively high standards in administration and telecommunications and in which deficiencies are mainly due to financial constraints.... The course of action the collaborators tend to advocate—that is rushing into joint ventures or management contracts—would merely stultify the venture and deprive it of an opportunity to develop itself and its self confidence. Such a choice would also reduce the opportunity for the local personnel and management to develop their skills and gain experience. (Mendis, 1989, p. 100)

For countries in which large unmet demand exists, Smith and Staple (1994) acknowledge that step-by-step reform has been the path taken by most countries that have implemented reform, but they argue that now:

incrementalism can dissipate public support for reform. Once expectations have been raised, only a bold initiative, one that makes a real break from the past, is likely to be acceptable. (p. xvi)

Again, none of their examples support this conclusion—Thailand has taken a bold initiative in licensing a consortium of private transnational firms to build and operate a million new lines—but the results are not yet in. In other Asian countries, such as China, South Korea, and Singapore, it is public-sector providers who boldly invested to rapidly increase telephone densities and advanced capacities:

> The histories of wireline and cellular radio service both suggest that a broadly available, mature telephone network is not a prerequisite to liberalizing market entry. In fact the opposite is true: licensing multiple service providers probably is the best way to accelerate the investment necessary to establish a more broadly based and mature national telecommunications network. (p. xiii)

Evidence from their sample countries does not support this claim either. The Asian countries that demonstrated the most profound increases in telephone densities were led by public sector operators. Some, such as Japan, eventually privatized and liberalized (1985), but others, such as Korea and Singapore, continue as state-owned enterprises. In situations where multiple licenses for cellular are overlaid on an inadequate public or private monopoly for wireline services, as in Mexico, the availability of the more expensive cellular service effectively stratifies the base of potential customers into those willing and able to pay the significantly higher price for cellular and those stuck waiting for wireline.

Telecommunications Sector Reform in Asia, like other important Bank publications, eschews discussion of government-led national industrial strategies. Given the importance of national industrial policies for East Asian economies (Amsden, 1994; Najmabadi et al., 1992; Singh, 1994)—including the "tigers," the recognizably "strategic" nature of the sector and the relationship between telecommunications policy and industrial policy should be explored by Bank economists. If, as Smith and Staples assert, it is essential "to have the unflagging commitment of the head of government," in order to reform the telecommunications sector, then it surely makes sense to try to design strategic complementarity between telecommunications reform and industrial policy. Bank evaluators criticize the Bank and advise a more thoughtful approach to industrial policy:

> In its general approach to industrial strategy, [the Bank] tends to assume that governments lack the ability to be selective, while in its project and subsector work it displays a more nuanced, realistic stance. The latter is what needs to be better articulated, and transposed to the general level: the ability of governments to be economically selective should be assessed on a case-by-case basis rather than assumed absent. (Najmabadi et al., 1992, p. 56)

Visions in the United States

During the 1980s, the United States put enormous pressure on the Bank to promote privatization. Pressure came from transnational corporations, Congress, and the U.S. administration: "Nationalized industries should be privatized" (Quayle, 1991, p. 700). Thus, part of the Bank's rhetoric can be heard as supplication to its Washington neighbors. The record in the telecommunications sector is diverse, as are Bank experiences in particular countries (Sullivan, 1992), but there is a definite trend toward privatization of telecommunications as a Bank activity. The difference between pre-1990 and post-1990 project objectives among samples of 30 loans reveals this trend. Of 21 completed projects from the 1970s and 1980s, only one included funding for restructuring or regualtory work among its project objectives or "institutional development components." Of nine ongoing projects funded during the late 1980s and early 1990s, a majority—six projects—included funding for restructuring or regulatory work.

Little is known about the consequences of privatization on service provision to the majority of communities in poor countries. The introduction of new services such as cellular telephones, fax, and e-mail make the lines-per-100-people statistic even more useless as a measure of distribution, for it hides the extent to which elites have more than one line—a wireline *and* a cellular phone, for example. Given how little is known about the distributional effects of privatization and the extent to which privatization is not an easily reversed process, the Bank and others should commit more of their research resources to evaluating the consequences of induced structural change for the majority of users and would-be users. To continue to push countries toward privatization without addressing these issues is reckless.

REFERENCES

Amsden, A. H. (1994). Why isn't the whole world experimenting with the East Asian model to develop?: Review of the World Bank's *The East Asian miracle: Economic growth and public policy. World Development, 22,* 627–633.

Antonelli, C. (1991). The diffusion of advanced telecommunications in developing countries. In OECD (Ed.), *Strategic industries in a global economy: Policy issues for the 1990s* (OECD International Futures Program). Paris: OECD.

Barbu, A. (1993, November 2). *The Bank's experience in the telecommunications sector and OED review* (Report No. 12445). Washington, DC: World Bank.

Blackman, C. (1994, January/February). Editorial: To have and have not. *Telecommunications policy 18* (No. 1, pp. 3–4).

Booz, Allen & Hamilton (1992, March). *Strengthening the telecommunications regulatory and analytical capabilities of the Directorate General of Posts and telecommunication (DGPT).* Government of Indonesia Ministry of Tourism, Posts and Telecommunications.

Communications and Development Team (CDT), Vice Presidency for Finance and Private Sector Development, the World Bank (1994, February 16). *Telecommunications sector background and Bank Group issues* (Joint World Bank/IFC/MIGA Seminar). Washington, DC: World Bank.

Danielian, N. R. (1939). *AT&T: The story of industrial conquest.* New York: Vanguard Press.

Gibbon, P. (1992). The World Bank and poverty, 1973–91. *Journal of Modern African Studies,* 30(2), 193–220.

Hills, J. (1993, July 4–7). *Economics as ideology: The World Bank and privatisation.* Paper presented at the International Association of Mass Communication Research Conference, Seoul, South Korea.

International Telecommunications Union (1991). *Financial resources for telecommunications projects in developing countries handbook.* Geneva: ITU.

International Telecommunications Union (1991, September). *Restructuring of telecommunications in developing countries: an empirical investigation with ITU's role in perspective.* Geneva: ITU.

Maitland Commission (1984). *The missing link.* Geneva: ITU.

Maitland, D. (Sir) (1992, November/December). *The missing link revisited* (Transnational Data and Communications Report).

Mendis, V. L. B. (1989). Phased privatization with proposed foreign participation: The Sri Lanka experience. In B. Wellenius (Ed.), *Restructuring and managing the telecommunications sector* (World Bank Symposium; pp. 99–106). Washington, DC: IBRD.

Najmabadi, F., Banerji, S., & Lall, S. (1992). *World Bank support for industrialization in Korea, India and Indonesia* (World Bank Operations Evaluation Study). Washington, DC: The World Bank.

Nulty, T. E. (1989). Emerging issues in world telecommunications. In B. Wellenius (Ed.), *Restructuring and managing the telecommunications sector* (World Bank Symposium; pp. 7–18). Washington, DC: IBRD.

Nulty, T. E., & Schneidewind, E. (1989). Regulatory policy for telecommunications. In B. Wellenius (Ed.), *Restructuring and managing the telecommunications sector* (World Bank Symposium; pp. 28–39). Washington, DC: IBRD.

Quayle, D. (1991, April 14). *Africa must develop its private sector.* Address before the African Development Bank, Abidjan, Cote d'Ivoire. (Department of State Dispatch, pp. 700–701).

Richter, W. (1990, May 18). Economic justification for telecommunications investment in developing countries. Lecture presented at the 10th ICC Executive Forum, *Global Imperatives for the '90s.* Neuss Duesseldorf.

Sahasakul, C., Thongpakde, N., & Kraisoraphong, K. (1989). *Lessons from the World Bank's experience of structural adjustment loans (SALs): A case study of Thailand.* Bangkok: Thailand Development Research Institute.

Saunders, R. J., Warford, J. J., & Wellenius, B. (1983). *Telecommunications and economic development.* Baltimore: Johns Hopkins University Press for the World Bank.

Singh, A. (1994, April 8). Close vs. "strategic" integration with the world economy and the "market-friendly approach to development" vs. an "industrial policy": A critique of the *World development report, 1991* and an alternative policy perspective (International Studies in Planning Seminar, Cornell University, Ithaca, New York).

Smith, P. L., & Staple, G. (1994). *Telecommunications sector reform in Asia: Toward a new pragmatism* (World Bank discussion paper No. 232). Washington, DC: The World Bank.

Sullivan, D. J. (1992). Extra-state actors and privatization in Egypt. In I. Harik & D. J. Sullivan (Eds.), *Privatization and liberalization in the Middle East.* Bloomington: Indiana University Press.

Telecommunications Journal (1992, July-August), p. 385.

Telephone Organization of Thailand [TOT]. (1981). *Annual Report, 1980.* Bangkok: TOT.

Telephone Organization of Thailand [TOT]. (1985). *Annual Report, 1984.* Bangkok: TOT.

Telephone Organization of Thailand [TOT]. (1993). *Annual Report, 1992.* Bangkok: TOT.

Wellenius, B. (1989). Beginnings of sector reform in the developing world. In B. Wellenius (Ed.), *Restructuring and managing the telecommunications sector* (World Bank Symposium; pp. 89–98). Washington, DC: IBRD.

Wellenius, B., et al. (1993). *Telecommunications: World Bank experience and strategy* (Discussion Paper 192). Washington, DC: The World Bank.

Wellenius, B. (1994a, May 23). Personal interview. Washington, DC.

Wellenius, B. (1994b, February 28–March 1). Financing telecommunications infrastructure—who wants to invest? In what, at what price? *Asia-Pacific telecommunications—a magnet for foreign investment* (pp. 14.1–14.5). Hong Kong: Financial Times Conferences.

World Bank (1991). *The challenge of development* (World Development Report 1991). Washington, DC: World Bank.

World Bank (1993). *The East Asian miracle.* New York: Oxford University Press.

World Telecommunications Advisory Council (1993). *Telecommunications Visions for the Future.* Geneva: ITU.

6

The Role of Domestic Capital in Latin America

Eduardo Barrera
University of Texas at El Paso

Four of the largest economies in Latin America went through the privatization of their respective telephone companies in a period of 4 years. Chile was the first to privatize the State-Owned Telecommunications Enterprise (SOTE) in 1988, with Mexico and Argentina doing the same in 1990 and with Venezuela the following year. Other countries in the region that privatized their SOTEs include Jamaica, Bermuda, St. Kitts, and Belize. Uruguay, Peru, Ecuador, Colombia, and Costa Rica are already planning to hand SOTEs over to private capital.[1] A common denominator in these processes has been the key role of foreign capital and the absence or the subordinated role of domestic capital.[2] Despite the commonalities in the timing and actors involved, these countries have very different political systems in terms of their centralization of power in the executive branch, the nationalist orientation, and the extent of support to capital accumulation.

The objectives of this analysis are (a) to identify the reasons that led Latin American countries to privatize their SOTEs; (b) to compare the political systems of four of the most important countries in the region and their relation to the privatization of SOTEs; (c) to examine the role

---

[1]The Panamanian government's attempt to privatize INTEL failed when the National Assembly rejected the draft law.

[2]Venezuela is the only country with a relatively lower direct foreign investment in telecommunications.

of domestic capital in the four cases; and (d) to assess the impact of the inclusion of workers as capitalist partners.

DEBT CRISIS

The privatization of SOTEs were not isolated actions by Latin American governments. Countries in the region started to privatize the economy after decades of buying or creating them. This change in political orientation can be seen in two ways: as the rise of neoconservatism and as an economic crisis. The first view would point out the coincidence of Reaganism and Thatcherism and how they were followed in Latin America by administrations that used a neoliberal economic framework. It privileges the ideological sphere and points out how policies of commodification, deregulation, and privatization can be adopted by intergovernmental organizations and exported by industrial powers (Mosco, 1989; Samarajiva, 1990). The second view would explain it as a consequence of the economic crisis that left Latin American governments not only with no surplus to redistribute but with foreign debts of historical proportions.

The 1980s debt crisis was actually the fourth in the history of Latin America. The previous crises were:

1. *The 1820s Debt Crisis.* Large European bond issues to finance the wars of independence were serviced by exports of gold and silver. This coincided with the European financial crisis of 1825–1826 and was renegotiated in the 1850s and 1860s by issuing new bonds.

2. *The 1870s Debt Crisis.* The construction of railroads in Latin America stopped after two decades because of the European Depression. Three solutions were simultaneously adopted by different countries: (a) Most countries rescheduled and negotiated new bonds, (b) Argentina, Brazil, and Chile increased their exports, and (c) After its foreign debt default in 1885, Costa Rica created the debt-equity swap, a mechanism through which the national railroad was sold to private investors.

3. *The 1930s Debt Crisis.* The Great Depression reduced the imports from Latin American countries. This crisis was extinguished by the mid-1940s by rescheduling, renegotiation, bond buyback, and issuing new long-term bonds.

The current Latin American debt crisis exploded in 1982 when several Latin American countries could not service the debt and its increasing interest. Among the debtors were countries that privatized their telephone companies within a decade: Argentina ($43.6 billion U.S.), Chile ($13 billion U.S.), Mexico ($86.1 billion U.S.), and Venezuela ($32 billion U.S.). Most analyses acknowledge that there is a relation between these two phenomena. The link has traditionally been viewed in three different ways:

1. *Streamlining the public sector.* The public sector was responsible for a large portion of the foreign debt. By 1979, the public debt in those four countries amounted to 41% in Argentina, 53% in Chile, 68% in Mexico, and 40% in Venezuela (Fig. 6.1). This debt originated in the loans obtained by these governments not only for the regular budget but for the buyout and maintenance of State Operated Enterprises (SOEs). The Latin American State became the main economic actor during the Import Substitution (1940s–1950s) and Stabilizing Development (1960s) phases, when it created many of the SOTEs or acquired them from the private sector because they were considered strategic or simply because they were not profitable and the state played the role of saving jobs and bailing out capitalists in trouble.

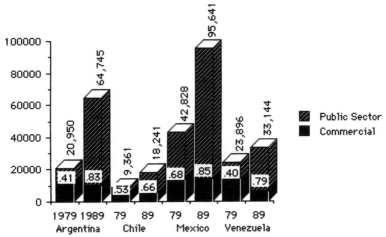

FIG. 6.1. Public sector debt of Latin America. Raw data from Grosse (1992, p. 67).

2. *Servicing the debt and its interest.* SOEs were sold and the money could be used to service the debt directly. In Mexico, the De la Madrid administration (1982–1988) initiated the sale of the 1,214 public enterprises that were both nonprofitable and nonstrategic under the Economic Solidarity Plan (PSE). By the end of the administration, 751 of them had been sold for $400 million U.S. In Argentina, the Alfonsín administration created the Commission 414 to privatize the SOEs but was not as successful as the Menem administration. The Pinochet regime in Chile preceded

these and the rest of Latin American countries in the overall liberalization of the economy, which included privatization, under the advice of a prominent monetarist economist from the United States, who replaced the U.S. "intelligence" advisers who supported him in the coup to support the democratically elected Allende administration.

3. *Investment needed to expand and modernize the infrastructure.* It would become more difficult to increase the telephone density and create the enhanced services demanded by large users such as the finance sector and multinational corporations (MNCs). In the case of Mexico, it was estimated in 1989 that Telmex needed to invest $14 billion by 1994, thus needing the collaboration of the private sector (Lerner, 1989).

After the collapse in 1982, all the actors involved were in disarray and governments could not meet the debt service or get new loans, while the credit institutions had a chaotic balance sheet. The Baker Plan presented at a 1985 meeting of the International Monetary Fund (IMF) was the first comprehensive package by the U.S. government to deal with the less developed countries (LDCs) debt. Ten out of all the countries involved in the plan were from Latin America. These countries were required to open their economy to foreign investment and privatize the SOEs. The Plan also called for $7 billion U.S. per year for three years from banks and $3 billion U.S. annually from industrial countries through the IMF and the World Bank.

The implementation of the Baker Plan failed because of the lack of response by the banks and strong opposition by some governments, particularly from Brazil and Peru. It was replaced in 1989 by the Brady Plan that promised, for the first time, to grant debt and debt service relief and included more specific steps, such as Debt Equity Swaps (DES),[3] debt-to-bonds swaps, debt buybacks, a three-year period for negotiations with banks and other agencies, and economic reforms contained in the Baker Plan.

The Brady Plan was first put in practice the same year (1989) in Mexico. This included a $40 billion U.S. swap of debt for new bonds and the reopening of the DES that was suspended in 1987. Venezuela and Costa Rica followed rapidly while other debtor nations adopted many of the policies and mechanisms.

---

[3]The credit institutions espoused the sale of debt paper for cash at a discount. The investors, generally MNCs, negotiated with the debtor's central bank the redemption of the document in exchange for an equity investment in the country at a value equal or lower than the debt redeemed. See Parhizgari (1992).

ED: 1 LINE SHORT

Although there is no doubt that the debt crisis was the source of the structural change that led Latin American countries to the liberalization of the economy, there are several ways to interpret the crisis in terms of its nature, origins, and possible solutions (see Table 6.1).

Governments, the IMF, the World Bank, and neoclassical economists hold the dominant view that argues that the debt crisis was the fault of the nations that became indebted. According to this approach the origins were the protectionist strategies of import substitution and the fiscal indiscipline of Keynesian states, coupled with corruption practices that resulted from that very indiscipline. The solution of the crisis would therefore lie on the Latin American states that should have to reverse those policies and impose a draconian austerity, streamline the state by privatizing SOEs, and deregulate open markets.

Many intellectuals and political parties in Latin America agreed with that explanation of the crisis but thought that the solution posed was too hard for the low-income sectors; they proposed the rescheduling of the debt, as well as less draconian measures that would permit economic growth, creating the conditions to service the debt.

Other sectors following the structuralist and dependency traditions of Latin America offered different explanations and solutions to the debt crisis. In these views, the crisis was a result of the deterioration of the terms of trade that did not allow Latin America to earn enough from its exports, coupled with the underdeveloped sector of capital goods that created technological dependency on the Center. This approach would propose the cancellation of the debt and a new flow of loans to peripheral countries in order to develop integrally (Furtado, 1984).

The French Regulation School is an approach that focuses on the development and crises of regimes of accumulation. The crisis is seen as the exhaustion of the global Fordist regime of accumulation consisting in a model of intensive production and intensive consumption. The former was based on assembly line mass production, whereas the latter consisted in the coding of minimum wage indexes and indirect wages through public services. Latin American countries had the variant of peripheral Fordism that did not allow intensive consumption by all sectors. The crisis was due to a decline in the rate of profit, and the global system would still be in the way of finding a new regime of accumulation. This is a nonnormative theory, so it does not offer alternatives (Lipietz, 1987, 1989).

A class-theoretical approach such as *Autonomia* rejects the view of the crisis as a mere financial one, as do capital theoretical approaches that explain the process as the outcome of the laws of motion of capital. Both leave class struggle in a marginal position where labor appears as a passive victim. The Autonomia approach is also known as "operaism," "worker-

140

TABLE 6.1
Theoretical Approaches to the Debt Crisis

Approach	Nature of Crisis	Origins	Solution
Neoclassical	Financial	Import substitution Keynesianism Corruption	Austerity Liberalization Privatization
Reformist	Financial	Underconsumption Growth	Rescheduling
Dependency	Trade	Unequal terms of trade	Cancellation Growth Redistribution
Regulation	Accumulation	Crisis of global Fordism	Nonnormative
Autonomy	Class struggle	1. Transfer of value from Western consumers to Western capital through OPEC 2. Failure to invest in nonspeculative enterprises and to limit nonoil consumption imports 3. Inability of debtors to have a trade surplus and increase in interest rates	Collective repudiation Self-valorization

ism," and the Italian New Left. Its most prominent scholars are Italians Mario Tronti and Antonio Negri as well as Harry Cleaver from the United States. Other analyses that label themselves class-theoretical are actually closer to the dependentistas. Petras and Brill (1988) is an example of an analysis that privileges not only capital but the nation as a unit of analysis. According to Harry Cleaver (1990) the root of the crisis was the power of labor that ruptured global capital accumulation. Class struggle would have forced capital to respond to their demands. Capital used the oil crisis of the 1970s to transfer value from Western consumers through petrodollars, which went to speculative ventures instead of productive activities. The relative failure to attack labor through unemployment and inflation forced a decrease in nonoil consumption imports from industrialized nations, creating a trade deficit in the Third World. This was followed by the monetarist reforms initiated by Paul Volcker in the United States and the antiinflationary policies of the IMF that called for a breakdown of "structural rigidities" such as trade unions and wage indexation. Debtor nations had to renegotiate the debt and obtain new credits, which required the approval of the IMF, which demanded cuts in state spending, subsidies, and the "denationalization and privatization to break the workers leverage with the State" (Cleaver, 1990, p. 31). This class-theoretical approach does not call for the cancellation of the debt but actually repudiates it on the grounds that it was a maneuver by capital for capital and against labor.

THE PRIVATIZATION OF SOTE IN LATIN AMERICA

Chile. The first Latin American country to privatize its SOTE was Chile in 1988 (Table 6.2). The *Empresa Nacional de Telecomunicaciones* (ENTEL) was created in 1964, nationalizing most basic telephony service providers. This schema was changed in the mid-1970s when the Pinochet regime introduced a free-market concept of government economic regulation. In 1985 the state started selling shares in ENTEL and the *Compañía de Teléfonos de Chile* (CTC).[4] In the following 2 years, the government sold 30% of ENTEL and 25% of CTC to employees and Chilean nationals. In August 1987, CORFO (the government holding company and development bank that held 80% of CTC's stock shares) announced the sale of 151 million common shares, Series A, which accounted for approximately 30% of the government share in the company. By January 1988, CORFO transferred the shares to Bond Corporation Chile, S.A.,[5] which offered the highest price: $114.8 million U.S. Throughout 1988, the Bond Corporation increased its participation in CTC capital stock, buying shares in various other offerings. By the end of the year, and after investing $285 million U.S., Bond was in control of 50% of CTC's subscribed capital stock. The Australia-based company sold its share to Telefonica International Chile, subsidiary of Telefónica de España, for $392 million U.S.[6] The state also reduced its share of ENTEL below 50% and sold all its shares the following year. The CTC was also the first LDC company to float shares in international stock-exchange markets, selling 110 million shares for $92.5 million U.S. in 1990. The purchase of CTC's controlling share by Telefónica created jurisdictional problems, however, because the Spanish company also had a considerable share in ENTEL. The Spanish common carrier acquired 20% of ENTEL's shares in an early privatization of the company. Additionally, Telefónica's official bank, the Bank of Santander, controlled 10% of ENTEL's voting shares. In this way, Telefónica not only had a strong regional presence[7] but was posing monopoly problems for the Chilean government. Chile's Preventive Commission ruled that Telefónica should keep its shares in only one company, selling its part in the other. But the Spanish entrepreneurs hoped to keep a stake in both companies, and they have appealed the ruling that would ultimately have to be resolved by the Supreme Court. The rest of the stock

---

[4]ENTEL provides long-distance service and CTC provides local service to 77% of the total number of lines.

[5]The corporation, which is part of a holding with interests in gold, beer, yachts, and finances, is owned by Australian business tycoon Alan Bond.

[6]In only twelve months Bond Corporation gained substantial profits from the firm's operation as well as earning $122 million U.S. in the sale.

[7]The company is already in Argentina, Venezuela, and Puerto Rico; it lost in Mexico and it is interested in Paraguay, Uruguay, and other Latin American markets.

TABLE 6.2
Timetable of Privatization and Related Events

	1985	1986	1987	1988	1989	1990	1991
Argentina			Swaps		$\dfrac{\text{Menem}}{\text{Plan}}$	SALE	
Chile	Swaps		Plan	SALE		Sale 2	
Mexico		Swaps	Proposal	Salinas	GATT	SALE	Stock
Venezuela			Swaps		$\dfrac{\text{Perez}}{\text{GATT}}$		SALE
USA	Baker Plan				Brady Plan		

is in the hands of the Bank of New York (16.6%), Pension Funds of Chile (13.8%), foreign funds (6.1%), employees (1.3%), and others (18.6%).

Argentina. Various foreign common carriers, including GTE, Bell-South, Nynex International, Bell Atlantic, Continental Telephone (U.S.), STET (Italy), Telefónica (Spain), Cable and Wireless (England), France Cable and Radio (France), and Siemens (Germany), showed interest in investing in the Argentine telecom sector. The finalists were three consortia, in each case headed by a foreign telephone company and a foreign bank: Bell Atlantic and Manufacturers Hanover; Telefónica and Citicorp; and STET (with France Cable and Radio) and J. P. Morgan. The sale of ENTEL in December 1990 resulted in the formal creation of four new telecommunications firms. The consortia Telecom Argentina and Telefónica Argentina provide basic telecommunications services nationally.[8] In their turn, they are owners of equal parts of the other two companies, Teleintar, an international long-distance service provider, and Startel, which offers "services open to competition," the most important of which are mobile radio telephone, mobile maritime radio, telex, and data transmission within the national boundaries of Argentina.[9] In the

---

[8]Telecom, in the north region, operates 5.8% of the telephones of Buenos Aires, and the total for the provinces of Catamarca, Córdoba, Corrientes, Chaco, Formosa, Entre Ríos, Jujuy, La Rioja, Misiones, Salta, Santa Fé, Santiago del Estero, and Tucumán. Telefónica, in the southern region, operates 94.2% of the lines of Buenos Aires and all of the lines in the provinces of Chubut, La Pampa, Mendoza, Neuquén, San Luis, San Juan, Santa Cruz, and Tierra del Fuego. The rest of the provinces are still served by CAT, which renewed its license and will initiate negotiations with the consortia to sell its part in the system. Telefónica and Telecom have licences to operate in the regions that CAT operates, but they are not obliged to do it if CAT provide services.

[9]Telecom bought 160,000 lines from CAT at a price of $67 million U.S.

case of Telecom, the shares purchased by the consortium were divided among their members according to the following scheme: 30% was awarded to STET and France Cable and Radio, the private companies, and 30% to the Perez Companc group, while Morgan Bank received 10%. For Telefónica, the proportions were different: 33% for Telefónica de España, the only operating firm; 57% for Citibank; and 10% for the Techint conglomerate.[10] Each of these consortia comprised at least one telephone company, an Argentine financial conglomerate, and a U.S. bank. In both PTEs, the U.S. bank was one of Argentina's principal foreign creditors;[11] the domestic financial group was one of the most important of those responsible for the state indebtedness; and the operating company was a foreign-state firm tied to various companies producing telecommunications equipment.[12] At the same time, these financial groups, in joint ventures with equipment companies (that were related to some of the groups involved in the consortium that now owns ENTEL) had been selling services and/or equipment to the state for many years.

There are three European companies involved: (a) Italtel, a subsidiary of STET, which is a long-time supplier of transmission equipment to ENTEL; (b) France Cable and Radio, associated with Alcatel, which, during the years of radical rule and with the support of the French government, tried to make itself into a third ENTEL equipment supplier; and (c) Sintel, a subsidiary of Telefónica Española, which sold ENTEL the ARPAC data transmission network. There were two main domestic investors in the PTEs: (a) Perez Companc, associated with NEC, which entered the Argentine market by selling ENTEL a digital network and later became one of the members of the duopoly that sold switching equipment to ENTEL; and (b) Techint, which besides being one of the main installation subcontractors for ENTEL, is working with Italtel and Teletra on the digitalization of the telephone network financed by a credit line of $135 million U.S., granted by the Italian government. This means, then, that both Telefónica and Telecom are associated through

---

[10]The groups that are led by Techint and Perez Companc consist of more than 50 companies, covering a wide variety of activities.

[11]Citibank leads the Committee of Argentine creditors, and Morgan Bank is one of the twenty most important international creditors of the country.

[12]More specifically: (a) STET is a firm owned by the Italian state. It integrates a holding of the Italian Reconstruction Institute (IRI), that operates in Italy 20 million lines, and has commercial operations in Holland, Germany, the United States, and Britain; (b) France Cables et Radio S.A., is under the control of the state-owned France Telecom, and it operates 28 million lines in France; and (c) Telefónica de España is controlled by the Spanish government (40% of shares), and it operates 11 million lines in Spain. The company has a very aggressive and expansionist policy in other European countries, as well as in the former Soviet Union and Latin America.

144

BARRERA

Italtel and Techint, respectively, on a project supplying transmission equipment.

Upon purchasing 60% of the shares of ENTEL, besides those assets of the company, these consortia acquired, for an undetermined period of time, the license rights to provide public telecommunications services (except for radio stations), monopolistic rights for 7 years (extendible to 10) to provide basic domestic service in their corresponding zones (the South for Telefónica and the North for Telecom),[13] and, above all, the rights for all types of international service through Teleintar.[14]

Mexico. The decision to reprivatize Telmex by President Carlos Salinas was the culmination of a "dance" that lasted for 2 years where various officials took turns in denying the upcoming sale of the telco.[15] The sources of this rejection were, alternately, the secretaries of communications, budget, and revenue; three successive directors of Telmex; the secretary general of the union (STRM), and the president himself. The initiative for the sale started at the VIII Corporative Planning Reunion in 1987, where Carlos Casasús, Under-Director of Strategic Planning, presented the proposal "Structural Change in Telmex." In May 1988, the then presidential candidate Salinas strengthened the rumors when he presided over a meeting titled "Challenges to Modernization: Modernization in the Industrial Sector, National Infrastructure, Telecommunications and Informatics" in Monterrey. The powerful local economic groups proposed the reprivatization of Telmex, to which the candidate answered by saying that there was a serious lag in telephone services and added that in general he agreed with the proposals. Parallel to that process was a media blitz against the telco where the private television consortium Televisa and the newspapers *Novedades* and *El Norte* were the most aggressive. The campaign was so effective that *The New York Times* and *Telephony* echoed it. Coincidentally, behind Televisa and *Novedades* was Rómulo O'Farrill, who was an important stockholder of the telco before 1972 and is now one of the partners of Carlos Slim. Something distinctive about that campaign was the contribution made by Telmex itself through full-page ads in several newspapers with national circulation announcing how they did not meet goals of installation, repair, and maintenance of an emergency plan.

---

[13]During this period, third parties are allowed to install point-to-point private lines, but they cannot resell. The permission to install the system is granted if Telecom or Telefónica do not provide the service in 180 days.
[14]Includes, besides telephony, data transmission, telex, and value-added services.
[15]Telmex started as two separate companies at the turn of the century, owned by ITT and Ericsson. They merged in 1940 and were bought by domestic capital in 1958. The Echeverría administration converted it into a SOTE in 1972 by acquiring 56% of the stock.

On September 18, 1989, Carlos Salinas de Gortari, President of Mexico, made public his decision to reprivatize Telmex. On December 9, 1991, the Mexican Secretary of Revenue announced that 20.4% of the social capital of Telmex had been sold for $1,757.6 million to the Carso Group, headed by Carlos Slim;[16] France Cable and Radio, subsidiary of France Telecomm; and Southwestern Bell. Only two other groups got to the final phase of the sale of this SOTE with 21 subsidiaries, ranging from publishing and construction companies to real estate firms.[17] The first was integrated by *Acciones y Valores de Mexico*, headed by Roberto Hernández; GTE Telephone Corporation; and *Telefónica de España*, whose bid was 4% lower. The other finalist was the Gentor Group with no foreign partners.[18] The second stage of the Telmex privatization was also a financial success for the Salinas administration. By mid-June 1991, the government had sold "L" type shares representing 16.5% of the company in foreign stock markets, for $2.27 billion U.S.[19] Mexico sold 1,745 million "L" type shares, which were offered in stock markets all over the world in the form of American Depository Shares (ADS), and cost $27.25 U.S. for each ADS.[20] Telephone workers and the controlling consortium also purchased Telmex shares. The union, using a credit of $325 million U.S. from the Mexican government, bought 187 million type "A" shares through *Nacional Financiera* (Nafin), which constitutes 4.4% of Telmex's capital. Carso

---

[16]At the time of the reprivatization, Carlos Slim ranked 54th among Mexican entrepreneurs and held stock in 43 firms including *Inversora Bursátil* (Inbursa), Sanborn's, Frisco, La Tabacalera Cigarrettes, Euzkadi Rubber Company, Loreto y Peña Pobre Paper Factories and Segumex.

[17]The subsidiaries of Telmex are Anuncios en Directorios, S.A.; Compañía de Teléfonos y Bienes Raíces, S.A. de C.V.; Construcciones Telefónicas Mexicanas, S.A. de C.V.; Canalizaciones Mexicanas, S.A. de C.V.; Construcciones y Canalizaciones, S.A. de C.V.; Alquiladora de Casas, S.A. de C.V.; Editorial Argos, S.A.; Fuerza y Clima, S.A.; Imprenta Nuevo Mundo, S.A.; Impulsora Mexicana de Telecomunicaciones, S.A.; Industrial Afiliada, S.A. de C.V.; Operadora Mercantil, S.A.; Renta de Equipo, S.A. de C.V.; Sercotel, S.A. de C.V.; Servicios y Supervisión, S.A. de C.V.; Teleconstructora, S.A.; Teléfonos del Noroeste, S.A. de C.V; and Radio Movil Dipsa, S.A. de C.V. The last operates Telcel, the cellular telephony service that holds the monopoly for B-band nationwide, whose competitors on A-band in all eight regional markets bid for the concessions paying up to $12 million U.S. plus 12% of federal participation. Shortly before its privatization, the telco acquired a number of circuits in submarine cables TAT-5, TAT-9, TAT-11, and PTAT-4, along with the Federal Microwave Network that it was leasing.

[18]Those who didn't make it to the final phase but participated formally in the earlier stages were Inverlat (a stockbroker firm) and a long list of foreign firms: NYNEX International, Bell Canada, Nippon Telephone and Telegraph (NTT), Cable and Wireless, Singapore Telecomm, United Telecommunications, U.S. Sprint, and Citibank.

[19]Another element that the financial reform of Telmex brought about was the creation of type "L" shares. These new "L" shares are nonvoting shares, and they are valued 2.5 times less than the traditional "A" Telmex shares.

[20]Each ADS contains 20 "L" type shares.

Group, Southwestern Bell, and France Cable and Radio had access to
5.1% of the capital through type "L" shares. In this way, Southwestern
Bell, for example, buys $467.3 million U.S. worth of "L" shares, doubling
its participation in Telmex. The sale of the last portion of government
stock (0.5%) took place in May 1994 through swaps in the European stock
market through Inbursa (Slim's own broker firm) and Merrill Lynch at
$550 million U.S.

Telmex will keep the monopoly over local service with periodical
reviews, but it will have to provide interconnection capabilities by 1996
when the long-distance service market will open to competition. Telmex
created a joint venture with Sprint and will compete with foreign carriers that are preparing to enter that market including the following projects:

- AT&T made an alliance with the Alfa Group from Monterrey after
  declining the joint venture with Telmex that many analysts expected.
- MCI formed a joint venture with Banamex, the largest domestic bank,
  and is already building a fiber-optic network starting with the three
  largest cities.
- British-based Ionica sold 15% of its stock to the Pulsar Group for
  $22.5 million U.S. and is preparing for this market.
- Cellular firms with conspicuous foreign investment in both bands
  are already bypassing Telmex nationally and internationally with
  their own digital microwave routes that may be converted to long-
  distance carriers.

Venezuela. In May 1991, eight MNCs acquired 40% of the stock of
Compañía Anónima Nacional Teléfonos de Venezuela (CANTV) for $1,885
million U.S. This group was headed by GTE and included AT&T and
Telefónica Española among the foreigners and the domestic companies
Compañía Anónima La Electricidad de Caracas and *Consorcio Inversionista
Mercantil* as the domestic companies. The Venezuela Investment Fund
retained 49% of the shares, while the employees ended up buying 11%
of the stock, which was in trust of Citibank.

Jamaica. Even before Chile, the Jamaican government started selling
shares of Telecommunications of Jamaica to Cable and Wireless and to
the public and employees. However, at that point the government
retained 82.7% of the stock, whereas the British telco got 9.4% and the
public 7.8%. By September 1988, the state held less than 50% of the stock
for the first time (40%), whereas Cable and Wireless increased their share
to 39%, the public 20.9%, and the employees got 2%. Ten months later
Cable and Wireless achieved an absolute majority with 59%, leaving the

state with only one fifth of the stock. In November 1990, the government of Jamaica sold the one fifth share to the British company, enabling it to arrive at its current 79% of the shares.

Cellular and Enhanced Services. Besides foreign participation in local and long-distance telephone services, Chile, like many other Latin American countries, has a considerable amount of foreign investment in its cellular telephone system. The market, which is growing fast, is divided among four companies: (a) CTC holds a license for nationwide services; (b) Cidcom de Telefonía Celular de Chile, owned by Pacific Telecom, holds licenses for Santiago and Valparaíso; (c) VTR Telecomunicaciones, owned by the British company Millicom and VTR Telecomunicaciones, holds a nationwide license that excludes Santiago; and (d) Telecom, constituted by Entel-Chile (33.3%), Télex-Chile (33.3%), and Motorola (33.3%), also holds a nationwide license that excludes Santiago.

The Argentinian ENTEL's privatization has considerably increased FDI in the national telecom market. However, foreign capital was already present in the early diversification of the industry, such as the development of cellular telephony and data services. In early 1988, the government opened to private investment the concession of cellular telephone services in the Buenos Aires area. By July 1988 a consortium led by BellSouth won the public bidding and was granted the license to operate what in 1989 came to be the first private cellular network in Latin America. The consortium, which had strong foreign participation and operated under the commercial name of Movicom, was constituted by BellSouth (31%); Motorola (25%); Citicorp (8%); and two local companies, Socma (16%) and BGH (20%). The government also planned to offer to private investors concessions of cellular systems throughout the country by the end of 1991.

In data and enhanced services, a fast-growing telecommunications sector, foreign financial companies are becoming partners of local entrepreneurs or installing their own services.[21] Besides Startel (owned by the new owners of Entel) the government granted concessions to five companies to operate data transmission at the domestic level: Impsat, Satelnet, and, most recently, Alcatel are the active ones; the other two companies, Keydata and Tecsel, hold licenses to operate but are currently out of the market. Among the active companies, Satelnet is a good example of joint venture trends with foreign capital participation. The company, originally owned by a local group, today includes the participation of the Italian

---

[21]According to studies carried out by the *Camara de Informática y Comunicaciones*, the Argentine data market has a potential for approximately $500 million U.S. a year (*El Cronista Comercial*, 1991).

financial group Banca Nazionale del Laboro. Other telecom services, such as electronic mail, are carried out in Argentina through U.S. companies such as Sprint Mail and MCI Mail.

Concessions for cellular telephony in Mexico had generated more than 140 applications to the SCT by 1989 (Ibarra, 1989). These applications come from a wide spectrum of entities, ranging from Telmex to small town businessmen. In order to assign concessions, the SCT divided the country into nine regions, two of which had been already operating: the Tijuana area, where DIPSA, one of the subsidiaries of Telmex was the sole concessionaire, and the Mexico City area, where the same firm and IUSACELL shared the market. The latter already had a solid customer base of SOS, the parent company owned by Alejo Peralta, a close friend of the Salinas clan. Peralta took advantage of technicalities in the wording of the 50-year concession given to SOS by the SCT in 1957 that gave him the monopoly in 27 cities for mobile telephony without considering future developments in telecommunications. Peralta avoided a big legal fight by keeping only the main market to exploit it with cellular technology instead of with the original radio communication still useful in low-density areas.

These franchises had a foreign investment component, as well as a public one. This pattern of ownership included a public component that was supposed to be inefficient and should not be involved with these types of enterprises according to the pro-privatization arguments (Table 6.3).

There are four companies offering satellite data transmission directly to the United States: Vitacom, Houston International Teleport (HIT), GTE Spacenet, and OTI. The first two have few users (e.g., Vitacom operates isolated plants of General Electric in Altamira and San Luis Potosí). HIT was the first international user of Morelos but remains with one or two clients. GTE is just trying to get into the market. Telecomm, the concessionaire of the Morelos System, may operate through commercial agents like SERSA. SERSA is linked to GeoComm in San Antonio and has a teleport in Ciudad Juárez. They have 18 MHz in Morelos I and had to bring the signal down in their teleport before sending it through MW across the border. The teleport has 45 Mbps. Their 20 users are concentrated in Coahuila, Tijuana, and Mexico City. The target market of SERSA is changing from the export-oriented maquiladora industry to large national VSAT networks.

These five cases of privatized telecom entities (PTEs) illustrate how the process of privatization in Latin America is characterized by the key role of foreign capital in the form of either telecommunication firms or financial institutions from Europe and the United States. Four types of domestic capital are present in this process: (a) holdings with a finance branch, (b) social funds, (c) the general public, and (d) labor.

TABLE 6.3
Franchises of Cellular Telephony in Mexico

Franchise	Region	States	Domestic Investors	Foreign Investors	Offer (U.S. M)[a]
Baja Celular Mexico (Bajacel)	Baja Cal.	Baja Cal. S. B. Cal.	Aurelio Lopez Grupo Canada	General Cellular Co.	1.4 6%
Movitel del Noroeste	NW	Sonora	Tamsa Ind. Bachoco Ind. Balco D.B.M.	McCaw Cellular Comm. Contel Cellular	4.5 4%
Telefonia Celular del Norte (Norcel)	North	Chihuahua Durango (Laguna)	Domos Int.	Motorola Centel	.75 4%
Celular de Telefonia	NE	Coahuila N. Leon Tamaulipas	Protexa	Millicom	2.78 12%
Comunicaciones Celulares de Occidente	West	Jalisco Michoacan Colima	Grupo Hermes Grupo Canada	BellSouth Racal	7.0 5%
Sistemas Telefonicos Portatiles Cellulares	Central	S.L. Potosi Aguascalientes Zacatecas Guanajuato Queretaro	Heraldo de Mex.	Bell Canada	6.9 5%

(Continued)

149

TABLE 6.3
(Continued)

Franchise	Region	States	Domestic Investors	Foreign Investors	Offer (U.S. M)[a]
Telecomunicaciones del Golfo	S. Gulf	Veracruz Puebla Tlaxcala Guerrero Oaxaca	IUSA G.M. Desarrollo Jorge Zapata	Bell Canada	11.6 5%
IUSACELL	Metro	Mexico City Mexico St. Morelos	IUSA	Mitsubishi	0
DIPSA	Metro	Mexico City Mexico St. Morelos	Telmex	NA	0

Note. Raw data from Ibarra (1990).
[a]The percentage below the offer is the share given to the Mexican government. Some changes in the ownership of cellular companies include the sale of BellSouth's share of the Western region franchise to IUSACELL and Motorola moving toward dominating the four northern franchises as part of its Iridium project of low earth-orbiting satellites.

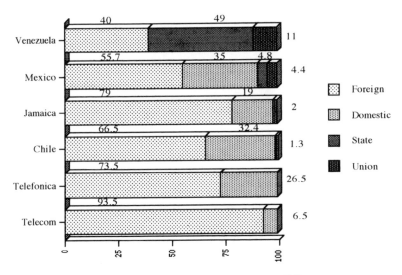

FIG. 6.2. Distribution of ownership in PTEs.

Argentina and Mexico are the only two cases where domestic private capital has a largely visible role with the participation of Gregorio Péres Companc and Carlos Slim, respectively. The two social funds are the Pension Funds of Chile and the Venezuela Investment Fund. The PTEs workforce is present in all cases except in Argentina (Fig. 6.2).

POLITICAL SYSTEMS IN LATIN AMERICA AND PRIVATIZATION

At first glance, the study of a phenomenon like the privatization of telecommunications in Latin America can be done as a block. It is true that all these countries went through a debt crisis, that they all privatized their SOTEs within 4 years, and that they killed the welfare state by espousing neoliberal economic policies in the recent past. The same actors are present in all cases: the international organizations (IMF, World Bank, and GATT), the U.S. government through the Baker and Brady plans, and the foreign large users that put on the external pressure to privatize. Internal forces were domestic large users and top management of SOTEs. Opposition parties and the labor unions of the telcos showed some resistance until the latter were coerced or coopted (Table 6.4). The historical academic construction of "Latin America" as a homogeneous whole is one of the reasons why it is somewhat surprising to find certain differences in the distribution of ownership of the PTEs and such profound differences in their political systems, which explain the former.

TABLE 6.4
Social Actors Influencing Privatization

	Sectoral	Nonsectoral
Foreign	ITU World Bank GATT	IMF Baker Plan Brady Plan MNCs
Domestic	Management Union	Financial sector Mass media Political parties

There are three dimensions of the political systems that help explain the different outcomes and pressures: the nationalist orientation, the capitalist orientation, and the centralization in the executive branch. The nationalism of a policy is not a matter of a binary classification but a matter of degree. Besides, the general set of policies may contain some that do not have coherence in their nationalism. However, individual policies or an abstract "factor" of the set of policies could be located in a spectrum that ranges from nonnationalistic to nationalistic for exploratory purposes. For example, a communication policy that reserves the exploitation of border telephone lines to Mexican citizens could be high in nationalism, whereas the policy of authorizing U.S. companies to exploit the domestic satellites could be low on the same spectrum. Despite the range of policies, each country has a general nationalist orientation. Before the privatization, Chile was the only country that had an open economy dating back to the coup by Pinochet. In Argentina, Mexico, and Venezuela, nationalism became institutionalized and codified in the legal framework during the Import Substitution phase. This may explain why Chile privatized its SOTEs even before the Brady Plan was proposed. It was after the Brady Plan was in operation that all these countries moved to a less nationalistic orientation (Fig. 6.3).

There are two levels of the capitalist character of national telecommunication policies: the commercial orientation of the telecommunications industry given by (a) the degree of participation of the private sector, and (b) the degree they protect or promote a high rate of capital accumulation. Typically a high degree of private participation corresponds to a high degree of capitalist orientation. However, a telecommunication service provided by the state does not mean it protects or promote capital accumulation to a lesser degree. A policy of investing vast resources in infrastructure for large users could be highly capitalistic, whereas giving priority to rural telephony or telemedicine is low on the same spectrum.

Bob Jessop (1982) stated that "the mediating institutions between political forces do not exist independently of the State: they are shaped in

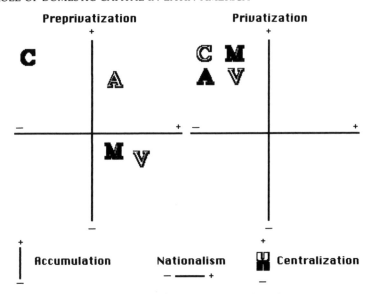

FIG. 6.3. Political systems in Latin America.

part through its forms of representation, its internal structure, and its forms of intervention" (p. 122).

State intervention can be in the form of (a) formal facilitation to maintain the general external conditions of capitalist production and implies a self-expanding and self-equilibrating capitalist production; (b) substantive facilitation to reproduce certain general conditions like labor-power and infrastructure; (c) formal support to alter the general external conditions of production and it is left to the market forces to determine whether these changes are exploited by free and autonomous economic agents; (d) substantive support in the direct allocation of particular conditions of production to particular economic agents rather than leaving it to the autonomous choice of market forces and can be in the form of licenses, monopolies, state credit, state sponsorship, and so on; and (e) direction to override the

> . . . formal freedom of economic agents and directs that they either act or refrain from acting in particular ways . . . (the State) intervenes to support, counteract, or modify them through restrictions on the formal autonomy or freedom . . . (it) may promote the substantive rationality of capitalism through recognition of the substantive interdependence among economic agents and promotion of their collective interest at the cost of their particular interests. (Jessop, 1982, p. 234)

Argentina, Mexico, and Venezuela did redirect their intervention and moved from formal and substantive facilitation to formal and substantive support. State interventions cannot be thought of as totally arbitrary activities because they are influenced by the demands and struggles of different sectors. The systems of representation that mediate class-specific and nonclass demands are *raison d'état*, parliamentarism, corporatism, clientelism, and pluralism. Pinochet's Chile was a clear example of the first category, while Mexico is a clear case of corporatism and Argentina and Venezuela have a more pluralist system. The importance of the centralization of power in the executive branch lies in the fact that economic liberalization tends to be incompatible with political competition. This explains why "Chile is a good example of a rather authoritarian government that understands what it has to do and has a much easier time at it. To some extent, that is Mexico's situation as well" (McCormack, 1992). On the other hand, the Alfonsin administration failed in an earlier attempt to privatize ENTEL because of the pressures of various social coalitions, while Menem was successful by insulating the state from societal pressures and concentrating power in his own hands.

THE STRANGE WAYS OF THE ESTRANGED
DOMESTIC CAPITAL

Despite the differences in political systems and the apparent blame of the public sector in piling up the foreign debt that led to economic liberalization, it was a form of nonconspicuous substantive support to a large extent. According to Petras and Brill (1988), foreign debt was being used to finance foreign investment and creating a new class: the transnational capitalists. They argue that large portions of the debt financed capital flight. It would have accounted for 98% in Argentina, 87% in Mexico, and the total amount in Venezuela (Fig. 6.4).

Domestic capital funnels its savings to international banks, which, in turn, lends capital to Latin American governments. These governments,

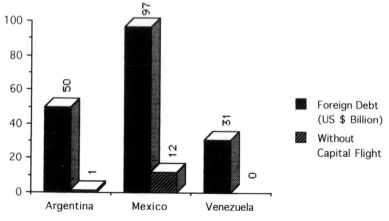

FIG. 6.4. Foreign debt and capital flight in Latin America.

in turn, lend to private capitalists. These mechanisms are used by capital to protect their savings, while the risks of the loans are guaranteed by the state. This circuit of Latin American transnational capital explains the nominal or real poor participation of domestic private capital, except in Mexico and to a lesser extent Argentina.

The two visible domestic capitalists are Gregorio Pérez Companc in Argentina and Carlos Slim in Mexico. There are some similarities like the sectoral diversification that includes a finance arm. Pérez Companc is the head of a clan that owns Banco Río and is the leader in oil production (PASA), telecommunications (Pecom Nec, Telecom, and Telefónica), construction (SADE), nuclear energy (Nuclear, Conuar, and Fae), data processing, metallurgy (San Eloy, Mellor Goodwin), agribusiness (Pecom Agra and Aguila Saint), supermarkets (Carrefour), and shopping malls (Alto Palermo and Paseo Recoleta) (Seoane & Martínez, 1993). Carlos Slim, the head of Grupo Carso, owns *Inversora Bursátil* (Inbursa) and holds stock in 43 firms, including Sanborn's, Frisco, La Tabacalera, Euzkadi Rubber Company, Loreto y Peña Pobre Paper Factories, and Segumex (see Fig. 6.5).

Another similarity is how they were accused of illegal practices. Pérez was accused of discriminatory provision of services denying long-distance access to the firm *Expresión y Medios* for teleconferencing, while granting the service to Television Española. Slim was accused by other stockbrokers of inside trading of Telmex through Inbursa. The last similarity is the rumors about being frontmen of more powerful interests. It has been rumored that Pérez is the frontman of the Opus Dei and even the Vatican, whereas Slim is rumored to be the frontman for President Carlos Salinas

FIG. 6.5. Circuit of Latin American transnational capital.

himself and/or Josep Córdoba Montoya, the French-born chief of staff until
early 1994, when he was sent to the Interamerican Development Bank after
the Chiapas revolt and after being linked by some sectors to the assassina-
tion of the ruling party's presidential candidate.

There is a difference that is as important as the similarities. While Pérez
belongs to an old dynasty and is the richest man in Argentina, Slim was a
nouveau rich who ranked 54th in Mexico. However, this difference was
eliminated when Slim became one of the four richest men in the world
along with Yoshiaki Tsutsumi, William Henry Gates III, and Warren
Edward Buffett. His fortune increased by 78% during 1993, amounting to
$6.6 billion U.S. The Argentine state was strengthening the old oligarchy,
while the Salinas administration was creating a young capitalist class. This
seems to be part of a restructuring of Mexico's system, where Salinas
momentarily stopped the growth of the old established groups. Salinas put
in jail the head of the Legorreta clan who owned the largest bank and did
not support the expansion efforts of Emilio Azcárraga's Televisa. The latter
was not allowed to bid for Telmex nor to participate in the design of the
Solidaridad satellites to be launched in 1994, since it needed transponders
for direct broadcasting. On the other hand, the Salinas administration
strengthened the Vargas group to compete with Televisa by granting them
concessions of wireless radio, MDMS, and special satellite services.[22]

The contrast between these two capitalists reveals the two strategies
of Argentina and Mexico: Menem wanted to close down the political
system and strengthen the oligarchy, whereas Salinas wanted to break
the old corporatist system while keeping the concentration of power in
the executive branch and realigning the political and economic forces.

THE SHORT-CIRCUIT OF LABOR: WORKERS
AS CAPITALISTS[23]

One of the fundamental actors in the privatization of SOTEs in Latin
America was labor. The general pattern in the region was that of workers
moving from a strong opposition to becoming partners in the deal.
Telephone workforce in Latin American countries is large both in absolute
and relative terms. Whereas Chile has 7,500 workers, Argentina and
Mexico have almost 50,000 each (Fig. 6.6).

The size of the workforce is a result of the state of the technology in
the region, which still has a lot of analog equipment. This situation results
in a relatively large number of workers for every thousand lines: 14.5 in
Argentina, 11 in Chile, and 9.5 in Mexico (Fig. 6.7).

---

[22]Televisa revived domestically at the end of the Salinas administration with more than
50 concessions for television and the marketing and decisive influence over the soccer
national team in the 1994 World Cup.

[23]This section is largely based on data, drafts and comments by Ben Alfa Petrazzini.

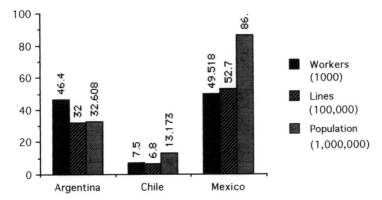

FIG. 6.6. Workers and lines in Latin American countries. Unpublished raw data from Petrazzini (1992).

In the telecommunications sector, a worker's acquisition of shares is present in all cases. In Mexico, the Salinas administration granted Telmex employees a low-interest loan of $325 million U.S. for the purchase of 187 million Type A shares. Through this acquisition Telmex employees control 4.4% of the company's capital. The Chilean government, through a creative financial mechanism, also brought workers into the sale of Compañía de Teléfonos de Chile. The government offered workers 50% of their severance payment in advance, with the condition that they would use 80% of it to buy shares of CTC. It was guaranteed also that the value of shares would not fall below their entitled severance payment at the moment of retirement. Through this offer, 84% of the company's workers bought 6.4%

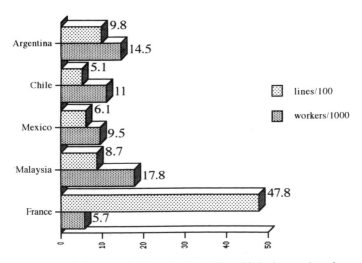

FIG. 6.7. Telephone and labor densities. Unpublished raw data from Petrazzini (1992).

158

BARRERA

of CTC shares. Two years later the government sold shares to public sector employees. With this purchase, by 1989, workers controlled 11.5% of the company. (Galal, 1992). However, this share has declined in relative terms as the company's capital stock grew. In October 1992, CTC workers controlled only 1.3% of the company's shares (Terol, 1993).

In Argentina, the Programa de Propiedad Participada (PPP) specifically required that 10% of a privatized company's stock go to its employees. Following the sale to new private corporate owners, 10% of ENTEL's shares were reserved for the companies employees. Yet, as of late 1992 no agreement had been reached regarding the specifics of how to transfer these shares. Despite PPP regulations that established that shares must be purchased at market value and financed by the new enterprises over a period of time, controversy has arisen regarding the current value of those shares. One of the main problems is that the value of the shares at the moment of ENTEL's sale had been radically altered by the sharp increase in the value of the company following privatization.

Workers' participation in the ownership of the privatized SOEs has a variety of possible consequences for the company, for labor, and for society as a whole. Some analysts argue that labor participation in ownership reduces the level of intrafirm conflicts, improves productivity, reduces costs by erasing theft and unbilling, and creates a corporate, entrepreneurial spirit in the labor force. Others, based on recent empirical data from 12 privatized SOEs, argue that productivity grew in eight of the 12 cases analyzed—not so much because of better management or a stimulated capitalist work force—but due to massive work force reduction (Galal, 1992). There is some evidence from the Chilean electricity sector that privatization has reduced losses—such as theft and unbilling—improving operational costs, which eventually can translate into reduced tariffs. However, it is not clear whether the reduction of losses is due to higher commitments of workers to the company or to a better performance of private managers in achieving this task. Finally, the level of labor conflict has diminished in most privatized SOTEs. Some see this as a consequence of the workers' unwillingness to take action against the company, which could jeopardize the image of the firm in the stock market, and subsequently affect the value of their stock.[24] However, it is not clear whether the source of stability is the new ownership in the company, higher salaries, the fear of dismissal, the weakness of unions, or other more general factors.

Labor's ownership of shares has granted workers two valuable assets: access to information about the company and power derived from the

---

[24]This indirect consequence of workers' ownership has been clearly identified: Some union leaders are advising workers not to accept shares from would-be-privatized SOEs. According to the head of the Congress of South African Trade Unions, Mr. Mashele, the institution is advising workers "not to accept these shares," because "workers who buy shares will be reluctant to take industrial action" (Wren, cited in Petrazzini, 1992).

threat of massive sale of shares in the stock market. In a period in which access to reliable and relevant information becomes crucial for any kind of bargaining with the company,[25] workers will have, due to their right as shareholders, the ability to request relevant information from management. Further, recent events have highlighted the potential impact that worker decisions regarding their shares may have in local and international stock markets. In Mexico the rumors spread on June 14, 1992, that workers were planning to sell their 315 million shares. This put such pressure on the stock market that Telmex shares dropped sharply in the following days. Only the formal announcement from the union's leadership about a majority decision by the workers to keep their shares and create a *fideicomiso* reversed the trend, allowing Telmex shares to recover their attractiveness in local and international capital markets. After 12 years of oppositional union politics followed by five years as the leader of the pet Union of the Salinas administration, Francisco Hernández Juárez has been forced by rank and file members to express some degree of dissent. The leader of the STRM was asked by members to negotiate a doubling of their wages just after they knew the scandalous increase of Slim's personal fortune (Acosta Cordova & Robles, 1994).

In social terms, it has been argued that employee ownership contributes to broaden and redistribute property in the country. This is scarcely true for most Latin American privatization. In Chile significant efforts were made to incorporate workers and small investors as shareholders of privatized SOEs, but in most other countries privatization led to higher levels and new forms of capital concentration. This seems to be the case in Argentina and Mexico.

FINAL CONSIDERATIONS

The generally low participation of domestic capital in the region points out three general considerations related to the traditional categories for analyzing the actors involved in these processes:

First, there is a need to redefine the notions of private and social capital. Whereas Chile is the country with the most liberal economy and Venezuela is the country that applied the IMF conditions with a shock treatment, there is no significant participation by private capitalists in either country. The domestic capital in these two countries is of a social nature with the participation of a pension and a public fund. Mexico and Argentina have the participation of two capitalists with diversified operations but with differences that reveal the different strategies of a political agenda. Menem wanted to concentrate power in the executive branch and strengthen the

---

[25]Asymmetric information on the market and the corporate business is one of the key concerns of regulators in the postprivatization period. See Helm and Yarrow (1988).

oligarchy, whereas Salinas was using the privatizations to restructure the political and economic forces while retaining power.

Second, there is a need to redefine the distinctions between foreign and domestic capital. The flight of Latin American capitalists and their integration in the circuit of transnational capital makes it difficult to differentiate among national lines. The joint ventures and the cartelization and globalization of firms will make the distinction increasingly difficult to establish. According to Gordon (1988), the best predictor for the relocation of the operations of MNCs were the economic advantages combined with a stable institutional climate. This climate consists of a state willing to attract foreign investment by creating infrastructure with an amortization period that is much longer than that of the plant and equipment of the firm. This could change if the development of infrastructure is privatized. Domestic capital will have the risk of expropriation and would rely on the state as guarantor of foreign credit. Foreign capital would have the long-term monopoly situation as the only guarantor. The risks are higher than before because of the social impact of the dismantling of the welfare state, a process linked to privatizations, resulting in 'IMF Riots' such as those in Venezuela and Peru.

Third, the need to redefine factors of production. The addition of the dimension of capital to labor short-circuits the latter by "widening in the differentials between those with the power to maintain or increase the real wage and those who could not. Such widening differentials tended, *ceteris paribus*, to decompose previous structures of power in the working class in favor of capitalist control" (Cleaver, 1990, p. 7).

APPENDIX A
Distribution of Shares in Telecom, Argentina

Company	Percentage of Shares
France Cable and Radio[a]	44.25
STET[b]	44.25
Compañía Naviera Perez Companc[c]	6.50
J. P. Morgan Bank	5.00

[a]France Cable and Radio is the international branch of the state-owned France Telecom, a powerful telecom conglomerate that provides monopoly services throughout France. The company also controls Alcatel, which is one of the main providers of telecommunications equipment in the Argentine market.

[b]STET, Societa Finanziaria Telefonica, is part of the Italian IRI (Institute for Industrial Reconstruction). The telecommunications holding company controls a variety of other enterprises related to the sector such as SIP, Italcable, Telespazio, Italtel, Sirti, Aet, Necsy, Siemens Data, and Cselt. The company also has indirect participation in Sic, Consultel, Accesa, Seva, SIRM, Telesoft, SSGRR, Data Spazio, and Teleo.

[c]Compañía Naviera Pérez Companc is one of Argentina's most powerful economic groups, and has a considerable share in Pecomnec, an Argentine–Japanese telecom equipment producer that had a close commercial relationship with Entel.

APPENDIX B
Distribution of Shares in Telefónica de Argentina

Company	Percentage of Shares
Citicorp Venture Capital	20.00
Banco Rio (Cayman)	14.56
Telefónica Internat. Holding[a]	10.00
Techint[b], S.A. (España)	8.31
Banco Central	7.04
Comercial del Plata	5.00
Banco Hispano Americano	5.00
Manufacturers Hanover	4.33
Bank of Tokyo	4.16
Bank of New York	4.16
Zurich Ltd.	4.16
APDT	4.03
Arab Banking Co.	3.41
Republic New York Financiera	1.50
Centrobanco (Panama)	1.42
Vanegas Ltd.	1.25
Banco Atlantico, S.A.	0.75
Bank of Nova Scotia	0.60
BFG	0.30

[a]Telefónica de España, the telecom operator of Cointel, is a society under Spanish law, with minority participation (34.1%) of the Spanish government, a distributed share among foreign investors (22%), and domestic capital (43.9%).

[b]Techint is an Italian multinational that provides telecommunications construction services (among many other commercial activities). It also has much experience in the Argentine market.

APPENDIX C
Distribution of Preferred Shares of Telefónica de Argentina

Company	Percentage of Shares
APDT	24.1
Manufacturers Hanover	15.5
The Bank of Tokyo	9.6
Union Bank of Switzerland	6.1
West LB International	5.4
First National Bank of Boston	5.1
Select Arcturus Holding	4.2
Generale Bank	4.0
Riobank Inter, Trust	3.9
Bank Fuer G.A.	2.9
Westdeutsche L.G.	2.9

(Continued)

APPENDIX C
(Continued)

Company	Percentage of Shares
Société Generale Panama	2.7
Swiss Bank Corporation	2.6
DG Bank Deutsche G.	1.7
Midland Bank PLC	1.7
Deutsche-Sudamericanische	1.3
Dresdner Bank A.	1.3
Banco Hispano Americano	1.1
N.V. Philips G.	1.1
Bank of Nova Scotia	1.0
Caterpillar Americas Co.	0.7
R N B and Co.	0.7
Republic New York Finanziaria	0.4

REFERENCES

Acosta Cordova, C., & Robles, M. (1994). Hernandez Juárez, sobre la enorme fortuna de Slim: "Tengo mucho respeto por Carlos, pero es una manifestación lamentable de lo que no debe ocurrir" [Hernandez Juarez about the huge fortune of Slim: "I respect Carlos very much, but it is a sad manifestation of something that should not happen"] *Proceso, 924*, 27–30.

Cleaver, H. (1990). Close the IMF, abolish debt and end development: A class analysis of the international debt crisis, *Capital and Class*, 17–50.

Furtado, C. (1984). *No to Recession and Unemployment: An examination of the Brazilian Economic Crisis* (London: Third World Foundation)

Galal, A. (1992, June 11–12). Chile: Background, CHILGENER, ENERSIS, Compañía de Teléfonos de Chile Paper presented at the World Bank Conference on the Welfare Consequences of Selling Public Enterprises, Washington, DC.

Gordon, D. M. (1988, March–April). The global economy: New edifice or crumbling foundations?" *New Left Review, 168*, 24–65.

Helm, D., & Yarrow, G. (1988). The regulation of utilities, *Oxford Review of Economic Policy, 2*.

Ibarra, M. E. (1990, March 26). Concesiones sin explicación otorgadas por la SCT. Los beneficiarios de la telefonía celular: Hank Rohn, Peralta, Gabriel Alarcón, Fernandez Hurtado, Aleman, Claudio X. González . . . [Franchises without explanation given by the SCT. The winners of cellular telephony: Hank Rohn, Peralta, Gabriel Alarcon, Fernandez Hurtado, Aleman, Claudio X. Gonzalez . . .]. *Proceso, 699*, 25–29.

Jessop, B. (1982). *The capitalist state: Marxist theories and methods*. New York: New York University Press.

Lerner, N. C. (1989, April 24). Mexico's development dilemma. *Telephony*, pp. 26–35.

Lipietz, A. (1987). *Mirages and miracles: The crises of global Fordism*. London: Verso.

Lipietz, A. (1989). Three crises: The metamorphoses of capitalism and the labour movement. In M. Gottdiener & N. Komninos (Eds.), *Capitalist development and crisis theory: Accumulation, regulation and spatial restructuring* (pp. 59–95). New York: St. Martin's Press.

McCormack, R. (1992). Latin America: Adjusting from illusion to substance. In R. Grosse (Ed.), *Private sector solutions to the Latin American debt problem* (pp. 177–181). New Brunswick, NJ: Transaction Publishers.

Mosco, V. (1989). *The pay-per society: Computers and communication in the information age.* Norwood, NJ: Ablex.

Parhizgari, A. M. (1992). Latin American debt-equity swaps. In A. Jorge & J. Salazar-Carrillo (Eds.), *The Latin American debt.* New York: St. Martin's Press.

Petras, J., & Brill, H. (1988). Latin America's transnational capitalists and the debt: A class-analysis perspective. *Development and Change, 18,* 179–201.

Petrazzini, B. A. (1992). The politics of telecom reform in developing countries. La Jolla, CA.: Mimeo.

Samarajiva, R. (1990). Towards a theoretical understanding of telecommunication policy in the third world. Paper presented at the 17th Congress of the International Association of Mass Communication Research, Lake Bled, Yugoslavia.

Seoane, M., & Martínez, O. (1993, January 3). Argentina, S.A. *Noticias*, pp. 76–87.

Terol, L. (1993, January). Development of the telecommunications sector in Chile and its role within the Pacific basin. Paper presented at the Pacific Telecommunications Conference, Hawaii.

7

Telecommunications Privatization and Capital Formation in the ASEAN

Meheroo Jussawalla
East–West Center, Honolulu

It is a rare occurrence in history when an innovation in economic and financial policy catches on as quickly and in so many different countries as privatization of publicly owned monopolies and deregulation of privately owned ones. Since the beginning of the 1980s this trend has spread across continents from affluent nations to low-income ones, representing a structural shift that favors greater reliance on market forces for efficient resource allocation. The central issue is control. Who will control the switches of information and who will be the beneficiaries of such control? The popularity of the privatization movement rests on the assumption that competition will better serve the users and protect them from predatory pricing. However, policies to deregulate have to contend with existing cultures, traditions, and laws and cannot be transplanted from one country to another. These considerations often take precedence over economic factors in the saga of deregulation of telecommunications in the Asia-Pacific region.

International telecommunications networks are largely a collective, cooperative, and pragmatic process even though there is fierce competition in the provision of services and equipment. There are contentious issues of standardization, interconnectivity of networks, allocation of the frequency spectrum, services trade, and equitable access to information resources that persist and have significant economic and social ramifications. An important change wrought by the electronic revolution is the reliance that society places on information. Information Technology (IT)

165

has altered the processes of production and trade. It is no longer necessary for the collocation of labor and capital for production to take effect just as it is not necessary to establish the right of presence in the country to which services are being exported (Jussawalla & Lee, 1992). Information has become a factor of production and an input that provides the competitive edge to business. A paradox that is emerging from the dynamism of information technology and its political impact is that on the one hand we demand a removal of technical and commercial barriers to the flow of information, and on the other we find regionalism taking shape in such organizations as the North American Free Trade Agreement (NAFTA) and the European Single Market. The Asian countries alarmed by this development are calling for the East Asian Economic Caucus (EAEC), whose chief proponent is Prime Minister Mahatir of Malaysia.

This chapter examines the cultural and political differences that permeate policy decisions in the telecommunications industry in the Association of South East Asian Nations (ASEAN) countries and compares these with the policies of the Asian Tigers (Korea, Taiwan, Hong Kong, and Singapore). The first section describes the existing and planned changes in telecommunications organizations, their infrastructures and policy dimensions. The second section deals with the drive for technological supremacy among the Asian Tigers leading to the next section that examines the role of multinational corporations in the capital formation and foreign direct investment in the countries under study. The last section deals with the mergers between the multinationals of the technologically advanced countries and the developing ones in the Asia-Pacific region. The conclusions analyze the future plans for institutional and technological collaboration among the countries of the region and their impact on the integration of global markets.

POLICY CHANGES IN TELECOMMUNICATIONS IN THE ASEAN

Privatization and liberalization policies diverge in the Asia-Pacific region, depending on political and cultural differences. Although the ASEAN countries have attempted to emulate the Japanese model, they are conscious of the fact that Japan's policy was the outcome of two conflicting views on the role of telecommunications in the macroeconomy, even as national priorities have dictated the approach to deregulation in the ASEAN (Jussawalla, 1990). In Japan, one view held by the Ministry of International Trade and Industry (MITI) was that telecommunications was part of the overall electronics industry and it should move toward strengthening that sector. The Ministry of Posts and Telecommunications held the view that there is a social use for these technologies and they should be sponsored for the growth of the information society, or *johoka shakai*. The heuristic value of the political and cultural determinants of any deregulation policy lies in the flexibility or otherwise of a nation's culture and the ability of its political institutions to make structural adjustments.

The United States practices and advocates a largely free-market economy. But the conditions under which a free-market economy works in the United States are not similar to conditions in other countries. Neither the U.S. nor the Japanese model is applicable to the ASEAN region. Even within the region there are disparities in the levels of economic growth and the aspirations of policymakers to advance toward export-oriented strategies of development and there are diverse political doctrines. For example, until recently Singapore espoused a fully centralized regime for its telecommunications organization (Singapore Telecoms) and its National Computer Board. Both delivered the most sophisticated IT to its citizens, while Malaysia opted for a mix of public and private sector operations (Hukill & Jussawalla, 1991).

However, the fact remains that telecommunications has become the thread holding modern society together, while it contributes to growth across a wide variety of sectors. It now constitutes an inextricable part of ASEAN's economic growth prospects. Multinational corporations have aided in this process of transformation by generating demand through their own telecommunications needs and thereby creating additional investment opportunities. Policymakers within ASEAN are becoming increasingly aware that deregulation by itself does not guarantee network expansion. It must be accompanied by organizational and structural changes for transition management from a monopolistic Post, Telephone, and Telegraph (PTT) to market competition for newer services. The implications of deregulation are twofold and interrelated. First, the restructuring of the industry has an impact on all other sectors that use its services. Second, liberalization policies must be compatible with global market trends for these countries to take advantage of new market opportunities. The estimated growth of the telecommunications equipment and services market is expected to grow from $113 billion in 1990 to $178 billion in 1995. This has resulted in growing competition from multinational vendors. But such competition varies with the country's individual developmental goals and consumer demand (Jussawalla, 1993).

Indonesia's plan for upgrading its network centers around the dispersal of basic services throughout the nation. In 1991 both AT&T and Nippon Electronic Corporation (NEC) competed for the contract of the planned expansion of 350,000 new telephone lines. In response to government and local pressures, Indonesia doubled its plan and divided the award equally between the two companies. The two companies that supply telecommunications and satellite services to Indonesia's 13,000 islands are Perumtel and PT Indosat. They plan to increase the digitalized exchanges and telephone lines by 1997 to reach a target of 7 million lines. Indonesia does not plan to privatize its statutory bodies controlling telecommunications services but may invite joint ventures to expand its value-added services.

Likewise Malaysia has also begun to develop its infrastructure through the privatization of its former PTT (Syarikat Telekom Malaysia) now called Telekom Malaysia. It sold the shares on the stock exchange of Kuala Lumpur in 1991. The regulatory body is Jabatan Telekom Malaysia. The privatization of its PTT provided a window of opportunity for new service suppliers from the private sector, such as Celcom and Atur 800, to compete for the cellular telephone market along with the former PTT. Currently Celcom has the largest share at 53%. Telekom Malaysia also provides both basic and international services and has planned to spend $4.8 billion U.S. in the next 5 years on new equipment. Additionally it has called for bids to provide $800 million U.S. worth of new digital telephone lines. Although foreign suppliers are being contacted for these bids, such large investments have also spawned the formation of indigenous companies like Sapura and Federal Cables to compete for telecommunications contracts. The Malaysian telephone industry is expected to grow by 12% over the next 5 years. The country is also planning on speeding up the introduction of the Integrated Services Digital Network (ISDN). For this purpose the domestic network is being digitalized. By the end of 1991, the transmission network was 56% digitalized and the switching network, 78%. Two other major projects are the laying of fiber-optic cable throughout the peninsula by 1995 and the laying of a similar cable linking Johor with Sarawak, covering 600 kilometers. Malaysia has also placed an order for its own domestic satellite, called Measat, with Hughes Communications with an advance payment of $250 million U.S.

Thailand is following its own path of infrastructure development. There are two state-owned monopolies: Telephone Organization of Thailand (TOT) and Communications Authority of Thailand (CAT). The former controls domestic communications and the latter, international. Privatization is not foreseen in the near future, although liberalization of new services is being attempted with the government's permission. In July 1992, the Thai government announced that it will contract for an additional 1 million telephones for rural areas. Loxley (Bangkok) won a major portion of the contract. As in Indonesia, Thailand's basic services are supplied under state subsidy due to the low per capita income of the region that does not meet the required income elasticity of demand necessary for supporting a commercially provided system for basic services, especially for the rural areas. Thailand is therefore turning to cellular telephones to overcome the low penetration ratio of telephones in the rural areas. In September 1991, Hitachi launched a cellular 900 telephone network for Bangkok and a domestic firm AMPS 800 is finding it difficult to accommodate all the demand in the countryside, which is indicative of the heightened desire for IT and access to information as a source for economic development.

Singapore is the "intelligent city" of Southeast Asia and has just launched a plan called IT2000 for furthering its networks and providing state-of-the-art technology to its users. All its achievements have been under a state-controlled statutory body. Singapore Telecoms provides domestic and international services along with data, fax, paging, and cellular services. It has developed its own videotext system called Teleview for its business users. Similarly the National Computer Board is responsible for the tremendous growth of computer services in the country including its computerized stock exchange, SIMEX, and TRADENET, which supports the entrepot position of the republic. However, the privatization trend has caught up with Singapore also, and the government has announced its intention of privatizing its monopoly and placing the shares of Singapore Telecom on the market in 1993.

ASIAN TIGERS TAKE ON WORLD MARKETS

The Asian Tigers have been the leaders in the region for promoting IT in their domestic economies and for using them as the lead sector for their export strategies. For some time Japan's blue chip companies had the global market for liquid crystal displays (LCD) all to themselves. But that picture has changed rapidly. Hong Kong's Varitronix is a leading exporter to Europe supplying custom-designed LCDs with sales growing at 28% a year for 5 years. The electronic producers of the NIEs have established many niches in global markets with innovative products rather than copying American technological products. By the late 1990s when the digital revolution hits the mass consumer markets, the Asian Tigers will be supplying the HDTVs and the Multimedia PCs ("Asia's High Tech Quest," 1992). In automated chip design they are also taking the lead. They are still not a major threat to Japan or the United States because they lack capital and basic science for innovations of technological systems. Even so, they have established niches in such areas as aerospace, software, robotics, and telecommunications. They have entered the Borderless World that Kenichi Ohmae described some years ago (1991).

Many of these companies are infused with the Confucian ethic, exploiting the rapidly growing markets of Southeast Asia that are experiencing a 10% growth rate and staggering wealth accumulation among its operators. Korea has laid great emphasis on investment in education in its seventh 5-year plan, which aims at generating greater production technology in the country. A fund has been created for the domestic production of equipment that includes information-intensive technology. The share of science and technology in the budget was 3% in 1991 and is expected to rise to 5% by 1996. The main aim of these policy measures is to increase the industrial competitiveness of the industrial sector in global markets. The telecommunications sector has been privatized, and Korea Telecoms is investing 4.2% of its revenues for the telecoms sector. Even Taiwan, which is not a self-sufficient economy, is a trade-oriented one. Its comparative advantage of cheap labor has been wiped out over the years, but it is fast developing its technological prowess in order to overcome the appreciation of the Taiwan dollar that came about because of its trade surpluses. For its telecommunications sector Taiwan has introduced a 6-year development plan from 1991 to 1996. The annual output of the industry is targeted to increase by 20% per year and consumer electronics production will increase by 30% per year.

If we look at the global computer industry we recognize the role of Intel and Sun Microsystems in building the fastest microprocessors that revolutionized the very foundations of the computer industry. But today it is the Taiwanese who make the lowest priced circuit boards that are used inside 65% of all PCs. It is Korea's Samsung that is controlling the market for dynamic random access memory (DRAM) chips that store 4 million bits of data. These details show that Asia will continue to dominate the low end of the market for telecommunications technology and can even succeed in the multimedia revolution as it proceeds to invade the telecommunications industry. One reason for the success of the Asian newly industrialized economies (NIEs) is that they have not aspired to innovate their own technologies but have been content with competing in the small-scale industrial sector with the cooperation of the multinational companies of the West. Taiwan has carefully conserved its foreign reserves of $88 billion, and its largest computer manufacturer, Acer, is grabbing Asian markets rather than those in the United States or Japan. This proves that technology is no longer an impediment to development. It can be bought or obtained through partnerships. What development needs is capital, distribution, and markets. In a survey of Asia's High Tech Quest conducted by *Business Week* in December 1992, it is estimated that Asia's own markets are the key to development in as much as the region plans to spend $1 trillion on telecommunications and power generation equipment over the next 10 years. In order to capture these markets telecom giants like AT&T, NEC, Motorola, Ericsson, and Fujitsu rush there and create labs and design centers, giving skills formation to local workers and providing partnerships for niche markets. More than 1,000 multinationals are based on the island of Singapore, pampered with tax breaks and subsidies and government-sponsored training programs for local workers. All this implies that the Asian NIEs are reshaping the technological balance of power in the world. They have demonstrated that it is no longer necessary for them to catch up with Japan, rather they can move into new fields like HDTV and multimedia consumer products and maintain their edge in global markets.

In 1987, Taiwan progressively lifted regulations in telecoms to suit its political imperatives. Competition has been introduced in the customer premises equipment sector and in domestic VANs. Only certain international value-added networks (VANs) are permitted, such as SITA, SWIFT, Reuters, and Associated Press. With liberalization have come videotext trials and ISDN in metropolitan centers. Taiwan has also taken the lead in imaging technology and in the manufacture of mobile telephones (Jussawalla, 1990).

Historically Cable and Wireless has been the principal supplier of equipment to Hong Kong. As part of Mercury PLC, the company has a separate subsidiary in Hong Kong. The Hong Kong Telephone Company is also associated with Cable and Wireless and holds the monopoly for basic domestic and international telephone services. The most important deregulated private-sector enterprise that has emerged from Hong Kong is the Asiasat system, which has enhanced the role of the colony as the gateway to China. It has also revolutionized television viewing throughout Asia by its wide footprint extending from Kabul to Hokaido. Hong Kong also has a large mobile communications sector that grew by 13% between 1991 and 1992. Cellular market competition is spearheaded by Pacific Link and Hong Kong Telecom, the latter having a 50% share of the cellular market. It is possible that this market will grow exponentially to reach the new urban centers sprouting in the southern part of China to replace the antiquated terrestrial system. It has been suggested that the legalized resale of international circuits will promote greater competition in Hong Kong and may result in more rational pricing of international telecommunications services in the whole of Southeast Asia (Mueller, 1992).

Many of these developments in the region's leading NIEs are an indicator of their resilience to change and their adaptability to introduce major structural shifts without causing any disequilibrium in the markets or causing market failures. The extensive capitalization that has taken place in these countries is related to the growth of multinational corporations that have contributed to foreign direct investment (FDI) and to transfer of technology. These corporations have either established subsidiaries in the region or have teamed up with local enterprises in creating joint ventures. In many cases the governments have introduced the buy, operate, and transfer (BOT) schemes for these ventures, which means that the foreign company will buy into the market, operate the new enterprise, and then transfer its ownership to the domestic counterpart.

THE GLOBAL FIRM AS ENTREPRENEUR

Transnational corporations have dominated the markets of Asia and Latin America for many decades, and their operations have been the subject of the vast literature developed around the theory of Dependencia. After the onslaught of the Information Revolution and the transformation it brought about in the developing countries, that theory lost much of its credence because NIEs have taken the maximum advantage of collaborative ventures and maintained their superiority in export markets worldwide. However, trade relationships are fast changing and even though the unresolved Uruguay Round of GATT talks is still grappling with greater free trade in services, a new trading system is launched under regionalism. Facing the reality of the European Single Market and NAFTA, the Asian countries are concerned about the future of free trade, and Malaysia's Prime Minister Mahatir has called for an East Asian Economic Group. This indicates that the future of the global firm is now uncertain, and Freidham predicts that it will soon become obsolete ("Survey of Multinationals," 1993). The reason for such a prediction is that we are witnessing an era of strategic alliances that cut across not only national borders but also industries. For example, the world's largest telecommunications firms have formed a group to provide a global network of fiber-optic submarine cable. A multinational alliance of independently owned firms can have a larger capital base from which to operate and can tap local markets anywhere in the world through their corporate relationships. Management experts call such an enterprise a "Virtual Corporation." The ASEAN countries may well enter into such partnerships in the future, but as of now they are content to invite foreign capital on their own terms and extend their infrastructures to their own specifications.

Contrary to what Vernon predicted in 1977, today's multinational company (MNC) is considered the embodiment of modernity and prospect of wealth, rich in technology, capital, and opportunities for skilled employment. The FDI generated by the MNCs has become so attractive to developing countries that even a country like India has reversed its decision of self-supporting growth and has opened its borders to MNCs. This political welcome may or may not be permanent, but for the time being, Southeast Asia is taking advantage of the vast sums of FDI flowing in from the tripolar investors based in Europe, the United States, and

Japan. The top 100 multinationals account for 40% to 50% of all cross-border assets, but there is no domination of total assets as yet. In his book *Twilight of Sovereignty*, Wriston (1992) described the power of telecommunications and of satellite technology that transmits data and pictures that transgress the political sovereignty of most nations. Emmott (1993) said that CNN matters more than ITT. It is not trade but telecommunications that is integrating the modern world, and in this sector multinationals are exploiting markets that have already integrated by setting up mergers as a way to economize on transaction costs. This trend in global capital flows increased during the 1980s, whereas the flow to the developing countries was declining because of the high debt servicing ratios that made policymakers in such countries cautious of FDI.

In the Southeast Asian region there are many joint ventures that have led to foreign investment and technology transfer. For example, Taiwan's Institute for Information Technology teamed up with Hewlett-Packard (HP) in 1992 for a $6 million joint venture for starting Open Systems Software, Inc. It is a venture that started the supply of software to Chinese language–dominated parts of Asia. HP will get 40% of the profits. Rather than setting up assembly plants, the large corporations in the United States and Japan are forging strategic alliances with their Asian partners to share their product designs and collaborate with local employees. The advantage for the Asean countries is that they are getting the benefits of latecomer access to dynamic technology without having to invest billions of dollars to develop it on their own. Likewise, Taiwan's First International Computer, Inc. has forged a relationship with Intel, Texas Instruments, Motorola, and Microsoft to become the world's largest producer of mother boards, which are the insides of PCs. Similarly, Intel and Hitachi are both transferring capital and technology to Malaysia, and Penang has become the world's second largest exporter of microchips in the world. In his work *Partnerships For Profits*, Lewis (1992) assessed that in the long run the Asian partners will emerge stronger than their foreign counterparts and become world-class exporters. As the process technology gets transferred, and the product cycle shrinks, rising prices and higher research and development (R&D) costs make these partnerships vital to survival in a globally integrated market. There are advantages on both sides. For example, Motorola is now able to move products from design to manufacture far more rapidly than before. In collaboration with Cal-Com in Taiwan, Motorola is producing pocket secretaries for a global market. Even Korea's Gold Star has obtained technology from Hitachi for semiconductors even as Matsushita has set up a subsidiary in Malaysia for television design. This frees up Japanese engineers in the home country to work on more sophisticated R&D. The Taiwan Semiconductor Manufacturing Company was formed 5 years ago as a collaboration between the government and Philips of Holland and has become a model for chip foundries in Korea and Singapore.

Singapore has taken advantage of these high-tech collaborations in research and has become a thriving research center not only in microbiology but in information technology. In 1992 Apple Computer opened the Apple ISS Research Center in Singapore jointly with the government-funded Institute of Systems Science and will spend $10 million to develop voice and handwriting recognition software for multimedia products in Asian languages. Singapore has also started a virtual reality theme park with government funding and Japanese designs that is probably the first of its kind in the world. It will cost the government $20 million.

In Korea there is an effort to reap the rewards of these corporate alliances. Daewoo has become a supplier to Boeing and Lockheed through technology transfer schemes for new aircraft parts, including computerized systems. In the Shah Alam Industrial Estate in Malaysia, Japanese and Malaysian engineers collaborate in turning out 1 million television sets a year for the global market. Malaysian engineers are designing 90% of the chassis for the television sets because qualified engineers are becoming scarce in Japan. Even China is permitting the influx of foreign technology for its stored program switches for its telephone exchanges. Ericsson of Sweden has a major share in collaborating with the Chinese PTT. At long last AT&T has been able to get permission from the government of China to start a research lab in Beijing for telecommunications and will take advantage of the vast market in that country for cellular and terrestrial telephone networks. Hewlett-Packard is training Chinese engineers to design integrators that are printed readouts of lab tests.

The benefits of all this activity of corporate alliances are enormous for both parties. The Asians are gaining access to software design and manufacturing technology that was once out of reach for them. In Penang alone there are 40 multinational companies that sponsor skills training centers for CAD/CAM and other basic electronics. This is called the Robin Hood policy of taking from the big guys and giving to the little ones. But not all collaborative enterprises are doing that. There are almost equal advantages for the MNCs in exploiting integrated markets and getting a foothold into Asia. The competition for Asian markets is so keen that such joint ventures reduce transaction costs and simultaneously assure profits from sales. Asian technological strength in the ASEAN and the Asian NIEs is rapidly advancing, and these countries are, in turn, setting up collaborative ventures in other LDCs so that further technology transfer and flows of capital are assured.

The new feature of this kind of capital flow is that it has begun to flow within the region itself from the NIEs to the Near NIEs. For example, when Samsung Electronics set up its export-oriented electronics plant in

Indonesia, no one believed that it would succeed. In October 1992, the factory set up by Samsung in southeast Jakarta started to export VCRs and audio players to the United States and Europe. Large corporations from Japan like Sony, Sanyo, and Matsushita have also established plants in Indonesia, so that the country is no longer reliant on oil and gas for exports. In 1992, its nonoil exports totaled $18 billion. The capital invested in electronics shows that Indonesia has reached a new level of sophistication in its industrial structure. Matsushita's joint venture, called Kotobuki Electronics, is to manufacture VCRs for export; its target is 1 million VCRs to the United States and Canada.

Innovations in computer and telecommunications research is also being applied in Asian centers for exports to American markets. One such venture is a joint operation between the Read-Rite Corporation of California and Sumitomo of Japan to manufacture heads of the hard disk drive that is fitted into PCs. Jointly they have invested $60 million in Asian countries for this technology's output. The heads are fabricated in Bangkok and assembled in Penang, giving employment to 5,000 workers in Asia.

Another example of intraregional capital flow is that of a joint venture between Singapore Telecoms and Perumtel of Indonesia to develop telecommunications infrastructure for Batam island as part of the growth triangle (Hukill, 1992). The Overseas Telecommunications Corporation of Australia has invested in earth stations in Hanoi and Ulan Bator that enable these remote areas to link up with Intelsat for their international telephones and television services. Partnerships within the region have resulted in the incredible success of Asiasat, which drew capital from Hutchison Whampoa based in Hong Kong, the China Investment Company, and Cable and Wireless. This satellite system has more demand for its transponders than it can meet with its current capacity. The further investment made for using the satellite by Hutch Vision for STAR TV has changed the audiences and their viewing habits across the whole of Asia. Likewise the collaborative international fiber-optic project has drawn investors from nine telecommunications entities: AT&T, CAT of Thailand, Telekom Malaysia, PT Indosat, International Telecommunications Development Corporation, Hong Kong Telecom, KDD, the Philippine Long Distance Company, and Singapore Telecoms. These nine companies have signed an agreement to lay and operate the APCN submarine network to link Southeast Asia with the rest of the Pacific countries.

Texas Instruments invested $1.5 billion in Asia in 1992 as a cost-efficient production base for its global operations. Asia accounts for 30% of the company's nonmilitary turnover. Similarly, in the television industry, advertisers are capitalizing on the growing middle class in Asia but have to cut across varied cultures. Even so the Asian market provided $9 billion in 1991 and is growing at 15% a year. STAR TV has provided the first pan-Asian network. This has led the Thai Shinwatra Corporation to not only start Thaisat but also to launch a television station in Indochina. However, all is not easy for Asia's businesses because the first Asian Business channel on television faces several hurdles even before it commences. Asia Business News (Singapore Private Ltd), along with TV New Zealand and the Hong Kong-based Business News Network jointly want to start the venture. The Asia Business News network will be based in Singapore but managed by TV New Zealand. The hurdles it faces arise from a variety of regulations in the countries that comprise the region along with language differences. But its success depends on the reception it receives from a niche audience that comprises the Asian businessman.

FUTURE TRENDS IN TELECOMMUNICATIONS IN THE PACIFIC RIM

The future of telecommunications development in the region depends to a great extent on the standard setting process for equipment in the global market. It is obvious that the ASEAN countries and the NIEs will not be willing to invest in equipment purchases that are not interoperable. Even the success of the ASEANET electronic network for Electronic Data Interchange (EDI) that was set up in 1991 is based on interconnectivity of the systems for the use of the member countries. The competition between proprietary and negotiated standards poses a threat to the growth of technology and its applications in the region. The recommendations of ITU's Study Group XVIII for digital network standards as provided by the CCITT have not yet gained full acceptance by the member countries leading to a tripolar battle for a global standard. Political interest divide the global standardization issue between European Telecommunications Standards Institute (ETSI) in Europe, Telecommunications Institute (TI) in the United States, and the Telecommunications Technology Committee (TTC) in Japan. For the Asia-Pacific region, the crucial issue is how to make the transition from technology-driven demand to a market-driven one based on user needs (Jussawalla & Steinour, 1993). Those developing countries in the region that face a shortage of hard currency find it difficult to purchase equipment in the lowest cost market if the equipment is not compatible thereby causing network inefficiency. To overcome this problem, Asia Pacific Economic Cooperation (APEC) has created a special subcommittee on telecommunications that is addressing the issue of EDI and standards. Without global homogenization it appears as if the future of the telecoms market in the region becomes tenuous. In order to counteract the problem of accounting rates disparities between developed and developing countries, the system of international communications is being revamped by a collaborative venture between the major operators like AT&T, KDD, British Telecom, Cable and Wireless, and Singapore Telecoms to arrive at negotiated call tariffs. This will speed up global communications in data, voice, and video without a multitude of settlement problems. Mobile communications is by far the most dynamic sector of the future telecom scene. Currently there are analogue systems based on incompatible standards, but with the introduction of digital cellular radio, the benefits will be widely attained. The market for mobile services in Asia has been estimated at $9 billion in 1991 and estimated to rise to $28 billion by the year 2000 (Stern, 1992). As such, wireless communications will be the largest growth area of the future in the Asia-Pacific region. The data transmission market alone is predicted to be worth $8 billion by the turn of the century.

Another major trend invading the developing countries is the widespread increase of access from projects like the Iridium and the Inmarsat project of linking the world with LEOS to provide cellular telephones in the remotest areas of the globe. The recommendations of the Maitland Commission made in 1984 are likely to be complied with a decade later if these projects materialize. So far, Motorola's Iridium project has obtained several collaborators from Japan, including Sony. In any case, the future trend for using low earth-orbiting satellites (LEOS) for low-cost voice and data services is a welcome one for countries such as China and India that have vast terrain to cover for thin route linkages. Cellular technology is likely to become the most ubiquitous since computing and communications can travel with the user. Although this trend is likely to greatly improve the networks of less developed countries (LDCs), it is tied up in the debate of spectrum availability. This is the problem that was keenly contested at World Administrative Radio Conference (WARC) '92 in Torremolinos. An agreement has been reached making the International Telecommunications Union the central institution through which member states will allocate the radio spectrum. With the abolition of the Consultative Committee on International Radio (CCIR) and the Consultative Committee on International Telephone and Telegraph (CCITT), all radio spectrum issues will be entrusted to the Radio Communications Sector, which will assume the responsibilities of the International Frequency Registration Board (IFRB) for administration of the radio spectrum. But as regional free-trade blocs appear, developing countries may face restrictions of markets and technologies that are so important to their prosperity (Helman, 1992).

Having traced the problems and potential for the rapid spread of networks in the Asia-Pacific region, we have seen the role of multinationals in a totally different light than in the past. We have assessed their contributions and their competitive strengths and weaknesses. They provide both risks and opportunities to the growth sectors of the Asian countries. Most of these countries have taken the maximum advantage of their presence and opened their borders to international cooperation. They have realized the benefits of international trade and have strengthened their export sectors to compete in regional and global markets. The trend for capital flows through FDI, and through technology transfer it has moved ahead in the ASEAN region; technology driven collaboration has induced economic development. Future trends indicate that with changing technology the countries of the region will invest in free trade and maximize the benefits for their populations. Governments will need to redefine their rules for competition and shift them from national to global levels. History shows that when cross-border alliances become too powerful or abuse their power as cartels they become unstable and dwindle. The global corporation is far from becoming obsolete as it enters the 21st century.

REFERENCES

Asia's high tech quest: A special survey report. (1992, Dec. 7). *Business Week*, pp. 126–130.
Emmott, W. (1993, March 27). Everybody's favorite monsters. Survey of multinationals. *The Economist*, pp. 5–15.
Freidheim, C. (1993, February 6). The global corporation: Obsolete so soon. *The Economist*, p. 69.
Helman, G. (1992, May–June). After WARC. *Transnational Data and Communications Report*, pp. 41–42.
Hukill, M., & Jussawalla, M. (1991). *Trends in policies for telecommunications infrastructure development and investment in the ASEAN countries*. (Research Report). Honolulu: East–West Center.
Hukill, M. (1992, May–June). ASEAN telecommunications development 1980–1989. *Transnational Data and Communication Report*, pp. 33–36.
Jussawalla, M. (1990). Cultural and political regulation of international telecommunications. In J. Pelton & C. Sterling (Eds.), *Proceedings of the National Telecommunications Forum*. Boulder: University of Colorado.
Jussawalla, M. (1990, Oct. 22). Privatizing telecommunications: National priorities dictate approach to deregulation. *Asian Wall Street Journal*, pp. 11–14.
Jussawalla, M. (1993). Telecommunications and regional interdependence in Southeast Asia. *The Fletcher Forum of World Affairs*, 17(1), 85–96.
Jussawalla, M., & Steinour, D. (1993, March). *Regional versus global standards for the Asia-Pacific countries*. Paper presented at the APEC meeting, Brisbane, Australia.
Jussawalla, M., & Lee, C. (1992, October). *Trade in services in the Asia-Pacific region: The telecommunications sector*. Paper presented to the Wharton Conference on Service Management, Philadelphia.
Lewis, J. (1992, December 7). Partnerships for profit (special report). *Business Week*, p. 131.
Mueller, M. (1992, June). *Telecommunications in Hong Kong after 1997*. Paper presented at the ITS Conference in Sophia Antipolis, France.
Ohmae, K. (1991). *The borderless world*. New York: Harper.
Stern, P. (1992, July–August). Asian telecommunications: The challenge of change. *Transnational Data and Communications Report*, pp. 5–6.
Wriston, W. (1992). *The twilight of sovereignty*. New York: Maxwell Macmillan.

8

The Changing Role of the State

Bella Mody
Lai-Si Tsui
Michigan State University

The telecommunications trade press has been full of news about state PTTs "going private." Much literature on the privatization of telecommunications has been descriptive and prescriptive (e.g., Ambrose, Hennemeyer, & Chapon, 1990; Roth, 1987; Wellenius, Stern, Nulty, & Stern, 1989). We are now at a stage where we can step back and use theory and history to conceptualize why and how different nation-states are rebalancing public and private investment (Mody, Tsui, & McCormick, 1993). Part II of this book focuses on the political forces in present-day developing country contexts that have influenced the increasing private sector participation in their telecom sectors. This context-analytical approach is essential to avoid looking at the increase in private capital investment as a disembodied "rational" natural transnational process that comes from nowhere. Previous chapters have examined global capitalism, locational decisions of U.S. foreign telecom operator-investors, the role of the World Bank in orchestrating the global financial integration process, and the different roles of domestic capital in Latin America and East Asia. This chapter focuses on the state. In-depth analysis of individual forces in each chapter is conducted against the background of cyclical global restructuring and state-market pendulum swings that are an integral part of the historical capital accumulation process.

What we are presently witnessing is private sector participation at the invitation of the state. Reorganization of the ownership of particular sectors of the state economy is an attempt to actually strengthen the state fiscally

179

at this point in time. If export-led growth does not occur, or profits do not materialize when good times return, the pendulum could swing back. We begin with an introductory section that presents the changing role of the state in developing country telecommunications systems. The next section focuses on different conceptualizations of the state, its relations with capital, and the nature of its own telecommunications bureaucracy. The final section looks at the implications for those persons who could continue to live outside the reach of telecommunications systems in a privatized scenario too.

INTRODUCTION

Until the mid-1980s, telecommunications in most parts of the periphery was organized as a state-owned enterprise or directly operated by the ministry with the rare municipal company (Colombia), private coopera- tive (Bolivia), or subscriber-owned enterprise (Lima, Peru). Since then, governments of industrialized countries at the center of the world business system and international lending agencies they influence, such as the IMF, the World Bank, and regional development banks, have been urging developing countries to "open up" their telecommunications (and other) sectors to investment by their own capitalists. Simultaneously, in some regions of Latin America and Asia, an indigenous capitalist class of varying strength has emerged, partly nurtured by the state's economic development initiatives, now ready to compete and collaborate with it in the telecommunications sector. Domestic broadcasters have been showing interest in radio wave technology. Developing country business users of telecommunications and even many government agencies are investigat- ing the new telecommunications technologies to help them bypass their own unresponsive, under-resourced, inefficient state telecommunications monopolies. In some cases, the state is "broke," due partly to internal reasons and partly to impossible external constraints.

History

The British company Cable and Wireless is probably the oldest transna- tional operator of telecommunications services. Started in 1868 as the West India and Panama Telegraph Company in the service of the British colonial government, Cable and Wireless has remained largely unchal- lenged in the Caribbean (Dunn, 1994) while earning 75% of its revenues from its Hong Kong operations.

The first global wave of private capital investment in telecommunica- tion came after World War I when U.S. banks and U.S. corporations began replacing Britain as the leader in world commerce (Sobel, 1982). Sugar broker Sosthenes Behn founded and developed the International Tele-

graph and Telephone Company (ITT) with the help of Morgan and
National City Bank of New York. "The International System," the international version of AT&T, started in Puerto Rico and Cuba. It then
acquired telephone companies in Spain, Mexico, Uruguay, Chile, and
Argentina in the 1920s and Peru and Turkey in the 1930s. It added a
major cable and telegraph operation and factories in many European
countries (Belgium, the United Kingdom, and Germany). ITT's strategy
was to stack the board in each country with its influential citizens. Still,
Latin American complaints about ITT's bribery of politicians and the
incompetence of the telephone system were widespread (Sampson, 1973).
ITT had no incentive to improve the system because many governments
insisted on keeping the rates down. A 1950 World Bank report on Cuba
was very critical of the ITT system.

     After World War II, many newly independent states in Asia and Africa
organized their telecommunications entities as state-owned enterprises.
After the market failure of the 1930s and the early successes of the Soviet
Union, the conventional wisdom in international lending agencies was
that markets were not as free, efficient, or reliable as they were made out
to be. Immediately after decolonization in Asia and Africa, domestic
capital was scarce. Both nationalists and international agencies like the
World Bank saw state enterprise as the chief route to economic development. As the tide of nationalism rose, countries resented the poor operation of their telephone systems by foreign-owned companies. Castro
expropriated the rundown system without compensation. In Brazil, ITT
was paid a total of $19.5 million. In Peru, they received $17.9 million. By
the late 1960s, ITT was operating telephone systems only in Puerto Rico,
the Virgin Islands, and Chile (acquired from the British in 1930). Desperate
during negotiations on nationalization of the Chilean system in the 1970s,
ITT used the CIA to induce economic chaos and instigate a military coup
to get an election outcome favorable to it.

     During the first decade of political independence in the Caribbean
(1960s), governments of the more developed countries encouraged national acquisition of a part of the monopoly shareholdings of Cable and
Wireless. Thus, government shares in the profitable international network
went up to 40% in Barbados and over 51% in Jamaica and Trinidad (Dunn,
1994). By the 1970s, the international private business of operating telephones versus manufacturing them was in retreat. The first wave of private
capital investment in telecommunication had receded, with a few exceptions (e.g., Cable and Wireless in Hong Kong and parts of the Caribbean,
and a public-private oligopoly in the Philippines). The experience of
imperfect markets was followed by the experience of imperfect state
intervention: Like the prior private owners of telecommunications in parts
of Latin America (ITT) and the present private owners in the Philippines,

the subsequent state telecommunication monopoly in the South Asian and African periphery have often provided little quantitative expansion of coverage or improvements in the quality of service. Customers with and without the ability to pay have both been ignored by the state bureaucracy. In most cases, neither equity nor efficiency have been achieved. This neglect of telecommunications continued unquestioned until the mid-1980s when the interests of foreign capital, domestic capital, and the state changed. The convergence of interests in the three power groups is relevant to the telecommunications sector in light of phenomenal advances in telecommunications technology and reductions in cost: Foreign capital in telecommunications in North America, Western Europe, and Japan who are confronting saturation of their domestic markets for basic services and national regulations that constrain their expansion into related sectors are now actively seeking foreign investment opportunities in underserviced areas of the Eastern European and traditional periphery. Foreign telephone companies are now in a position to service the intenational communication needs of their domestic individual and corporate clients.

The second wave of private capital investment in telecommunications started in the mid-1980s, with Thatcher's Britain and then Japan. Private investment in operating telecoms has begun to be big business again. The transfer to private ownership of state-run telecommunications entities in the periphery began in Chile in 1987. Cable and Wireless moved from 9% of the total equity in the new Telecommunications of Jamaica in 1987 to 79% in 1982. In 1994, Cable and Wireless has exclusive licenses and uncontested markets in 15 Caribbean countries; in 9 of these, Cable and Wireless is a 100% monopoly operator of both domestic and overseas services (Dunn, 1994). In 1990, Booz, Allen, and Hamilton estimated that 9 million lines were privatized in Argentina, Mexico, and New Zealand, followed by Venezuela in 1991. Potential privatizations in Latin America alone are estimated at $50 billion (O'Neill, 1991). The trade press estimates that an additional 26 countries with 95 million lines are expected to look for private capital in the next 2 years (Dixon, 1991). Although this is a substantial part of the global market, many countries in the periphery of the world business system are wary: A 1991 ITU Centre for Telecommunications Development survey found that 77 % of its developing country respondents had initiated some form of restructuring other than privatization (Westendoerpf, 1991).

THE STATE

This section first outlines different conceptualizations of the nature of the state in the developing countries at the periphery of the world business system. It then presents diverse patterns of state-capital relations and a

range of statist logics underlying invitations to private capital to participate in the telecommunications sector in different national settings. We
then investigate the quality of state bureaucratic capability that has contributed to state divestment.

Neoclassical economists assume a passive pluralist state that is acted
upon by interest groups with equal access to its benefits. A nonpolitically
organized society coexists with an ideal liberal state, completely neutral
with regard to distribution. This pluralist model views economics and
politics as separable spheres. When development economists experienced
the problems of state capitalism, they decided it was easier to work
through imperfect markets and dismissed the state. Virtually all industrialized countries have benefited from state intervention and continue
to do so, the United States and Japan included. Clearly, a much more
sophisticated political economy is required that acknowledges the empirical reality of the state, qualified as it is by social forces (Migdal, Kohli,
& Shue, 1994).

In the past, Marxist social scientists conceptualized the state as subor-
dinate to the interests of private capital. Writing in the context of Western
Europe in the 1960s and 1970s, Domhoff (1967, 1979) and Miliband (1969)
saw the state as an instrument of private capital rather than as an adver-
sary. Poulantzas (1978) perceived the state as structurally supportive of
the capitalist social order. The state was relatively autonomous from
particular capital groups only because its interest was protecting the
long-term interests of the capitalist class as a whole. Writing in the context
of Latin America, the first wave of dependency theorists (Baran 1957,
Frank 1967) saw their state as fundamentally different from Japanese,
North American, or Western European states at the heart of the capitalist
system: The dependent state was relatively autonomous with respect to
weak national capital, but it could only be an instrument organized to
meet the needs of transnational capital. The argument of the second wave
of *dependistas* was that, under certain conditions, the state could take
initiatives and forge alliances to harness the dynamics of global capitalism
for their own development (Cardozo & Faletto, 1979). Consistent with
later dependency theorists, world system theorists (Wallerstein 1979)
emphasized the role of the state in carving out a space for indigenous
economic and political forces to move the nation to an intermediate
semiperipheral position within the world economy.

Early dependendency theorists predicted that the initiative for private
investment in state telecommunications entities would usually be taken
by predatory First World capital looking for overseas markets. The early
reality in this second wave of private capital investment in telecommunications has consisted of the peripheral state actively soliciting capital
from willing and able foreign investors waiting impatiently for the gates

184

MODY AND TSUI

to open. The state is offering what could be its most profitable sector (and in some cases, the most profitable parts of this sector) to private capital, often with little attention to regulatory mechanisms or requirements to service the total population. Low state capability, in the telecommunications sector in particular, and in national trade and fiscal management in general, combined with alternative models of telecommunications supply from the United Kingdom and Japan privatizations have provided windows of opportunity for private capital. The scale of capital and the level of technical knowledge required to manufacture advanced telecommunications equipment or provide telecommunications services is not available in many developing countries, leading to dependence of the sector on foreign firms as sole proprietors in some cases and joint venture partners in most others.

Following Evans (1979), we maintain that decisions on ownership of telecommunications flow from national and international accommodations between members of the ruling triple alliance of state, national capital, and foreign capital. With statist theorists Evans (1985), Skocpol (1985), and Deyo (1987), we hold that the state is a central actor with survival and consolidation interests of its own. In the past, it was military and administrative power that guided the expansion of states. In contemporary capitalist society, particularly since the Cold War, the political power of a state is increasingly determined by its economic wealth: The regime of capital is at the center of the stage. Therefore, it is self-interest that drives the state to support and advance the accumulation of national capital. A large part of the modern state's activity consists in guiding, advising, encouraging, and subsidizing national capital toward what Wallerstein (1979) calls profitable chains of production and distribution that cut across old political (national sovereignty-based) boundaries. Appelbaum and Henderson (1992) have shown how newly industrializing countries (NICs) have been led by "developmental states" that sought legitimacy through economic growth policies. State strategies to support national economic development range from encouragement of domestic industrial competitiveness through subsidies for R&D and provision of infrastructure such as roads, telecommunication, and education, to financing of basic research that will have spin-offs for the economy. Thus, capital development fits in with its own interests. After all, the state's ability to tax presupposes the existence of a thriving economy. This is not to say that the state promotes capital accumulation by following the dictates of the largest capitalist group in the country all or even some of the time. This strategy may actually not be in the state's interest. We frequently find one part of the state allying itself with one capitalist or class interest in competition with another part of the state in coalition with different class interests. Several states have played a very effective developmental role, albeit qualified by the interests of international capital and domestic capital (Cardoso & Faletto, 1979; Evans, 1979, 1985). Perforce, the national development path it pursues varies in unity and coherence, given state-capital relations, its internal fragmentation, and its absolute size.

Patterns of State Capital Relations

The prevailing configurations of state capital relations in each state will depend on the colonial and postcolonial histories of the specific country. The following regional patterns are presented to demonstrate the diversity within the periphery that constrains generalizations. Similar diversity holds within each of these regional groupings. The comparative examination of regional contexts cannot present a comprehensive overview but it can help understand the different reasons why countries as diverse as the United Kingdom and Malaysia would both use the same ownership strategy.

Despite mounting periodic challenges, the Latin American state has generally cooperated with foreign capital to promote capital accumulation. Although it speaks in the name of the "national interest," the pattern of accumulation it has promoted has actually advanced the interests of big domestic capital, leading to spectacularly uneven development. Big national capital has actually "privatized" or "captured" the state in its own interest. Thus, the labor party in Brazil talks of "deprivatizing" the state, that is, ridding it of penetration and control by a narrow range of big private interests. Gini coefficients in Argentina, Brazil, El Salvador, Mexico, Panama, Peru, and Puerto Rico show increasing inequality in income distribution (Cardozo & Helwege, 1992). Entrenched poverty and drug-trade violence threatens the social and political order. With a political leadership in disarray, basic economic survival is the primary national goal.

East Asian states have dictated terms to foreign capital rather than being used by it, due to initial development conditions, domestic politics, domestic ideological harmony, and geopolitical factors when compared to the Latin American state (Evans, 1987; Koo, 1987). The role of the state in providing credit, channeling investment into particular industries, subsidizing export products, protecting the domestic market, and attracting new technologies has been more important than the role of foreign capital in contributing to political and economic development (Haggard & Cheng, 1987). Their national goal is economic and technological enhancement.

Southeast Asian states like Thailand, Indonesia, and the Philippines, with weak domestic capital holdings (like Latin America) have been influenced by foreign capital more than East Asian states. Their goal is infrastructure development.

African economies are characterized by very low domestic capital (much of it Asian Indian and Lebanese) and continually declining foreign direct investment. The state in most African countries is comparatively uncontested in its reign over constantly shrinking resources. The recent colonial legacy of an economy dominated by foreign firms has resulted in an African state hostile to what it sees as the continuation of an exploitative neocolonial presence that constrains its national security. The large number of job-providing but money-losing, unprofitable public sector enterprises are being recognized as economic drains that need to be divested.

We have very briefly examined the diverse configurations of state power in the periphery of the world capitalist system in relational perspective to understand the state's invitation to private capital. We now look at the state's reasons.

Statist Logic in Inviting Private Sector Participation

If the nation-state in the periphery had been able to efficiently use its revenues to develop a national infrastructure (in collaboration with domestic capital where it existed) that met the needs of other government agencies and private businesses (not to mention rural and residential users), there would have been limited underserved areas of the world for foreign capital to penetrate. For historical colonial and postcolonial reasons, this has been the case only in East Asia. States in South and Southeast Asia, Latin America, and the Caribbean are now inviting private capital to invest in different parts of their telecommunications infrastructure. The objective is to accommodate internal pressures from domestic users in commerce, industry, government, and their growing telephony-demanding middle class on the one hand and the external interests of foreign capital and its national and international supporters on the other. Based on individual conditions, the state's logic has varied from paying off the national debt through the sale of a profitable sector (e.g., Jamaica, Mexico), to financing essential infrastructure development (e.g., Eastern Europe, Indonesia), to infrastructure enhancement for international competitiveness (e.g., Singapore), and combinations of the foregoing. We discuss two distinct statist scenarios in recent openings to private capital in developing countries in terms of pace and motivation: the East Asian case and the Latin American case.

The East Asian case is generally characterized by relatively efficient state telecommunications bureaucracies, clear national and sectoral enhancement goals, low domestic capital, and clear-headed, unhurried leadership. Singapore's state-guided capitalism has determined that an infusion of private capital into its very profitable state telecommunications corporation will help to stimulate its stock market. The goal is to further develop Singapore as a financial center so it can take Hong Kong's place after its reunification with the People's Republic of China in 1997. To ensure that all the stocks were not controlled by foreign capital, and that the public got a chance to buy shares, there was a distinct domestic offering: Government provided S$200 for every adult Singaporean with an active Central Provident fund to buy the utility's shares ("Singapore Awards," 1993). It is expected that a privatized Singapore Telecoms will face less resistance in its attempts to buy parts of other national telecommunications sectors (e.g., mobile phones in India) than might a public state corporation of a sovereign government. Malaysia's goal is to reduce the dominance of Chinese capital by increasing the participation of Malays (and foreign capital, if necessary) in domestic share equity. Foreign capital is only allowed to set up joint ventures with *Bhumiputra* (sons of the soil) due to this national attempt to redistribute wealth by race (Hukill & Jussawalla, 1989). With a relatively efficient ministry-run state telecommunications service in Taiwan, and no large domestic capital groups wanting to capture the telecommunications sector, the Taiwanese state faces no great pressure to privatize. In 1990, it did privatize value-added services. Provision of basic services are expected to be offered by an autonomous corporation in the next few years. Unlike the Mexican case where opening up to the market was a result of the state's past failures, the primary impetus for Korea's economic liberalization was the state's continuing success at economic development. In addition to domestic pressure from *chaebols* (domestic conglomerates) and worker unions, international pressures from the United States increased in the mid-1980s. The growing U.S. trade deficit prompted the Reagan administration to press Japanese and Korean markets to open up to U.S. goods and financial institutions (Hamilton & Kim, 1993). This led to the Korean state's July 1990 announcement of a gradual plan to introduce competition into basic, long-distance, international, and mobile markets then monopolized by state corporations.

The World Bank has played a major role over the last few years in pushing restructuring and opening up the telecommunications markets in both China (0.05 lines per 100 population) and India (0.8 lines per 100 population). In May 1994, both states initiated major changes, with China taking a major step toward granting autonomy to its dominant state-owned service provider and India introducing competition in basic services.

Private capital investment in Latin America and the Caribbean has been encouraged by a state in economic distress, an inefficient state telecommunications bureaucracy, willing and able foreign capital looking for unsaturated markets, and big national capital groups interested in joint ventures. The hasty pace of telecommunication privatization in Argentina, Mexico, and Venezuela has been likened to the distress sale of family jewels, parting with a cash cow, or giving away the goose that will continue to lay golden eggs. The granting of monopoly rights to make the deal more attractive to investors and the lack of regulatory protection for consumers is an indication of state desperation, because several states are competing with each other to attract foreign capital investment. Thus, telecommunications privatization has been a matter of economic survival in Latin America compared to economic enhancement as in East Asia. Although the Latin American state is in a strong position vis-à-vis national capital, foreign capital plays the dominant role in the ruling alliance. After years of fighting the Spanish colonial presence, some states in South and Central America have invited Spanish capital (namely, Telefonica de Espana) to do for it what this state-owned firm has been unable to do in its domestic market at home. Other sources of foreign capital that Latin American states have successfully courted are France Telecom, STET, Southwestern Bell, Bell Atlantic, Ameritech, GTE, and AT&T.

In no country has privatization been carried as far as in Chile after 1975. The absolute control of dictator Pinochet and the political faith of the country's Chicago-trained economists that a decentralized privatized economy integrated into the world system is morally and economically superior to state intervention were primary reasons. Nevertheless, the process took 10 years in Chile.

Unlike other East and Southeast Asian scenarios, the Philippines is an embellishment on the statist privatization scenarios described in the preceding section. In 1993, President Ramos succeeded in breaking the long-running near-monopoly of the private Philippines Lond Distance Telephone Company (PLDT). Given that the power of the state is much weaker vis-à-vis domestic capital in the Philippines than in Latin America, this exercise of state autonomy and political will is noteworthy. When Ramon Cojuangco engineered the purchase of a major stake in the PLDT from GTE in 1967, it was seen as a triumph of Filipino nationalism. Of the 105 million shares, Filipinos own 91.37%. The voting majority is held by a cast of U.S. investment bankers and portfolio managers that changes every day, an amorphous group that cannot be organized to influence the company in any way. This overseas diffusion of common stock has allowed the Cojuangco family and associates Alfonso Yuchangco and Antonio Meer to maintain their grip over PLDT. This elite *comprador* class owes much to Douglas MacArthur who squandered a perfect opportunity to break up the ruling oligarchies and to foster a working democracy. As a result, PLDT remains under the control of an effete corrupt aristocracy who has a cozy relationship with the judiciary, the Marcos family (who had shares in the company), and the Aquino family. The Cojuangco family held the country and its telecommunications system hostage through its network of patronage within the executive branch, the legislature, and its control of *The Manila Chronicle*, one of the largest national newspapers

("Philippines Phone Monopoly Assailed," 1993). It is little wonder that many states are wary about giving additional power (through telecommunications ownership) to some of their own large domestic capital groups.

In Brazil, Roberto Marinho of TV Globo dominates his cellular joint ventures with NEC; potential competitors are more afraid of his political power and market power than they are of NEC. As an extramarket decision maker on which capital group it wants to encourage and who it can manage, the Mexican state decided against allowing its large national communication conglomerate to grow larger: It excluded Televisa from bidding on the privatization of Telmex. In December 1990, Grupo Carso, a mining, manufacturing, and tobacco firm, was allowed to buy a controlling interest along with Southwestern Bell and France Telecom. National capital in large states (e.g., Thailand, Brazil, South Korea, and India), are participating actively in joint ventures in equipment manufacture and service provision.

Producing telecommunications equipment and providing telecommunications services are capital-intensive undertakings. Asian and Latin American states with large domestic capital firms are capable of jointly investing in their telecommunications sectors with foreign firms and institutional investors. Although national capital may be more amenable to state control, many recognize that so-called "domestic" capital has no nationality. The primary interest of domestic capital may actually be transnational rather than local. Also, their focus may be monopoly-seeking security maximization rather than risk-taking profit-maximization.

Whereas Nigeria and Ghana have initiated private sector participation in their telecommunications sectors, some smaller, younger nation-states in parts of Africa presently express concerns about allowing private investment into a state security-related sector. This is also the case in Taiwan where the military assigns all radio frequencies. In Thailand, the military opposed conventional approaches to privatization of telecommunications on this ground, preferring to grant concessions to private industry to build, transfer, and operate lines.

Bureaucracy

In addition to its relationship with capital, the capability of the state sector also depends on the bureaucracy and its independence from other social groups and sectional interests. Evans (1985) and Hamilton (1992) showed how strong effective state interventions require coherent bureaucracies autonomous from dominant social interests. They indicate that, unlike Europe's former colonies, the postcolonial administrations that Japan left behind in Korea and Taiwan were effective state organizations that helped integrate their countries into the world capitalist system to their own *national* advantage. South Korea and India have similar levels of state participation in their national economies; the difference is that in South Korea, government intervention is used to discipline the private sector, whereas in India, government is not insulated from sectional interests and cossets domestic industry rather than forcing it to be efficient. Amsden (1989) identified the central role of bureaucracies in late-industrializing countries (Japan and the NICs) in designing a "catch-up" strategy based on learning and adapting foreign technology and processes under licensing agreements.

Private capital guided by a military bureaucratic developmental state explains how Korea's telecommunications sector succeeded in achieving a 35% penetration rate that ranked the country 11th in the world in 1992. This was less than 40 years after its system was destroyed by wars. In 1981, telecommunications operations were decentralized from the Ministry of Communication. In 1983, a step-by-step plan was implemented to encourage management efficiency in state-owned enterprises as part of national economic development policies. Korea Telecommunications is now a domestic public-private corporation. Competition was introduced in 1990, resulting in major reductions in tariffs. Dacom is now a privately owned domestic competitor. There are several foreign and domestic joint ventures in the mobile and value-added markets (Kim & Ro, 1993).

East Asian telecommunications bureaucracies in Singapore, Taiwan, and South Korea have earned a reputation for goal-oriented efficiency in comparison with other countries in the Third World. Their unity of national purpose rather than partiality to sectarian interests may have been caused by the need for concerted action against the perceived communist threat in the early years of nationhood. Established before domestic capital grew or international capital penetrated them, the East Asian bureaucrat was not involved in comprador capital relations. Successful land reforms meant there were no negotiations with agricultural landlord interests. The city states of Singapore and Hong Kong had no backward rural constituency to serve (Koo, 1987). While bureaucrats in telecommunications entities around the world are fighting the possibility of being restructured as private sector entities, bureaucrats in Singapore Telecoms International are busy investing in joint ventures all over the world to achieve their own new global market penetration goals.

Many public telecommunications operators in South Asia and Africa are authoritarian bureaucracies interested in consolidating their own power, frequently leftovers from the colonial era. The colonial inheritance includes rigid centralization, obsessive legalism, authoritarianism, and administration by clerks due to the frequent transfer of officers. After national independence, without a palpable enemy on their soils to organize against, administration of telecommunications (and other public sectors) consisted of job creation for political survival of the ruling regime rather than efficiency. Appointments continue to be based on individual favors and on class and ethnic interests.

Foreign investors frequently make comparisons between the quick decisions their proposals receive in *authoritarian* bureaucratic states in East Asia, Southeast Asia, and in China as against the delays and misuse of authority in *democratic* bureaucracies like India. India and China may be counted among the most restrictive and largest centralized telecommunications bureaucracies in the world. India's state telecommunications monopoly has over 460,000 employees and only 7 million lines (0.8 per 100) managed from Delhi through circulars, directives, orders, and 14 volumes of rule books. Although China installed 7 million new main lines in 1 year (1993–1994), it took India more than 50 years of independence to reach this stage. The public sector work culture in India emphasizes hierarchy, procedures, precedents, and audit requirements rather than customer service and connectivity characteristic of the new technology. Although private capital investment is eagerly sought by politicians in the ruling party and senior civil servants, several dozen private sector proposals have been collecting dust on the desks of junior administrative and technical staff who have not received instructions on how to process the applications in the new liberalized regime. Telecoms in India was separated from Posts only in 1985. A year later, two supposedly autonomously state corporations were set up, one to handle metropolitan traffic and the other to manage international traffic. Employees in the parent Department of Telecommunications (DOT) went on strike when their former colleagues received a raise based on the performance of their new autonomous corporation, given that DOT wages have traditionally been based on seniority, rather than performance. Without the required training in market management that requires fair competitive bidding procedures transparent to all, the DOT bureaucracy's clumsy attempts at selecting cellular system operators and paging system providers landed it in court.

In 1994, China had 589,000 employees, 0.05 main lines per 100 population, and a reputation for vicious bureaucratic fighting between competing government agencies. The almost-monopoly service provider, the Directorate General of Telecoms (DGT) was recently separated from its parent ministry to make it commercially autonomous, signaling a first step toward corporate autonomy. Then, there are over 200 private networks owned by official agencies of the state (e.g., the army, the railways, the electronics ministry, the banking system) eager to sell their excess capacity. Additionally, two competitive service providers were approved in 1994 too, one directly and the other indirectly controlled by the ambitious electronics ministry that would like to expand into service provision in a big way (Xu, 1994).

The state telecommunications bureaucracies in many Latin American countries (e.g., Brazil under military rule) were remarkably efficient in the 1960s and 1970s. Exhausted by the inflation and debt situation in the 1980s, the Brazilian state is now overwhelmed by the resurgence of vocal political interests following the return to civilian rule after 1985.

In many states, attempts are being made to improve the efficiency and financial autonomy of telecommunications bureaucracies through the creation of some form of decentralized public sector undertaking or corporation autonomous from the government, (e.g., Morocco, Tanzania, China). These steps could be interpreted as attempts to make the public monopoly work and thus keep privatization at bay for as long as possible. These are also first steps preparatory to privatization (e.g., Singapore, Sri Lanka).

Unions

Once antagonists, state bureaucrats and union members in telecommunications entities considering privatization now find themselves united in their opposition to the "rationalization" of management and labor proposed in this private profit-maximizing organizational system. Both bureaucrats and workers realize state divestiture is the fall of their empire: They fear staff reductions, new work conditions, and "creamskimming" of revenues by the assignment of value-added services to a separate profit-making entity. British Telecommunications has a 7-year plan to slash its peak of 245,000 workers by 55% by 1997. The Telecommunications Corporation of New Zealand made massive layoffs. Privatization was accompanied by a hiring freeze in Chile. In Argentina, early voluntary retirements reduced 10% of the 45,000 employees.

Early attempts at privatization made no provision for officials and workers who would be declared redundant as the bloated bureaucracy was cut down to size. In countries that have subsequently considered this option, the mere mention of privatization has led to union opposition leading to reduction in the proposed scale of privatization in a few (Colombia, Uruguay, Costa Rica) and even cancellations (postponements?) in some (Greece, Thailand, Sri Lanka). The privatization of the Mexican telecommunications system Telmex might not have been possible had it not been for the dramatic demise of organized labor within government policymaking that peaked during the 1980s (Davis, 1992), the transfer of all 50,000 workers to the privatized entity, and the provision of a limited number of shares to employees. Argentina's conversion to a private duopoly took place at a time when the once strong labor movement was low: The head of the telecommunications workers union was coopted by the offer of the job of Secretary of Communication (Petrazzini, 1993). Malaysia offered its telecommunications employees early retirement, continuation in the Department of Telecommunications that was now the regulator, or a 5-year job guarantee in the privatized entity. Any changes in wage and labor laws in South Korea will have to contend with the fact that about 92% of workers are now members of the Korea Telecommunications Trade Union, one of the largest in the country (Kim & Jin, 1993). NTT Japan improved labor conditions for employees by lifting civil service salary caps and allowing wage levels to be set through collective bargaining on the basis of actual business performance. Thus, wage rates have been higher for NTT employees in the first 6 years in the privatized entity in comparison to comparable government enterprises and public corporations (Takano, 1992). There were no layoffs; workers were transferred to new NTT-controlled companies on the same terms and conditions. When Bangladesh's telephones were restructured as a government-owned autonomous corporation, employees were given the option of transferring to more protected government departments ("Number Unobtainable," 1990). Thus, "sweeteners" such as distribution of shares to workers and wage increases before divestment have been used to placate unions. Petrazzini (1993) concludes that while jobs have been protected in most cases of privatization of telecommunications in developing countries, labor laws have definitely changed in Argentina and Chile (on tenure, promotion, strikes, union membership, retirement, and health), rolling back labor's successes after years of struggle. In industrialized countries, a process of negotiation between management and labor has tended to minimize problems with regulators and speed the process of modernization and expansion (International Labor Organization, 1993).

CONCLUDING REMARKS

The preceding sections have shown how different developing country telecom systems have strategically taken recourse to private capital to improve their capability in various ways, albeit at the loss of some autonomy. State-capital relations are definitely being transformed, as evidenced by the state's need and willingness to use private capital to adjust to internal and external pressures it faces. The state versus market balance has tilted in favor of the market in many countries.

The market-based approach to national economic development benefited those with capital and purchasing power in the 1960s. In the 1970s and 1980s, many states in the periphery claimed they were going to benefit those without capital and purchasing power, but they primarily redistributed benefits to themselves self-servingly, and thus helped the expansion of a middle-class bureaucracy. The state did little to benefit the underprivileged, and now in the 1990s, its reliance on private capital to meet the state's capital accumulation needs (distinct from the public interest) could primarily benefit those who have the ability to pay.

What are the consequences for the excluded? What services will trickle down to the working-class majority, both urban and rural, small users of the telecommunications system who have limited purchasing power and influence with private capital or the state, potential users who did not benefit from the state as owner, and may not benefit from the private capitalist either, unless special regulations are designed and enforced? Because private capital investment is motivated by profits, less profitable rural and residential users are perceived as irrelevant drains on the system in the absence of universal service policies and measurable goals.

Developing country-states have offered monopoly rights to private investors even though the research indicates that it is competition rather than a public to private change in ownership that energizes efficiency and growth (Cook & Kirkpatrick, 1988). In cases of complete state divestiture and total privatization, the state has gained from a fresh infusion of capital (Mexico, Chile, Venezuela) or cancelled some amount of capital debt (Argentina)—all one-time benefits. The state has lost its goose that laid the golden eggs.

Most telecommunications markets are oligopolistic. The theoretical justification for relying on market signals to promote entrepreneurship and capital accumulation fades in such noncompetitive settings; state regulation becomes necessary. In recent conversions of the state telecommunications monopoly to a private monopoly, regulatory oversight has been reduced to a minimum to make the sale attractive to buyers, thus diminishing the state's ability to align telecommunications policies with national economic and social policies. The IMF, the World Bank, and potential investors are eager to ensure that there is an objective autonomous regulator, clear entry and exit regulations, and guarantees of profit repatriation. These are reasonable economic regulations because capital investment is not a philanthropic activity.

Because the state's goals in democratic regimes include meeting the needs of the largest numbers of voters, it is necessary for the state to also initiate social regulation in terms of universal service at an affordable cost. Unless the conditions of the franchise require public call offices in remote areas and slums, they will not be installed. Telmex is complying with the service conditions that require that all villages with a population of 500 be provided with some form of basic telephony. However, private sector participation implies cost-based pricing: The cost of local calls (the only calls made by the poor except in an emergency) have gone up in the United Kingdom, the United States, Malaysia, and Mexico as cross-subsidies are lifted. Middle-class and business users, but not the urban and rural working classes, benefit by a reduction in long-distance rates. Large users migrate to special value-added networks, leaving the working-class majority to pay the costs of running the expensive local loop in the absence of strong regulation. Without strong social regulation, private sector participation in telecommunications could represent a return to the trickle-down approach that benefited those with capital and purchasing power in the 1960s: The gaps between the rich and the poor grew larger. This could result in electoral defeats for the ruling party and a swing back to state ownership and populist policies.

No developing country telecom sector has adequate domestic capital and modern know-how to be able to afford to exclude foreign direct investment by U.S. and European telecommunications firms. The positive relationship between foreign direct investment and industrialization has been emphasized in Latin America (e.g., Frobel, 1980). With the exception of Singapore and Malaysia, the significance of foreign direct investment to national development in Japan, South Korea, Taiwan, and Hong Kong has been limited. Where economic development did occur, it was significant only when it stimulated linkages with local production firms and/or when it resulted in higher value-added production. In many parts of the Third World, neither significant linkages nor shifts to higher value-added production has resulted (Appelbaum & Henderson, 1992). This is another area for state policy and regulation.

What are possible outcomes under the current private sector participation scenarios? We suspect this is a time of growth rather than a time of distribution. This is a time of accelerated unequal development when the international and domestic voice, data, and video needs of large and medium investors and users in urban areas will be better met; the intra-rural and rural-to-rural needs of remote rural voters could continue to be neglected. This is a time when national identity could increasingly become identified and eclipsed by the corporate identities of foreign and domestic private capital. Like company towns, there are now company countries too, dominated by ventures with large multinational firms, corporate states themselves. State regulations are necessary to protect its information networks against a volatile capital base controlled by stateless unelected offshore fund managers and trade managers whose actions will directly affect the health of this sector. Whether this second wave of private sector investment—some call it a tide—will be better for the quality, quantity, and accessibility of telecommunications service will depend on the class interests of the particular state as regulator.

A majority of the top 40 public telephone operators with revenues greater than $1.5 billion U.S. are now privately owned, claims the ITU.

The majority of the next 40 are state-owned (Tarjanne, 1994), but this is changing. The tripartite role of the state in telecom is shifting in emphasis. When it was policy maker, regulator, and operator, it emphasized operations and neglected regulation. The state telecom entity that has now divested itself of some or all of its operational responsibilities is now required to make its policies and regulations transparent for private operators. The state telecom system that has opened its doors to competitive operators of basic, value-added, and mobile services in addition to running its previous operation (e.g., India) has increased its work load to also emphasize policy and regulation.

We maintain that the boundary conditions for bureaucrats, politicians, and regulators themselves are set by the national and local context of class and ethnicity. The challenge is not more or less regulation by the state but choosing the right set of economic development and social equity policies and regulations, that is, "getting the policies right" (see chapters 13 and 14, this volume, on regulation). Whether or not this will happen does not depend on training in policy and regulatory economics alone.

REFERENCES

Ambrose, W. W., Hennemeyer, P. R., & Chapon, J. (1990). *Privatizing telecommunications systems* (Discussion Paper 10). Washington, DC: The World Bank.

Amsden, A. H. (1989). *Asia's next giant: South Korea and late industrialization.* New York: Oxford University Press.

Appelbaum, R., & Henderson, J. (1992). *States and development in the Asian Pacific rim.* Newbury Park: Sage.

Baran, P. (1957). *The political economy of growth.* New York: Monthly Review Press.

Cardoso, E., & Faletto, E. (1979). *Dependency and development in Latin America.* Berkeley: University of California Press.

Cardozo, E., & Helwege, A. (1992). Below the line: Poverty in Latin America. *World Development, 20*(1), 19–37.

Cook P., & Kirkpatrick, C. (1988). *Privatization in less developed countries.* Brighton, UK: Wheatsheaf Books.

Davis, D. E. (1992). Mexico's new politics: Changing perspectives on free trade. *World Policy Journal, 19*(4), 655–671.

Dixon, H. (1991, October). The sleeping giants awaken. *Financial Times, 3*(1), 12–13.

Deyo, F. (Ed.). (1987). *The political economy of the new Asian industrialism.* Ithaca, NY: Cornell University Press.

Domhoff, G. W. (1967). *Who rules America?* Englewood Cliffs, NJ: Prentice Hall.

Domhoff, G. W. (1979). *The process of ruling class domination in America.* New York: Vintage Books.

Dunn, H. S. (1994, July 3–8). *Transnational corporations, the IMF and communications policy-making in the Caribbean: Implications of a Cable and Wireless/TOJ controversy in Jamaica.* Paper presented at the Conference of the International Association of Mass Communication Research, Seoul, South Korea.

Evans, P. B. (1979). *Dependent development: The alliance of multinational, state and local capital in Brazil.* Princeton, NJ: Princeton University Press.

Evans, P. B. (1985). Transnational linkages and the economic role of the state: An analysis of developing and industrialized nations in the post-World War II world. In P. B. Evans, D. Rueschmeyer, & T. Skocpol (Eds.), *Bringing the state back in.* New York: Cambridge University Press.

Evans, P. B. (1987). Class, state and dependence in East Asia: Lessons for Latin Americanists. In F. Deyo (Ed.), *The political economy of the new Asian industrialism.* Ithaca, NY: Cornell University Press.

Frank, A. G. (1967). *Capitalism and underdevelopment in Latin America.* New York: Monthly Review Press.

Frobel, F., Heinrichs, J., & Kreye, O. (1980). *The new international division of labor.* Cambridge, UK: Cambridge University Press.

Haggard, S., & Cheng, T. (1987). State and foreign capital in East Asian NICs. In F. Deyo (Ed.), *The political economy of the new Asian industrialism.* Ithaca, NY: Cornell University Press.

Hamilton, C. (1992). Can the rest of Asia emulate the NICs? In C. K. Wilber & K. P. Jameson (Eds.), *The political economy of development and underdevelopment.* New York: McGraw Hill.

Hamilton, N., & Kim, E. M. (1993). Economics and political liberalisation in South Korea and Mexico. *Third World Quarterly, 14*(1).

Harris, N. (1991). *City, class and state: Social and economic change in the Third World.* London: I.B. Taurus.

Hukill, M., & Jussawalla, M. (1989, Spring). Telecommunications policies and markets in the ASEAN countries. *Columbia Journal of World Business,* pp. 43–57.

International Labor Organization. (1993). *Telecommunications services: Negotiating structural and technical change.* Geneva: Author.

Kim, E., & Jin, Y. (1993). The evolution of telecommunications policy in the Republic of Korea. *International Review of Comparative Public Policy, 5,* 231–256.

Kim, J. C., & Ro, T. S. (1993, September–October). Current policy issues in the Korean telecommunications industry. *Telecommunications Policy.*

Koo, H. (1987). The interplay of state, social class and world system in East Asian development: The cases of South Korea and Taiwan. In F. Deyo (Ed.), *The political economy of the new Asian industrialism* (pp. 165–181). Ithaca, NY: Cornell University Press.

Lee tells Manila it needs to expedite economic reforms. (1992, Nov. 23). *Asian Wall Street Journal,* p. 14.

Mexico reaches for new telecommunications heights. (1992, February 3). *Telephony,* pp. 22–23, 26, 28.

Migdal, J., Kohli, A., & Shue, V. (1994). *State power and social forces.* New York: Oxford University Press.

Miliband, R. (1969). *The state in capitalist society.* New York: Basic Books.

Mody, B., Tsui, L. S., & McCormick, P. (1993). Telecommunications privatization in the periphery: Adjusting the private-public balance. *International Review of Comparative Public Policy, 5,* 257–274.

Mody, B. (1985). First World technologies in Third World contexts. In E. M. Rogers & F. Balle (Eds.), *The communication revolution in North America and Western Europe.* Norwood, NJ: Ablex.

Mody, B. (1987). Contextual analysis of the adoption of a communication technology: The case of satellites in India. *Telematics and Informatics, 4*(2), 150–158.

Number unobtainable: Dhaka unveils ambitious plans for phone network. (1990, June 21). *The Far Eastern Economic Review,* pp. 74–75.

O'Neill, J. (1991, May). *Foreign operator desires in telco privatizations: Their relationship to government realities.* Paper presented at the World Bank-ITU CTD Seminar on Implementing Reforms in the Telecommunications Sector, Washington, DC.

198   MODY AND TSUI

Petrazzini, B. A. (1993). *The politics of telecommunications reform in developing countries.* Unpublished doctoral dissertation, University of California, San Diego.
Philippines phone monopoly assailed by critics. (1993, February 11). *Far Eastern Economic Review,* p. 55.
Poulantzas, N. A. (1978). *Classes in contemporary capitalism.* London: Verso.
Ramos seems to push Philippine court reform. (1993, Feb. 15). *Asian Wall Street Journal,* p. 1.
Roth, G. (1987). *Telecommunications: The private provision of public services in developing countries.* New York: Oxford University Press.
Samarajiwa, R. (1992). *Telecommunications restructuring in Sri Lanka.* IAMCR paper, Brazil.
Sampson, A. (1973). *The sovereign state of ITT.* New York: Stein & Day.
Singapore awards telelecoms underwriting. (1993, March 18). *Far Eastern Economic Review,* p. 67.
Skocpol, T. (1979). *State and social revolutions.* Cambridge: Cambridge University Press.
Sobel, R. (1982). *ITT: The management of opportunity.* New York: Truman Talley Books.
Southwestern Bell's move. (1993, Nov. 22). *New York Times,* Real Estate section, p. 32.
Tarjanne, P. (1994, January). *The missing link: Still missing?* Paper presented at the Pacific Telecommunications Conference, Honolulu.
Takano, Y. (1992). *NTT privatization study: Experience of Japan and lessons for developing countries.* Washington, DC: The World Bank.
Wallerstein, I. (1979). *The capitalist world system.* Cambridge: Cambridge University Press.
Wellenius, B. W., Stern, P., Nulty, T., & Stern, R. (1989). *Restructuring and managing the telecommunications sector.* Washington, DC: The World Bank.
Westendoerpf, D. (1991). *Restructuring of telecommunications in developing countries.* Paper presented at the World Bank–ITU CTD seminar on implementing reforms in the telecommunications sector. Washington, DC: The World Bank.
Xu, A. (1994, July). *Perspectives and issues on China's telecommunications institutional reform.* Paper presented at the International Association of Mass Communications Research, Seoul, South Korea.

III

CASE STUDIES

9

Privatization of Telecommunications: Lessons From the Philippines

Alexandrina Benedicto Wolf
The Ohio State University

Gerald Sussman
Emerson College
Portland State University

The Philippines represents an exception to the pattern of statist types of Third World telecommunications systems. From its origins, telephone service in the country, as in the United States, has emphasized regulated, private sector control. This chapter argues that: (a) problems attributed to either state-owned or privatized systems are not necessarily inherent in the technical structure of the telecommunications system per se; (b) privatized systems of telecommunications ownership do not necessarily alter common characteristics of the infrastructure (e.g., lack of universal service, high tariffs, heavy indebtedness, insufficient investment in distributed plant and services, political manipulation and corruption, foreign equity or supply controls, etc.); and (c) the success or failure of telecommunications systems cannot be isolated from the national and international political and economic framework in which the systems operate.

The study of the complex nature of the Philippine telecommunications sector provides a useful antidote to the simplistic advocacy of private ownership as a panacea for the aforementioned problems. Despite its allegiance to privatized ownership since its formation under American colonialism as well as large private and public investment in the late 1970s and early 1980s, the Philippine telephone system remains among the world's least developed with less than 2 telephones per 100 people, in comparison with Thailand (13), Indonesia (3), and Malaysia (7). Other states in the region, such as Singapore and South Korea, which, until recently, pursued state-ownership approaches have among the highest telephone access (density) ratios in the Third World, 38 and 33, respectively. The Philippines' major telephone operator, Philippine Long Distance Telephone Company (PLDT) has had limited reach under both American (GTE until 1967) and domestic management. PLDT still has about 50% foreign equity control. As one observer has noted, the Philippines state "is more a product of external forces . . . than of internal factors" and even in more recent years, foreign investment has determined the course of its debt-driven economic growth (Riedinger, 1994, pp. 2, 7).

In this chapter, the following questions are raised: (a) What factors led to the Philippines developing and retaining private ownership of its telecommunications system? and (b) How has PLDT performed under a private form of control?

Private Sector Participation

Much of the recent discourse in economic development has focused on the role of telecommunications, the need for developed and less developed countries to pursue structural changes in their telecommunications policy to make them more competitive in the world economy, and the need to shift from state-owned systems (PTTs) to liberalized, deregulated, or privatized telecommunication systems.

State-owned telecommunication systems have been widely blamed for problems in the sector, including inattention to and underinvestment in telecommunications infrastructure, failure to adopt new technological standards, long waiting lists, and unfulfilled demand from the business sector. Vertical and horizontal integration in services and equipment and diversion of profits from telecommunications to subsidize other government operations is another perceived shortcoming of state-owned systems. The use of outdated equipment is said to make many telecommunication systems unresponsive to the growing needs of the domestic and international business community, protecting inefficient and bureaucratic management and maintaining artificial tariff structures.

On the other hand, arguments for state-owned and state-controlled telecommunication systems have focused on the natural monopoly characteristics of the system and question the willingness of privately owned systems to provide service and be accountable to rural and less profitable areas. Exaggerated attention to business users, it is argued, results in social and economic distortions. Roth (1987), however, believes that private ownership induces more private investments in telecommunications, allows investments to be managed efficiently without political intervention, and provides service prices based on actual costs. Private ownership of telecommunication systems also frees the government to undertake less profitable activities that are not in the interest of private ownership to pursue.

Increasing pressure has been brought to bear on PTTs as a result of economic and political instability; demands by national and transnational business users for newer, expanded, and integrated services; and by the World Bank and the International Monetary Fund, which seek to limit government spending and encourage multinational trade and investment. Deregulation and privatization in the United States and Britain have led the way to a similar, though less extensive, trend in other parts of the world. The Philippines was already on board.

THE PHILIPPINE CONTEXT

The Philippines, long mentored by the United States in its private-sector orientation to economic and telecommunication development, represents a complex example of national growth strategy. Made up of more than 7,000 islands, about 800 inhabited, and 65 million people, the Philippines in many ways has had a difficult transition to political and economic self-rule. Under President Ferdinand Marcos (1965–1986), the country suffered deteriorating social and economic conditions that left it impoverished, unstable, and deeply in debt. Public health, one of the only sectors to improve before the dictatorship period (1972–1986), had registered by the mid-1980s the highest rates of preventable disease occurrence in the western Pacific, had the lowest caloric intake, suffered real-income erosion so badly that 79% of preschool children were suffering from malnutrition by 1987, and had infant mortality rates two to four times higher than other Southeast Asian countries (Crone, 1993).

In 1989, 70% of Filipinos lived under the official poverty line of $540 annual income (Aquino, 1989; Sussman, 1991b). The wealthiest 12.9% of the population had 45.5% of national income. Per capita GNP rose only slightly during Corazon Aquino's presidency (1986–1992), from $709.20 U.S. in 1988 to $723.30 U.S. in 1989. The Philippines continues to be plagued by slow growth, low levels of investment, extremely low wages, and high levels of rural and labor discontent.

Although geographically located in Southeast Asia, the Philippines is regarded in some respects as outside the Asian mainstream. It is the only Roman Catholic country in the region, has an overlay of both Latin American and North American cultural characteristics, and has a political system patterned after that of the United States. At a time when most Asian governments were reluctant to entrust their countries' economic future to private enterprise and the United States had been reluctant to assist state capitalist or socialist economies, the Philippines stood as a showcase of American-style development. The Philippines has maintained a free-market economy in which foreign capital, especially American and more recently Japanese, has been able to freely operate. The United States remains the Philippines' most important international political, economic, and sociocultural conduit (Montgomery, 1962; Myrdal, 1968; Owen, 1983; Thompson, 1991).

The Political and Economic Framework

To place the present analysis in perspective, it is important to briefly review the historical development of the Philippines, particularly its economic and political framework. Technological or economic structures do not diffuse mechanistically but rather depend on a host of variables, including historical linkages; social class structure; government policy; allocation of resources; goals and capabilities of institutions and individual actors; cultural, psychological, and ideological orientation and practices; leadership; and core political dynamics. Technology does not operate in a vacuum but rather in a complex matrix of social and personal relationships.

Colonial History. Over the past century, the Philippines has evolved from a European statist to the North American private enterprise concept. From the Spanish era to the present, the Philippines has been integrated into the world market mainly through it supply of raw materials for manufacture and consumption in the West and more recently in other industrialized Asian nations. Under Spanish (1565–1898), American (1898–1946), and Japanese (1941–1944) rule (Table 9.1), the economic, political, and social structures of the Philippines were molded in ways that left the country dependent upon powerful foreign states, a legacy relevant to the present era.

The manner in which the Philippines achieved its independence suggests that the end of formal colonialism did not break established commercial, cultural, and political ties with the former colonizer. Although educated under American institutions, the intelligentsia were divided in their attachments to maintaining American economic and military linkages to their country, which gave rise to varying articulations of nationalism and economic independence. Under the Spanish colonization, the dominant domestic economic and social classes, the *hacenderos*, based their power mainly on land ownership and agricultural (e.g., sugar) production. With the growth of foreign trade, especially after the Suez Canal opening (1869), the merchant and landowning classes became more powerful. Following the transfer of political control from Spain to the United States and after passage of Congress's Payne–Aldrich Free Trade Act in 1909, sugar and coconut products began to dominate the economy. To this day, economic and political power is vested in the hands of a small number of powerful family dynasties and coalitions, with 10% of Philippine households, domi-

TABLE 9.1
Significant Periods in Philippine History

1565–1898:	Three centuries of Spanish (divide and conquer) colonialization of the Philippines. Revolutionary struggle against Spanish rule began in 1870, initially calling for secularization of Catholic parishes. Educated Filipinos, led by Jose Rizal, joined the struggle in 1892. In the same year, Andres Bonifacio, representing the working class, set up an underground organization called the Katipunan. Armed revolution began August 26, 1896.
1898–1901:	American conquest and assimilation. American dominance was established and principles of constitutional governance were laid down.
1902:	The Philippine Organic Act, adopted by the U.S. Congress, declared the end of the Filipino–American War and established full-fledged civil government in the colony.
1909:	The Payne–Aldrich tariff passed by the U.S. Congress gave Americans protected trade status in the Philippines and ushered in the era of economic "special relations."
1934:	Tydings–McDuffie Act signed into law that provided for Philippine independence to take effect on July 4, 1946. Independence Act set a 10-year transitional period during which the Philippines was designated a commonwealth with limited autonomy.
1941–1944:	The Pacific War and Japanese occupation of the Philippines. Philippine government in exile; puppet government established by the Japanese in Manila.
1944–1947:	Americans reconquer the Philippines. United States emerges as a dominant power in the Asian-Pacific Region. U.S. policies and interests overshadow interests for Philippine sovereignty. Bell Trade Act restored colonial economic ties: parity amendment to the Philippine Constitution established national treatment for Americans in exploitation of natural resources and operation of public utilities. U.S. military bases given a 99-year lease as a condition for foreign aid (subsequently revised).
1955–1965:	Period of Philippine economic nationalism. The Laurel–Langley Agreement revised the terms of trade and was followed by entrepreneurial initiatives in Philippine legislation.
1972–1986:	Martial law and dictatorship. Period of economic instability, loss in productivity, rising inflation, and large foreign debt (about $28 billion). Martial law technically ends in 1981, rule by decree continues until 1986.
1986 to date:	Post-Marcos era: economic and political restructuring.

Note. From Friend (1987); Pomeroy (1993).

nated by Spanish and Chinese *mestizos* (i.e., of mixed ancestry), holding 32.1% of household expenditures (Riedinger, 1994).

During the Spanish and American regimes, economic development was characterized by "progressive alienization." Ownership of productive assets outside the subsistence agriculture sector was dominated by Westerners, Chinese, Japanese, and other Asian minorities. When the United States consolidated power in 1901, its Philippine Commission urged Congress to provide liberal access to investors in Philippine public lands and mineral and forest resources; to empower the insular government to

extend favorable terms in chartering corporations and granting franchises
to investors proposing to build railroads, electric utilities, and communication systems; to authorize the sale of bonds in the United States to
provide funds needed to erect an appropriate infrastructure of roads,
portworks, and other public improvements supporting economic development; and to grant the colony certain nonreciprocal tariff preferences
to ensure a lucrative and expanding market for insular exports.

By 1914, direct American investment in the Philippines accounted for
roughly 80% of total foreign investment. Some of the direct investment
went into public utilities, but the great bulk went into extractive industries
and plantations (Myrdal, 1968). American investment in public utilities,
including transportation facilities, accounted for an additional 22% of
American direct investment in the 1930s. The concentration of American
investment in export production and processing and public utilities, including telephone, at the end of the colonial period was extreme, as these
activities counted for more than four fifths of all American direct investment (Golay, 1961).

Postcolonial Era. The Philippines in the postwar period gradually
shifted from traditional (agricultural) to modern sector dominance.
Foreign reserves started to drain by the late 1940s, inducing the Philippine
government to install import and exchange controls in 1950 and promote
industrialization through import substitution. The greatest beneficiaries
of the manufacturing and import substitution boom continued to be
American investors and their Philippine–Chinese counterparts who
enjoyed preferential treatment accorded by the Laurel–Langley Act in
licensing, patent protection, royalties, and the repatriation of dividends.
Excessive import substitution also discouraged the export of manufactured goods and the growth of a domestic investment-goods industry.

The ratification of the parity amendment in 1946 Philippine Trade Act
established national treatment for Americans in the exploitation of natural
resources and the operation of public utilities and other infringements
on Philippine sovereignty (Golay, 1961). The U.S. Congress also specified
that compensation for war damage would be contingent upon implementation of the parity clause of the Bell Trade Act. This was based on the
premise that it would result in a massive inflow of foreign investment,
which would accelerate rehabilitation and economic development. However, development of the Philippine economy was modest, and American
investment was limited compared to those in other countries (e.g., Japan,
South Korea, Taiwan). American investment in Cuba in 1930, with a third
of the Philippines' population, was eight times larger.

The search for capital to finance export-oriented industries leads to
heavy reliance on foreign investors, particularly transnational corporations, because of their control of affluent country markets, their access to

risk capital, and their experience with the requisite technology (Snow, 1983). In the Philippines, export-oriented industrialization (EOI) grew rapidly in the 1970s, particularly in garments and electronics, with the United States still the largest single foreign investor in both areas. On the eve of martial law, the dominant economic position of the United States in the new export-oriented economy aroused fervent opposition to "special relations," a persistent theme of Filipino nationalists since independence. Yet, each postwar president sought to increase external, especially American, sources of investment, loans, economic and military aid, and foreign exchange. It is incumbent upon Philippine presidents, according to former president Diosdado Macapagal (1961–1965), "to obtain the support of the American government or at least not antagonize it in their bid for the Presidency" (Lindsey, 1987, p. 230).

With a substantial electoral war chest and the financial backing of his wife's wealthy family, Marcos was perhaps less directly beholden to the United States than his predecessors for electoral support, although he had built his political capital on his reputation as an allied guerilla leader during the war with Japan. When he became president (1965), he showed his fidelity to U.S. policy interests by sending military forces to Vietnam (1966) and (illegally) contributing between $1 million and $2 million to Nixon's presidential campaigns in 1968 and 1972 (Bonner; 1987; Karnow, 1989). In reciprocation, Nixon endorsed Marcos's seizure of state power in September 1972. Marcos used his unprecedented powers in the presidential palace to approve many new state enterprises, most run by relatives or secret business partners, including many in the telecommunications sector.

In the 1970s, the shift from import substitution to export-oriented industrialization and the declaration of martial law (1972) created new pressures to lure foreign investment and at the same time suppress nationalist and radical opposition. *Business International* reported that the foreign business community in the Philippines welcomed martial law as "the best thing that ever happened to the country" (Lindsey, 1987, p. 231). The creation of export processing zones and other new areas for foreign investment would require a more sophisticated system of telecommunications to link the EPZs and urban industrial parks to Manila and to other world centers of capital. Telecommunications expansion in the 1970s and since has been in essence a project of global capital integration.

Under Marcos, the "free enterprise" policies of his predecessor, Macapagal (who abandoned exchange and import controls), shifted toward a more active state role in promoting modern sector development. State-run companies became a vehicle for rewarding the president's family and allies (so-called "crony capitalism") and grossly enriching the first family with ownership or management portfolios in potentially lucrative enter-

prises, including oil, automobile, airline, banking, agribusiness, and telecommunication businesses. In sectors where transnational corporations
were prohibited from owning land, as in agriculture, companies, such as
Dole, bypassed the restriction through leasing or joint venture arrangements. The main direction of state economic policy was toward establishing export-oriented industrialization, which helped change the Philippines from a seller of primary commodities to a seller of cheap labor
power (Snow, 1983). By 1986, a vast network of palace-affiliated industries, many of them already bankrupt, fell under the sweep of the "people
power revolution," which restored power to the premartial law elite (the
"old oligarchs") and led to government sequestration of the Marcos and
crony holdings.

In Philippine business, as in politics, control is typically vested in
wealthy, extended family or kinship groups and coalitions of such groups,
rendering elections as a means of contesting intra-elite economic, rather
than ideological, conflict. The end of Marcos' rule in 1986 represented a
dramatic and more democratic form of elite circulation rather than a
"revolution," as the landless and urban poor continued to suffer the
ravages of poverty, low wages, high unemployment and underemploy-
ment, political neglect, and human rights abuses inflicted by the entrenched and restored ruling elite under Aquino and later Fidel Ramos
(1992–).

Early in her administration, Aquino spoke of the need for economic
reconstruction and supported several programs toward that end. However, the initiative that emerged was essentially short term, focusing on
dismantling the Marcos monopolies, selling off crony and state enterprises, and starting a public works program to absorb some of the un-
employed labor pool (Hooley, 1991). The new president yielded to pressures from the United States, the World Bank, IMF, and conservatives
within her inner circle for liberalization, easing rules for foreign investment, deregulating several economic sectors, reducing tariffs and government spending, opening state enterprises to foreign debt–equity
swaps, and ridding her cabinet of social democrats—but not without
resistance from former political allies, nationalist elements in government,
organized labor, and the left. From the far right, she faced destabilization
efforts with half a dozen coup attempts. By the end of her term, most of
the business monopolies remained, many with new management, and
one that survived with its ownership intact was PLDT.

In 1987, the Philippine government also embarked on an ambitious and
controversial program of privatizing 135 of the 296 government-owned or
government-controlled corporations. This covered two main areas: state-
owned corporations and nonperforming assets that ranged from cars to
private jets, financial accounts, and even whole projects. One of these is

Philippine Airlines, which was subsequently taken over by a consortium headed by Antonio Cojuangco, president of PLDT (and a cousin of Corazon Aquino). Other companies to be privatized included Petron Corporation (oil), National Steel Corporation, and Manila Electric Company, from which the government hoped to raise at least $283 million U.S.

THE PHILIPPINE TELECOMMUNICATIONS SECTOR

Early Growth and Development

Telecommunications services were present in the Philippines as early as 1867, with the installation of telegraph lines between Manila and Corregidor and later to several provinces in Luzon and the Visayas. In 1878, the British opened Manila to Hong Kong via international cable, which Admiral Dewey, during the Spanish–American War, severed to cut Spanish ties with Madrid. The U.S. Army Signal Corps would eventually lay 1,500 miles of cables and landlines between all the major Philippine islands for military applications, later for commercial uses, as private American interests organized telegraph and telephone services. In 1905, Americans organized the Philippine Islands Telephone and Telegraph Company, and in 1928, the Philippine Long Distance Telephone Company, under American control, was given a 50-year franchise by the colonial Philippine legislature. Almost completely destroyed during World War II, PLDT did not recover its prewar scale until 1953. Telephone services and other communication industries have remained in the private sector from the outset of American rule.

During the 1930s, PLDT acquired the assets and franchises of other telephone companies and initiated long-distance services to some parts of the country. By 1940, it had installed approximately 28,600 telephones, many of which were destroyed during the war. By the 1950s, other small telephone companies were established outside Metro Manila and were interconnected to PLDT.

In 1956, GTE bought out the British Columbia Telephone Company, which then owned PLDT, and acquired the Philippine company in the deal. In 1964, the first high-quality voice circuit system was introduced in the Philippines with the installation of the Philippine-Guam coaxial submarine cable, owned and operated by PLDT. By 1968, PLDT had more than 180,000 telephone lines in service and had become a regulated national and international service monopoly, similar to AT&T.

Telecommunication development during this phase got a major boost from the existing U.S. military operations in the country. In 1966, the Defense Communication Agency arranged for intermediary INTELSAT

210

WOLF AND SUSSMAN

voice channels to be set up in the Philippines to augment the U.S. military's expanding communication infrastructure war in Vietnam. Marcos complied by setting up Philcomsat, a satellite corporation, nominally run by the government but in fact managed by Clark Air Force Base officials and GTE executives. The U.S. military–GTE group also began planning a regional satellite consortium called EASAT that was to include countries in the ASEAN nations. With the end of the Vietnam war, the Marcos government took control of Philcomsat and additional communication facilities in order to set up an integrated telecommunications system for commercial exploitation and for presidential command and control (Sussman, 1987).

Another shift toward Filipinization of telecommunications occurred when GTE began divestment of its holdings in PLDT. (By 1990, 40% of PLDT equity was still foreign-owned.) Following his proclamation of martial law, Marcos issued a presidential decree permitting American corporations to continue to occupy land under a lease limited to 25 years, renewable for a second 25 years, and that no action would be taken to change existing parity rights held by Americans. Marcos utilized his dictatorial power to repress the demands of nationalist intellectuals and politicians, socialists, peasant and labor leaders, and grassroots organizers, while courting the international business and financial community, U.S. government representives, the U.S. military, and other foreign influentials. The divestiture of GTE from PLDT in 1967, the last wholly owned American utility enterprise, resulted from the conjuncture of nationalist pressures, domestic business demands, and economic pragmatism.

First, the Public Service Commission rejected applications for rate increases submitted by the Manila Electric Company and the Philippine Long Distance Telephone Company as long as they were American-owned. This affected the utility services provided and generated customer complaints and negative publicity, which contributed to the decision of the American owners to sell out to Filipinos. Second, because of the scheduled 1974 expiration of the Laurel–Langley agreement, which would end preferential and parity rights granted to Americans, GTE began to transfer its PLDT stocks to Filipino ownership, in the meantime working out a 15-year (1966–1981) supply contract with PLDT and an anticipated major role in the planned East Asian Satellite Consortium. Other foreign companies were disinvesting to reduce U.S. equity to 40%. Marcos himself would become a major silent partner in PLDT via his offshore holding company, Prime Holdings, through which he manipulated the telephone company's stock (Sussman, 1991b).

Third, in 1976–1977, the U.S. Securities and Exchange Commission exposed illegal GTE payoffs in 27 countries, including the Cojuangco-led group, who were given interest-free secret commissions, credits, and uncollected loans worth $4.5 million, plus a 40% interest in GTE-Philippines as conditions for favorable tax treatment and $20 million in purchases of GTE telephone equipment. As a result of the investigation, GTE agreed to sell to the Cojuangco group its $15 million in common and preferred PLDT stock and terminate its supply contract. Although this ended formal American monopoly control in Philippine telephony, PLDT found other transnational partners, transferring its business ties to the German-owned Siemens Corporation and other TNCs. The leading suppliers of telecommunications equipment to the Philippines by the early 1980s were West Germany, Sweden, France, the United Kingdom, and Japan (Sussman, 1982).

Telecommunications in the 1990s

As a result of easy licensing in the past, the Philippines has had a multitude of telecommunication entities, most of which are small local or regional companies. Critics of the system have argued that the quality and extent of services and facilities are inadequate to cover the whole country; existing networks are insufficient to meet basic communication needs; and that the telecommunications sector is uncoordinated, fragmented, and highly segmented by service, technology, and geography. Service coverage represents only about 16% of the total land area of the country, with a skewed concentration of telephone facilities in Metro Manila (Fig. 9.1). There are still 13 provinces, including three subprovinces and eight cities without telephone service, and 198 towns do not have telegraph service.

Over 50 other private telephone companies that essentially provide local telephone service in limited areas. Of the 226 local telephone exchanges in the country, 25 are still manual; about 73 exchanges are step-by-step Strowger switches, and the rest use a variety of crossbar switches. PLDT operates some 20 electronic switches: 73% are in Luzon, 19% in the Visayas, and 8% in Mindanao. The government-owned Bureau of Telecommunications (BUTEL) provides telephone service to less profitable areas in northern Luzon under contract to PLDT. In effect, state-owned entities provide limited service in areas where private capital has not found it profitable to do so.

PLDT dominates the telecommunications sector, with over 94% of total telephone traffic and until recently with its monopoly in long-distance service. It is the biggest private firm in the country, with assets listed at $1.4 billion U.S. in 1991. It has 1.1 million telephones in service but with a backlog (800,000) of unserved applications almost as large. Foreign shareholders control more than 50% of common stock, and the Cojuangco family wields an estimated 11%.

The domestic long-distance network serves about 157 cities and reaches out to 249 islands where no local toll exchange exists. Long-distance traffic

The Philippines

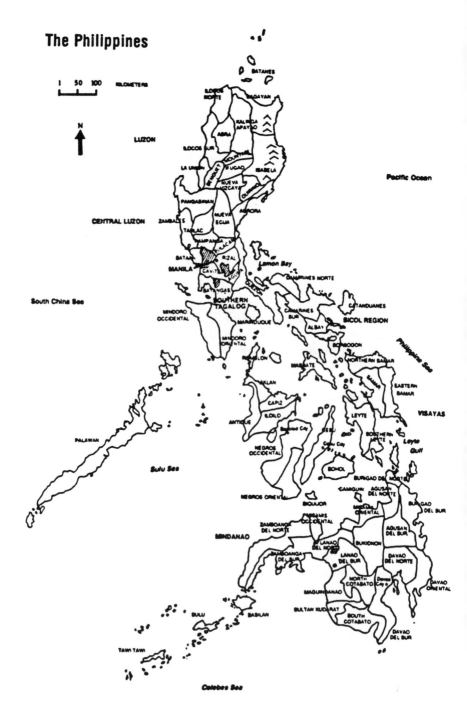

FIG. 9.1. Map of the Philippines showing major provincial capitals and other large cities. National teledensity is 1.02/100 people (5.73/100 in Metro Manila and 2.25/100 in other urban areas). Service coverage represents only about 16% of the total land area, with 13 provinces (3 subprovinces, 8 cities, and 198 towns) without telephone or telegraph service.

is still handled mainly on an operator-assisted basis, and domestic direct dialing is available only in larger cities. There are 2,113 telegraph stations in the country, 1,921 located outside Metro Manila, although the expected 24-hour message delivery time is not often met because of limited traffic and time-sharing of backbone networks. Moreover, many of these stations are frequently not in operation. The telex data network, with 12,860 lines, is relatively efficient, simply because its 2,254 business subscribers are currently limited to major cities where facilities are available.

Record carrier service is provided by BUTEL and eight private companies. Global-Mackay Cable and Radio Corporation (GMRC), Philippine Global Communications Inc. (PHILCOM), Eastern Telecommunications Philippines Inc. (ETPI), and Capital Wireless International provide international service. Globe Mackay is associated with ITT; Eastern Telecommunication is associated with Cable and Wireless. Philippine Telegraph and Telephone Corporation (PTTC), Radio Communications of the Philippines, Inc. (RCPI), Capitol Wireless, Inc. (CAPWIRE), and Universal Telecommunications Services, Inc. provide domestic service.

Demand for telecommunication services remains high in the Philippines. Although both government-owned and private communication facilities extend throughout the country, historically the service provided has not coped effectively with demand. For example in 1969–1970 the 1,428 cities and municipalities in the country had access to only 1,346 radio/telegraph/telephone stations. The country did not have full long-distance service until 1967, and by 1970 there was only one telephone per 2,300 people (Burley, 1973). Metro Manila, with 12.7% of the nation's population, has 70% of the telephones in service. Even the booming Cebu area in the Visayas region has only 7% of the lines (Aquino, 1989; Tiglao, 1992).

National teledensity for the Philippines is 1.1 per 100 people (5.73 per 100 in the Metro Manila area and 2.25 per 100 people in other urban areas). In comparison, Malaysia has 6.5 per 100 people, Thailand has 13 per 100, Indonesia has 3.1 per 100, South Korea has 33, and Singapore has 38. In the Philippine case, a long-standing strong commitment to privatization in the sector has not fulfilled growth expectations.

Even though current demand for telephone service far exceeds supply, private telephone companies like the PLDT operate on short-term profit incentives, concentrating their investments almost exclusively in the main business centers (meeting only 12% of demand in the secondary cities). As a result, people desperate for service connection are willing to pay black market prices for telephones at 300% to 700% of normal subscriber installation fees. Weak state intervention has encouraged illegal wiring alternatives and corruption of service practices.

214   WOLF AND SUSSMAN

FACTORS AFFECTING GROWTH
AND DEVELOPMENT OF PHILIPPINE
TELECOMMUNICATIONS

Analysis of the growth and development of Philippine telecommunications can not be isolated from the complex realities that exist in the
country, such as the transition from colonial to postcolonial rule, the
interplay between postcolonial intervention and the country's relatively
weak state apparatuses, and the general condition of underdevelopment.
This section focuses on important factors that have played key roles:
transnational investments, IMF and World Bank lending, and a weak
regulatory framework for telecommunications.

Foreign Investments and Lending

Lending in the telecommunications sector was conditional upon World
Bank–imposed rules that required Third World countries to establish or
maintain monopoly local telephone service providers and centralized
regulatory bodies (Broad, 1988; Sussman, 1991a). Externally directed policies in this sector became evident in the 1950s from proposals of the
International Cooperation Administration (predecessor of the U.S. Agency
for International Development) and the International Telecommunications
Union, and Marcos responded in 1969 by setting up a reorganization
committee that recommended full control of the telecommunications system by the private sector. In 1973, his government created an FCC-type
communications board. This institutional framework was important to and
in tandem with the export and foreign investment strategy of the regime
(Sussman, 1991a).

Historically, the United States has intervened in Philippine politics,
openly or covertly, on behalf of American business and pro-American
business policies to protect its privileged market access, evident in the
conditions attached to the U.S./International Monetary Fund stabilization
fund totaling almost $300 million: devaluation of the peso against the
dollar, abolition of import controls and exchange licensing, extension of
incentives to foreign capital, and tightening of domestic credit.

Although the economic theory of markets suggests that an expansion
of available information, together with enhanced and improved telecommunications, favors efficient decision making and extension of markets
across geographic and industry boundaries, in reality, only the largest
national and transnational corporations and government agencies have
the need for and the ability to take full advantage of these new opportunities. "In fact, smaller firms and countries are likely to find themselves
disadvantaged because of the new technological developments.... For

traditional, simpler communication requirements, such as basic telephone service, the new upgraded systems will serve quite well, but at substantially increased costs to small users" (Melody, 1991, p. 29). The introduction of telex, facsimile, advanced phone services, satellite, and new submarine cable systems in the Philippines in the 1970s and 1980s served mainly the large banking and business community and the international record carriers that serviced their information requirements.

Foreign-financed telecommunications tends to widen the gaps in communication access between rich and poor and between urban and rural areas. Many national planners and international development experts see telecommunications as a luxury for wealthy, urban business executives and are not convinced that telecommunication projects would benefit the rural poor. As technology introduces more specialized services, users become more fragmented. If newer technologies are affordable largely by the affluent in developed countries, the poor in the LDCs will be priced out of the market entirely (Hudson, 1983, 1984; Jussawalla, 1983). This is particularly evident in the Philippines, where state-of-the-art telecommunication facilities are concentrated in metropolitan business districts, and most rural areas are still without even basic telephone service. Mody (1993) found that "the typical privatized telecommunication scenario is neglect of rural communities, price increases for domestic users, price cuts for international clients, and cost-cutting on all fronts: employment reduction, wage freezes, equipment, R&D" (p. 262). Transnational users on the other hand continue to have better channels and services. This is a trend that has been encouraged by the World Bank's emphasis on ability-to-pay principles in telecommunication services and on the private monopoly ownership of telephone operations. In 1982, in response to World Bank recommendations, PLDT increased telephone interconnection fees to $415 dollars for residential private connection and $305 for party line (when annual per capita income was $783), as a way of pricing out "inefficient" users (Sussman, 1991a).

Samarajiva and Shields (1989) make the point in their analysis of the Philippines telecommunication investment and debt relationship that there are no built-in incentives for policymakers to include the marginalized in their policy decisions unless political actors within and without government force the policy process to respond to social concerns. Despite the rhetoric describing telecommunications as a factor for growth and development in the Philippines, social objectives have been traditionally overlooked in the policy agenda. In the historically privatized Philippine context, the state has largely escaped accountability, and where the state took a more active role (the Marcos dictatorship), cronyism and wholesale embezzlement were primary features of the system.

Broad (1988) stated:

Under the weight of the World Bank's and the International Monetary Fund's successful policy influence from 1979 to 1982 . . . [Filipino] nationalists lost every foothold of influence on policy formulation as transnationalists assumed hegemonic control of all industries. Within the private sector, economic nationalist factions whose enterprises depended on domestic markets were decimated as a class. As the industrial sector policy changes left an ever more concentrated industrial sector in their wake, so, too, universal banking undermined all but a small circle of banks. (p. 13)

Resistance was overcome mainly through corps of transnationalist allies and Western-educated technocrats occupying key government positions.

Marcos and his technocrats were also advocates of the developmentalist approach and of communication diffusion theory, using their leverage under martial law to centralize control of the mass media and other telecommunication services. Marcos saw telecommunications as one of the primary generators of the country's business environment, with emphasis on foreign turnkey investments and export-oriented industries, particularly electronics (Sussman, 1982). According to one critical scholar, the same community of transnational users of telematics has been responsible for the deregulatory policy changes in the United States, demanding expanded and integrated telecommunications services free of borders or barriers. "These major corporate users require global rebuilding of telecommunications network facilities just as they earlier pushed for domestic upgrading of the network" (Schiller, 1982, p. 101).

Political and Regulatory Protection

Two key departments are involved in the telecommunications policy of the Philippines. The Department of Transportation and Communications (DOTC) is responsible for the planning of telecommunications systems in the country. It implements its plans and policies through the National Telecommunications Commission (NTC), which is responsible for the supervision, regulation, and control of all telecommunications services. These include issues involving allocation and assignment of radio frequencies, spectrum-usage monitoring, and regulation and rate setting for telecommunication services. The 1984 National Telephone Program called for an integrated digital national network utilizing domestic and foreign investment. Advisers to the DOTC and NTC include the international record carriers and the World Bank, but there is virtually no representation from public interest groups.

Despite the lack of universal access, inefficient and poor service, and unresponsiveness to the needs of the rural sector, PLDT has maintained its dominant position in the sector through political patronage during the Marcos, Aquino, and Ramos administrations. Ramon Cojuangco's and his family successors' privileged relationship with Marcos and Aquino (Ramon's cousin) allowed him to thwart competitors. Competition in the airline and telecommunications industries require congressionally mandated franchises, which have been difficult to obtain.

The Justice Department blocked the sale of an existing government-owned telephone network covering northern Luzon and another system in central Luzon to Digitel, which had won the bid, because the firm failed to secure a franchise from Congress. The two systems would have made available 80,000 telephone lines by the end of 1995. Such a contract would have helped Digitel expand its telephone system into a national network independent of and, in the future, able to compete with PLDT. It might also have offered a yardstick for evaluating the efficiency of PLDT's operation. In 1991, threatened competition from ETPI, Digitel's sister firm, succeeded in forcing a 40% cut in PLDT's international toll rates. As a junior partner of the British transnational, Cable and Wireless, the joint venture expects to compete with PLDT and its technical partner, AT&T.

Extended families form the pillars of domestic capital in the Philippines. With the privatization of the Philippine Airlines in March 1992, Antonio Cojuangco (Ramon's son) was also installed as chairman of the airline. Heading the consortium (composed of the Zobel family who control Ayala Corporation; the Soriano family who control San Miguel Corporation of which another Cojuangco is also a major stockholder; George Co, owner of Equitable Banking; and the Retirement and Separation Benefits System, the military pension fund) that paid $370 million U.S. for a 67% shareholding of the airline monopoly, Cojuangco pledged his family's share in PLDT as collateral for most of the $212 million U.S. needed in financing Aeropartners. Other family holdings include Bank of Commerce; Manila Chronicle Corp., a newspaper publishing firm; First Pacific Land and Steniel Manufacturing Corp., a real estate firm with Hong Kong and Indonesian partners; Landmark Corporation, Makati department store; Sapphire Securities, stock brokerage; and Piltel, a telephone subsidiary of PLDT and the country's leading supplier of mobile telephones. Cojuangco has parlayed the power of PLDT to expand into real estate, media, financial services, and aviation.

PLDT has been more adept at stifling competition than in meeting demand for more telephones. Aquino's first secretary of transport and communications, Reynaldo Reyes, was replaced when he forced down the company's long-distance charges and raised its tax rates. He was replaced by Pete Prado, who took a far less aggressive stance toward PLDT. In mid-1991, the government awarded PLDT with a new 25-year national telecommunications franchise, despite the fact that its current franchise was not due to expire for 14 years.

In early 1993, however, President Ramos set out to deregulate the telephone industry and issued two orders that could make a dent in PLDT's monopoly position. These orders mandated interconnection among telecommunication networks and required international gateway operators and cellular companies to operate regular phone systems and install basic services in urban and rural areas. This means cellular operators have to install 400,000 land lines in 5 years, whereas international gateway operators have to install 300,000 lines. The National Telecommunications Commission expects to see 3.5 million lines in the next 5 years. With the 1.5 million lines to be installed under PLDT's "zero backlog" program, this will result in 5 million new telephone lines. NTC has given approval to five cellular telephone firms and five companies operating international gateways (Tiglao, 1993).

Since then, PLDT has installed some 172,000 new telephones in Metro Manila, and with its "zero backlog" program it plans to install 970,258 new telephone lines by 1996. Installing new lines, however, requires financing of $2.8 billion U.S. Total investment for all new lines is estimated at $6.5 billion U.S. over the next 5 years. Capital requirements will likely induce PLDT to concentrate on business districts and services with higher rates of return.

PLDT's competitors come from the country's wealthiest families, who have taken on "global partners with state-of-the-art expertise in telecommunications" (Tiglao, 1994). Among these are the Lopezes, who own the country's biggest broadcast network and control Manila Electric; shopping mall and manufacturing magnate John Gokongwei, who had already teamed up with the Lopezes in banking and power generation ventures, and the Ayala conglomerate (see Table 9.2).

TABLE 9.2
PLDT's Rivals

Firm	Controlling Filipino Owners	Foreign Partners (Nationality)
Globe Telecoms	Ayala Corp.	Singapore Telecom
Digital Telecoms	JG Summit (Gokongwei's holding firm)	Cable and Wireless (British)
International Communications	Benpres (Lopez' holding company)	Telstra, Ltd. (Australian)
Isla Telecom	Delgado family	Shinawatra Grp. (Thailand)
Smart Communications	10 investors, mostly company executives	First Pacific Group (Indonesian)
Bell Telecom	Ortigas, Maramba, and Puyat-Reyes	BellSouth (US)

Note. Revision of material from *Far Eastern Economic Review*, April 7, 1994, p. 52.

Expansion Problems

In 1981, the groundwork for the Philippine Master Plan for Development was laid. The first phase of the program called for the installation of a nationwide long-line transmission system equipped to handle data communications. Initially addressing the telephone system, the plan projected satisfaction of 60% of the total demand by 1990, which has yet to be met.

PLDT's funding shortage is about P2 billion (Philippine pesos) annually. The company's current expansion program is being paid out of retained earnings, a P4 billion commercial paper program, and a $125 million U.S. World Bank loan. This has beleaguered PLDT's expansion plans. However, this is also largely due to Cojuangco's reluctance to issue equity that would diminish his hold on the company and dilute its earnings. Despite building debt that has come close to breaching the allowable debt–equity ratio, the company has not issued any new common stock since 1987. Current debt–equity ratio stands at 1.6:1.

Since the moratorium on Philippine foreign borrowings in 1983, the company also has had difficulty in obtaining approval for foreign borrowings. Investment into telecommunications infrastructure averaged .35% of gross domestic product (GDP) for developing countries and .90% for developed countries. In the case of the Philippines, less than .30% of GDP was invested into telecommunications in 1980–1981. This represented 1% of the total infrastructure budget.

The Philippines external debt was $26.2 billion U.S. in 1985 and increased to $28 billion U.S. by 1988. The direct contribution of telecommunications investment to the debt burden was negligible and was dwarfed by the effects of oil imports and other factors (Samarajiva & Shields, 1989). However, there are indirect and intangible costs, sometimes difficult to quantify, that can be associated with telecommunications investment and that may contribute to a country's debt burden. Advanced information infrastructure that is biased toward the transnational community and world markets facilitates easier transborder data and financial transactions, which encourage repatriation and expatriation of savings and reduce foreign exchange holdings. It also reduces information costs of locating foreign goods and services, encouraging imports and trade and current account deficits. Information technologies, such as Landsat, may also accelerate borrowing for the extraction and export of natural resources and encourage transnationals to engage in transfer pricing, thereby reducing the government's tax receipts.

Advanced communication technology gives transnational capital an added advantage in peripheral economies, even as it allows for decentralized management. As a New York Chase Manhattan Bank executive commented on Brazil's debt situation in the late 1970s: "We know more about Brazil's economy than its own government" (cited in Gonzalez, 1988, p. 29).

SUMMARY

The Philippines has had private ownership of its telecommunications sector for nearly a century, but the putative social benefits of such a system have yet to accrue for most Filipinos (see Table 9.3). Privatization has not been an appropriate alternative for the communication-related problems that are faced by traditional state-owned and state-controlled telecommunications systems, some of which have made significant improvement in public service delivery through heavy state intervention or direct ownership. Equally important is the complex relationship between the historical and present-day industrial, political, economic, and social structures and actors, which eschews reductionist private ownership solutions.

As the Philippine experience shows, privately owned systems do not have any incentive to provide service to the poor and the marginalized. The current expansion plans illustrate the lack of responsiveness of private systems to service less profitable areas in need of basic telephone service and the propensity to direct investment toward highly industri-

TABLE 9.3
Summary of Private Ownership

Promises of Privatization	Philippine Reality
Provide services based on actual cost.	Interconnection and service fees remain high. Both local and long-distance rates remain high as well.
Efficient management without political intervention.	PLDT has always enjoyed a "favored" position and political protection. Although Ramos vows to deregulate the industry and open up competition, how the new players will fare remains to be seen. We may simply be seeing a new dog on an old collar.
Increase domestic and foreign investment in telecommunication.	PLDT has used its profits from the telephone industry to finance other ventures. Investment and service have also been concentrated in Metro Manila and transnational users. Service and connection in rural areas remain poor or nonexistent.
Raise capital to pay off national debt.	The contribution of telecom investment to the debt burden was negligible. PLDT's expansion programs require a $125 million U.S. World Bank loan.
Provide universal service.	As Philippine experience shows, privately owned companies do not have any incentive to provide service to unprofitable areas. Metro Manila transnational and business users have state-of-the-art service; however, basic service is lacking in most rural areas.

alized and profitable areas. A marked disparity exists between the communication services available to the large transnational corporations and domestic businesses in the metropolitan cities of the Philippines and those found in remote provinces, smaller cities, and rural areas in terms of access, quality, and efficiency.

Although the most modern telecommunications facilities are found in Metro Manila, most Filipinos await access to the most basic telephone service. Following World Bank advice, the government's and private sector's expansion plans focus on the capability to provide digital technology for data communication needs rather than basic service to remote, unconnected areas. And although the National Telecommunication Commission has mandated provision of telephone lines to unprofitable and rural areas by new entrants in the cellular and long-distance markets, it remains to be seen whether there will be adequate regulatory control and implementation. PLDT's rivals also come from the country's wealthiest and most influential families, a cornerstone of Philippine politics that government is unable to restrain. To the extent that they pose a threat to PLDT, the competitors' own political leverage may lead to the same kind of government protectionism that PLDT has enjoyed in the past.

Although privatization theoretically fosters competition for basic local, regional, and long-distance services, this has not occurred in the Philippines. Foreign ownership until 1967 (equity control to the present), government-supported monopoly, and weak regulation have favored one major operator. Even though the monopoly on long-distance service has been broken, it has not altered the functional monopoly that caters to business users rather than to broader planning needs of the country. Nationwide competition for basic local and regional service has not developed simply because it offers less promising returns on investment by private operators.

Telecommunications has an important role to play in Philippine development. The experience of privatization of that sector, however, has not been encouraging. The Philippines has one of the lowest rates in Asia of access to basic and affordable telecommunication services, yet it has perhaps the oldest private system in the periphery and was introduced to modern communication systems under the guidance of American colonial rule. And although democracy has, with the exception of the martial law period, long roots in the country, it is an elitist form of democracy that has failed to break with the oligarchic and even semi-feudal traditions of the past. Ironically, it was during the Marcos dictatorship period that the country underwent the widest expansion of telecommunications—new submarine cables, domestic and international satellite, digital telephone switching, telex, data and national, and all-color television broadcasting were introduced.

However, the expansion of telecommunications was predicated on providing a broad communication and information infrastructure for foreign and domestic business investment, export development and defense of the regime by way of a panoptical system of crony and palace wealth accumulation, propaganda, surveillance, and pacification through broadcast entertainment. Mutually beneficial as it was, neither foreign patrons nor the ruling elite challenged the assumptions of its design and did little to apply telecommunications toward rural, educational, social, or labor-intensive development. The energies of radical and ousted elite opposition were focused on gaining or recovering power, rather than pushing for social reforms. Until a multiclass political, economic, social, and communication consensus is reached—and the faction-ridden politics of the present period do not suggest that is happening—telecommunication policy will remain an elite enclave. For the Philippines, privatization has never been a panacea for making telecommunications an instrument of economic growth and social development, and, unlike Japan or South Korea, neither has the state been sufficiently independent of transnational capital to do any better.

ACKNOWLEDGMENTS

The authors wish to express their appreciation for comments offered by Consuelo Campbell, Jill Hills, and Rohan Samarajiva.

REFERENCES

Aquino, T. (1989, Spring). Philippine telecommunications in the Asia-Pacific Region. *Columbia Journal of World Business*, 73–81.
Bonner, R. (1987). *Waltzing with a dictator*. New York: Times Books.
Broad, R. (1988). *Unequal alliance: The World Bank, the International Monetary Fund, and the Philippines*. Berkeley: University of California Press.
Burley, T. M. (1973). *The Philippines: An economic and social geography*. London: G. Bell & Sons.
Crone, D. (1993). States, elites, and social welfare in Southeast Asia. *World Development*, 21(1), 55–66.
Friend, T. (1987). The Yellow Revolution: Its mixed historical legacy. In C. H. Lande (Ed.), *Rebuilding a nation*. Washington, DC: The Washington Institute Press.
Golay, F. (1961). *The Philippines: Public policy and national economic development*. Ithaca, NY: Cornell University Press.
Gonzalez, I. M. (1988, March). Scenarios of the communication revolution. *Philippines Communication Journal*, 22–30.

Hooley, R. (1991). Economic developments in the Philippines. In T. Robinson (Ed.), *Democracy and development in East Asia: Taiwan, South Korea, and the Philippines* (pp. 193–212). Washington, DC: The AEI Press.

Hudson, H. (1983). The role of telecommunications in development: A synthesis of current research. In O. Gandy, P. Espinosa, & J. Ordover (Eds.), *Proceedings from the Tenth Annual Telecommunications Policy Research Conference* (pp. 291–308). Norwood, NJ: Ablex.

Hudson, H. (1984). *When telephones reach the village.* Norwood, NJ: Ablex.

Jussawalla, M. (1983). The economics of telecommunication infrastructure in Third World countries. In O. Gandy, P. Espinosa, & J. Ordover (Eds.), *Proceedings from the Tenth Annual Telecommunications Policy Research Conference* (pp. 309–320). Norwood, NJ: Ablex.

Karnow, S. (1989). *In our image: America's empire in the Philippines.* New York: Random House.

Lindsey, C. W. (1987). Foreign investment in the Philippines. In D. B. Schirmer & S. R. Shalom (Eds.), *The Philippines reader: A history of colonialism, neocolonialism, dictatorship and resistance.* Boston: South End Press.

Melody, W. (1991). The information society: The transnational economic context and its implications. In G. Sussman & J. Lent (Eds.), *Transnational communications: Wiring the Third World* (pp. 27–41). Newbury Park, CA: Sage Publications.

Mody, B. (1993). Telecommunication privatization in the periphery: Adjusting the public-private balance. *International Review of Comparative Public Policy, 5,* 257–274.

Montgomery, J. (1962). *The politics of foreign aid: American experience in Southeast Asia.* New York: Praeger.

Myrdal, G. (1968). *Asian drama: An inquiry into the poverty of nations (Vols. I–III).* New York: Pantheon.

Owen, N. (1983). Philippine–American economic interactions: A matter of magnitude. In N. Owen (Ed.), *The Philippine economy and the United States: Studies in past and present interactions* (pp. 177–208). Ann Arbor, MI: Center for South and Southeast Asian Studies.

Pomeroy, W. (1993). *The Philippines: Colonialism, collaboration, and resistance.* New York: International Publishers.

Riedinger, J. (1994). *Political economy of private sector development in the Philippines.* Unpublished manuscript.

Roth, G. (1987). Telecommunications. In *The private provision of public services in developing countries.* New York: Oxford University Press.

Samarajiva, R., & Shields, P. (1989, January). Debt and telecom investments in the Third World: The Philippines case. *Media Development.*

Schiller, D. (1982). *Telematics and government.* Norwood, NJ: Ablex.

Snow, R. (1983). Export-oriented industrialization, the international division of labor, and the rise of the subcontract bourgeoisie in the Philippines. In N. Owen (Ed.), *The Philippine economy and the United States: Studies in past and present interactions* (pp. 77–108). Ann Arbor, MI: Center for South and Southeast Asian Studies.

Sussman, G. (1982). Telecommunications technology: Transnationalizing the new Philippine information order. *Media, Culture and Society, 4,* 377–390.

Sussman, G. (1987, Spring). Banking on telecommunications: The World Bank in the Philippines. *Journal of Communication,* 90–105.

Sussman, G. (1991a). Telecommunications and transnational integration: The World Bank in the Philippines. In G. Sussman & J. Lent (Eds.), *Transnational communications: Wiring the Third World* (pp. 42–65). Newbury Park, CA: Sage Publications.

Sussman, G. (1991b). The transnationalization of Philippine telecommunications: Postcolonial continuities. In G. Sussman & J. Lent (Eds.), *Transnational communications: Wiring the Third World* (pp. 125–149). Newbury Park, CA: Sage Publications.

Tiglao, R. (1992, February 27). Politics on the line: Philippine telephone deal blocked by controversy. *Far Eastern Economic Review,* pp. 64–65.

Tiglao, R. (1993, October 14). Wake-up call. *Far Eastern Economic Review,* p. 63.

Tiglao, R. (1994, April 7). A matter of willpower. *Far Eastern Economic Review*, pp. 52–54.

Thompson, W. S. (1991). The Philippines in the international environment. In T. Robinson (Ed.), *Democracy and the development in East Asia: Taiwan, South Korea, and the Philippines* (pp. 235–242). Washington, DC: The AEI Press.

Telecommunications Restructuring: The Experience of Eight Countries

Joseph D. Straubhaar
Brigham Young University

Patricia K. McCormick
Johannes M. Bauer
Consuelo Campbell
Michigan State University

The ongoing telecommunications reform processes, although frequently aiming toward deregulation, liberalization, and privatization, nevertheless show a great diversity in the specific rationale for and design of measures. This is a reflection of the fact that the relation between institutional frameworks and the performance of a sector is more subtle than what some of the fashionable policy discourse seems to suggest (Bauer, chapter 12, this volume). However, practical policy may be unable to choose from among all theoretically possible solutions and may be restricted to approaches feasible under the particular circumstances. Not only are policy choices constrained by the socioeconomic base of a country, but they are also constrained and shaped by the historical experience of nations as well as the status quo ante of reform. Furthermore, they are linked to the "visions" of appropriate reforms held and promoted by specific interest groups in support of different proposals.

Viewed in such a broad perspective, telecommunications policy may or may not evolve along a path leading to more efficient organization. Institutional change may result in increased efficiency, redistribution of costs and benefits from the given arrangement of telecommunications, redistribution of the chances and opportunities to participate in the future expansion of the industry, or a combination of these possibilities (Bauer, chapter 12, this volume).

The countries selected for review in this chapter illustrate many of these points. All of them have embarked on reforms of the organizational

TABLE 10.1
Basic Economic Indicators

	GDP/ Capita 1992 $	GDP Growth 1980–1992	Average Inflation Rate 1980–1992	Terms of Trade (1987 = 100)	Foreign Trade Quota	External Debt as % of GDP
Argentina	6,050	−0.9	402.3	110	11.8	30.3
Brazil	2,770	0.4	370.2	108	16.4	31.2
Chile	2,730	3.7	20.5	118	46.4	48.9
Hungary	2,970	0.2	11.7	102	61.8	65.0
Jamaica	1,340	0.2	21.5	96	86.8	131.7
Korea, Rep.	6,790	8.5	5.9	106	53.3	14.2
Mexico	3,470	−0.2	62.4	120	22.8	34.1
Philippines	770	−1.0	14.1	105	48.1	56.8
Sri Lanka	540	2.6	11.0	90	67.9	41.0

Note. Sources: World Bank, IMF, UNESCO.

framework of telecommunications. However, they started from very different historical, political, and economic contexts. For example, in an economic perspective, the countries differ in terms of their real economic income (as measured by GDP/capita), their sustained growth rate during the past decades, as well as their ties to the global market through exports and imports (see Table 10.1). The countries also differ significantly in terms of their social and living conditions. Table 10.2 summarizes key indicators including population and population growth, literacy rate, and sanitary conditions. Furthermore, the countries differ with respect to their

TABLE 10.2
Basic Socioeconomic Indicators

	Population 1992 (million)	Population Growth 1960–1992	Literacy Rate 1992	Infant Mortality[a]	Life Expectancy
Argentina	33.1	2.0	95.0	29	71
Brazil	153.9	5.9	81.0	57	66
Chile	13.6	2.6	93.0	17	72
Hungary	10.3	0.1	n.a.	15	69
Jamaica	2.4	0.2	98.0	14	74
Korea, Rep.	43.7	2.2	96.0	13	71
Mexico	85.0	4.3	87.0	35	70
Philippines	64.3	4.0	90.0	40	65
Sri Lanka	17.4	3.5	88.0	18	72

Note. Sources: World Bank, UNESCO, IMF.
[a]Deaths per 1,000 live births.

TABLE 10.3
Basic Telecommunications Indicators

	Main Lines/100 Population	Growth Rate (ML/100 Population) 1983–1992	Waiting Time (years)	Telex Connections per 10,000 (1992)	Data Connections per 10,000 (1992)	Telecom Investment as % of GDCF[a]
Argentina	11.1	4.2	3.1	3.0	0.7	1.9
Brazil	6.8	4.3	0.8	9.0	3.9	4.0
Chile	8.9	10.2	1.7	5.0	2.6	5.5
Hungary	12.5	7.9	6.0	3.0	3.5	5.1
Jamaica	6.8	10.1	5.1	2.0	0.8	16.6
Korea, Rep.	36.3	13.1	0	2.0	9.5	2.9
Mexico	7.5	6.1	1.0	3.0	18.5	3.4
Philippines	1.0	2.1	>10.0	0.5	0.1	3.6
Sri Lanka	0.8	5.6	9.6	1.0	0.1	0.9

Note. Source: ITU (1994).
[a]Gross domestic capital formation (GDCF).

political situation and institutional framework, their technological resources, their integration into the global economy, and so on. These
conditions impacted on the development of the basic telecommunications
infrastructure (see Table 10.3) and shaped the specific course, speed, and
success of the reforms adopted. The next section reviews the main reform
steps in eight countries from South America, Asia, the Pacific Region,
and Eastern Europe that illustrate this diversity (for a discussion of the
Philippines, see chapter 9). Another section presents the main reasons,
goals, and effects of restructuring. This is followed by analyses of the
process of restructuring and the selection and design of regulatory regimes. We conclude with a review of the tentative experience with telecommunications restructuring.

A SYNOPSIS OF TELECOMMUNICATIONS
RESTRUCTURING PROCESSES AND APPROACHES

Argentina

Until 1990, the Argentine telecommunications system was operated by
Empresa Nacional de Telecommunicaciones (ENTEL), a state-owned monopoly. In 1990–1991 ENTEL was split and privatized as two separate
regional monopolies, Telefónica de Argentina in the south and Telecom
Argentina in the north, with Buenos Aires divided between them. In
1990–1991, 60% of the new entities was sold to foreign companies. Tele-

fónica de Argentina was sold to Telefónica de España and Telecom de Argentina became a subsidiary of Société Nortel, which is 100% owned by a consortium consisting of France Télécom (32.5%), STET of Italy (32.5%), the banking firm J. P. Morgan, and the Argentine company Perez Companc. The remaining 40% were sold in 1991–1992 to employees (10%) and to the public (30%). The international monopoly operator, Teleintar, is owned jointly by Telefónica de Argentina (50%) and Telecom Argentina (50%). Several private providers of satellite services have been licensed (e.g., IMPSAT S.A., Sateliatal, S.A., Satelnet, S.A.) and some 300 cooperatives provide local telephone service. Several cellular licenses have been granted in the metro Buenos Aires area and in the provinces. No foreign ownership restrictions exist for cellular operators.

Until 1997, with a possible 3 year extension, basic services and networks are provided on a monopoly basis by the two main operators and Telintar. Value-added services, data, private networks, cellular, and some satellite services are open to competitive provision. However, Telintar, jointly operated by the two regional telephone monopolies, holds a dominant position in the VAS and data services markets. Cellular services are operated on a duopoly basis.

The Telecommunications Law of 1972, modified to permit private ownership and competition, still applies. Prior to privatization, rate-of-return regulation for telecommunications tariffs was replaced by price-cap regulation with rates indexed to the U.S. consumer price index. Lack of interconnection regulation created difficulty for licensed carriers to access bottleneck facilities of monopoly carriers. Regulatory tasks are vested with the semi-autonomous agency, Comisión Nacional de Telecommunicaciones (CNT).

Brazil

Telecommunicaçoes Brasileiras S.A. (TELEBRÁS) is a mainly state-owned joint-stock holding company, the parent company of 27 operating companies, one in each state, and the Empresa Brasileira de Telecommunicaçoes (EMBRATEL), the interstate and international long distance company. Established in the late 1960s, TELEBRÁS has been relatively successful in consolidating more than 1,000 scattered telephone operations and, during the 1970s, in building up a modern telecommunications infrastructure. A policy of high prices for telecommunications service allowed a strategy of concerted investment and rapid network expansion. However, during the 1980s, under the overall deteriorating economic conditions of Brazil, this reform process came to a halt as price controls for telecommunications services restricted the ability of TELEBRÁS to fully recover its cost and generate continued surpluses for investment purposes (Wellenius, 1994). Brazil's policy of industrial autarky resulted in high equipment cost (Roche, 1990). The highly centralized organizational structure of TELEBRÁS limited the managerial and financial flexibility of the operating companies and further slowed down expansion. As a result, large waiting lists for basic telephone services developed, service quality declined, and new services were introduced only slowly.

In spite of the poor record during the 1980s, Brazil has been hesitant to push for liberalization of telecommunications and has put off privatization (which would require constitutional reform), although the topic is still debated and policy could change under the new president, elected in 1994. TELEBRÁS still holds a monopoly for basic domestic and international telephone services. However, several measures were undertaken to alleviate the constraints on the business community. During 1990, several ministerial decrees allowed users to construct and operate their own networks and enabled private providers to offer value-added services, satellite, data, and cellular service in partial competition with the publicly switched network of TELEBRÁS. In addition, new ways to finance the expansion of the telecommunications infrastructure, such as the authorization of communities and developers to build local facilities for interconnection with the publicly switched telephone network, were introduced. Telecommunications policy is formulated and implemented by the Secretaría Nacional de Communicaciones (CNC) under the Ministry of Infrastructure.

Chile

The telecommunications restructuring process in Chile began in the mid-1970s. Until then, telecommunications services were mainly provided by two public enterprises and several small private companies. Compañía de Teléfonos de Chile (CTC), originating from mergers between a large number of private local telephone companies that were consolidated in 1930 into one company owned by International Telegraph and Telephone Corporation (ITT) and, in 1971, taken over by the government, provided local telephone service to about 95% of the subscribers. Empresa Nacional de Telecommunicaciones S.A. (ENTEL), the second largest, publicly owned company established in 1964, provided long distance and international services. In 1975, as the powerful Chilean military regime started to embark on a policy of deregulation for national development, the comprehensive telecommunications monopoly was also gradually reduced, first with the introduction of resale of telephone lines and the interconnection of terminal equipment purchased from independent suppliers (Melo, 1994).

Beginning in 1982, the government started selling its shares in telecommunications companies, first in the smaller telephone companies in southern Chile, and, in 1986, in CTC and ENTEL. This process was completed by mid-1991, when the government held no more significant ownership in any telecommunications company. In the privatization process, Telefónica de España had become the largest single investor in Chilean telecommunications with a share of 43.6% in CTC and 20% in ENTEL. The remaining shares are held by foreign and domestic banks, pension funds, and smaller private investors.

The process of liberalization and privatization has opened the telecommunications market considerably. Several companies have begun to provide basic telecommunications services, especially long-distance and international services, although CTC and ENTEL are by far the dominant carriers. The antitrust authorities have given CTC and ENTEL permission to enter each other's market. Although considered generally successful, the reform process in Chile has also shown weaknesses. For instance, the expansion of telephone services to rural areas has literally come to a halt. Furthermore, the powers of the regulatory agency, the Subsecretaría de Telecommunicaciones (SUBTEL) under the Ministry of Transport and Telecommunications since 1977, are weak and make the enforcement of open access conditions to the bottleneck facilities of CTC or an efficient control of prices in the remaining monopoly areas difficult. At the time of writing, the regulatory structure is under further review.

Hungary

Before World War II, the expansion of the Hungarian telecommunications system, provided by an integrated PTT (Magyár Posta), was more or less in line with its economic development. After 1948, under Communist rule, Magyár Posta continued to provide postal, telephone, and broadcast transmission services on an integrated basis. Prices were set by the state and investment funds were awarded by the National Planning Office. National planning emphasized "material" industries in manufacturing (Whitlock & Nyevrikel, 1992) resulting in a telecommunications network that predominantly served key industries (and party officials) but placed low priority on residential and agricultural customers (Heller, 1994). In 1968, Hungary gradually embarked on a course of liberal economic reform leading to the emergence of a more dynamic economic climate. However, imports of advanced technology were impeded by the CoCom restrictions, starving the country of needed technology.

The current restructuring process began in 1985 with the separation of the PTT from the Ministry of Transport and Communication and its reestablishment as a stand-alone government entity with both operational and regulatory functions. In 1989, these functions were separated and MATAV, the Hungarian Telecommunications Company (HTC), as well as two regulatory agencies, one responsible for spectrum management and one for technical issues, were founded under the Ministry of Transport, Communication, and Water Management. The ensuing political debate focused on the creation of conditions that would support the deployment of universal service. A basic framework was finally created with the Telecommunications Law of 1992, which provided for partial privatization of HTC and created a unique, more liberal environment for Hungarian telecommunications, not unlike the framework for Finnish telecommunications.

In 1993, after some political struggles, 30.3% of HTC was sold to German Telekom and Ameritech and further privatization of up to 49% was authorized (Kiss, 1994). The 1992 Law also provided for the creation of 56 regions, each served by a concessionary, and potentially in competition with HTC. In February 1994, 25 of these local franchises were awarded by the government. Long distance and international services are provided by HTC but the monopoly will be subject to review toward the end of the decade. The provision of public-switched, mobile, and paging services will require a government license but all other services have been liberalized and can be offered subject to transparent, minimal technical requirements. If local governments are not satisfied with the services provided, they have the right to initiate a public bidding for concessions. Concessions also specify an obligation to serve and regulate the quality of service targets. Tariffs will be subject to price-cap regulation and a Telecommunications Fund was established to subsidize service to disadvantaged areas (Heller, 1994). Conditions for interconnection in the increasingly pluralistic environment will be specified in a government decree. Overall, Hungary has implemented a unique set of reforms that are largely compatible with the rules of the European Union.

Jamaica

Until 1987, telecommunication services in Jamaica were provided by two companies, Jamaica Telephone Company (JTC, for local service) and Jamaica International Telecommunications (Jamintel, for long distance and international services). In 1987, Telecommunications of Jamaica (TOJ) was incorporated as the holding company for JTC and Jamintel (McCormick, 1993). After the People's National Party (PNP) under Prime Minister Manley sold the remaining government share, TOJ is now owned by Cable and Wireless (79%) and public shareholders (21%). Jamaica Digiport International (JDI), a joint venture between JOT (30%), Cable & Wireless (35%), and AT&T (35%), offers international private line services for Jamaica's free trade zones. TOJ holds a 25-year license for the exclusive provision of all basic services including leased lines. In the common law framework of Jamaica with its respect for property rights, the licensing model, which cannot be altered unilaterally and is enforced by the courts, was preferred by the main players over a legalistic model.

South Korea

By the mid-1950s, due to wars and struggles with occupying forces, 80% of Korea's telecommunications system had been destroyed. Despite increasing recognition of the importance of an efficient telecommunications infrastructure and the inclusion of telecommunications into Korea's consecutive government-led Five Year Economic Development Plans which were introduced in 1962, the sector evolved only slowly and was characterized by significant shortcomings until the late 1970s (Park, 1993; Kim & Jin, 1993; Kim & Ro, 1993). The pace of renewal and expansion accelerated in the 1980s with the separation of Korean Telecom (KT) from the Ministry of Communications (MOC) and the establishment of DACOM (jointly owned by public and private firms) for the provision of data communications services based on lines leased from KT. At the same time, and to provide a stimulus to the Korean electronic industry, the government allowed customers to purchase terminal equipment from competitors of KT. In 1991, DACOM received a license to provide international services in competition to KT. In the same year, the Public Telecommunications Business Law reclassified telecommunications service providers into three categories that are subject to different regulatory regimes.

According to the new provisions, all carriers are classified as either network service providers (NSP) or value-added service providers (VSP). NSPs own telecommunications network facilities, whereas VSPs provide services based on lines leased from NSPs. Network Service Providers are further split into General Service Providers (GSP) and Specific Service Providers (SSP). The most stringent regulatory oversight is established for GSPs that offer voice services and other basic services based on a designation from the Ministry of Communications. SSPs offer, for instance, mobile communications and port communications and need MOC approval before being allowed to enter the market. SPs are subject to light regulation and only need to register with MOC. As of 1992, KT and DACOM were licensed as GSPs, 12 companies as SSPs, and more than 60 enjoyed the status of VSPs.

During the 1980s and early 1990s, Korea was successful in developing policies to secure universal service (Kim & Lee, 1991) and to finance the rapid expansion of its telecommunications network largely through internal funds (Kim, 1992). Currently, the partial privatization of KT, which became a joint stock company in 1991, is pending. For its seventh Five Year Economic Development Plan (1992–1997), the government aims at reaching the economic income level as well as a sophistication of its telecommunications infrastructure comparable to industrialized countries.

Mexico

Until the 1940s, telephone and telegraph service in Mexico was provided by subsidiaries of ITT from the United States and Ericsson from Sweden. In 1941, ITT merged its operations with Ericsson, leading to the incorporation of a joint company, Teléfonos de México S.A. (TELMEX), in 1947. A group of Mexican entrepreneurs acquired interest in ITT and Ericsson, and in 1971 the Mexican government acquired 51% of the capital stock. Responsible for its own budget, TELMEX was under the control of the Ministry of Communications and Transport (Secretaria de Communicaciones y Transportes, SCT). Until 1989, SCT operated the services reserved for the state by the constitution, such as satellite and telegraph services, in competition with TELMEX. To end this ambiguous situation, a para-state agency, Telecommunicaciones de México (TELECOMM) was founded in 1989 and entrusted with the operations of the services removed from the SCT, which was restricted to regulatory functions.

At the end of the 1980s, after a period of dismal economic development, President Salinas placed high emphasis on the modernization of the Mexican economy. Telecommunications was seen as an indispensable infrastructure in this process of modernization and included in a broad program of privatization. A majority of shares was sold to a consortium consisting of Grupo Carso of Mexico (10.4%), Southwestern Bell International of the U.S. (10.1%), and France Télécom of France (5%; according to the concession, a majority of shares needs to remain in the hands of nationals).

TELMEX holds a monopoly for basic services until 1996. It is subject to price-cap regulation and has to meet certain quality of service targets as well as minimum network expansion requirements. Under the terms of its concession, TELMEX was to invest $8 billion U.S. until 1994 and expected to reduce the urban-rural discrepancy in service availability. Regional duopolies are established for mobile services. TELMEX competes in each region via a subsidiary, TELCEL. Private networks, terminal equipment, value-added services, and information service markets are liberalized.

Sri Lanka

Since 1980, telecommunications in Sri Lanka has undergone major institutional reform, although an effort to privatize the Department of Telecommunications in the mid-1980s was unsuccessful. Before 1980, telecommunications was operated by an all-encompassing, unregulated, government-owned monopoly (Samarajiva, 1993). In 1980, the Department of Telecommunications was separated from the Post Office. In 1984, a presidential committee issued recommendations including the privatization of telecommunications and a more open competitive environment. Due to opposition from trade unions and internal political crises in the wake of the Indo–Sri Lankan agreement of 1987, the privatization process was abandoned. However, in 1991, a comprehensive reform was initiated regarding the introduction of regulation, the transfer of the assets, liabilities, and so on, of the Department of Telecommunications to Sri Lanka Telecom (SLT), a newly created, fully government-owned statutory corporation, and the more concerted introduction of competition.

REASONS, GOALS, AND PROCESSES OF PRIVATIZATION

Governments have embraced the policy of privatization for a variety of reasons and goals, some of which stem from the very creation of the public sector. Many developing countries began creating state-owned enterprises and nationalizing companies, including utility companies, after World War II. Through the early 1970s, the expansion of the public enterprise sector was encouraged and supported with loans and credit extended by bilateral and multilateral agencies. Markedly slow economic growth in the early 1980s and a rapid increase in international interest rates in the late 1970s and early 1980s, however, made it increasingly difficult for governments to service these debts. The fiscal crisis prompted a reassessment of governments' economic policies and state-owned enterprises in the developing countries and in some Eastern European countries that had also begun to borrow from Western banks.

Economic and Administrative Efficiency Goals

The reassessment of the public sector has resulted in a widely shared conviction that its size should be reduced. It is argued that the administrative capabilities of governments were overextended and the performance of the public sector has been poor. The inefficiency of the public sector is associated with bureaucratic failure and political interference (Hemming & Mansoor, 1987). The public enterprise is an instrument for political patronage in many developing countries.

Problems with a state-owned and operated telecommunications monopoly are not unlike those of other state-owned enterprises or, for that matter, private monopolies. They are often inflexible, subject to political interferences, and have little incentive to provide efficient operations, quality service, or responsiveness to customer needs. This lack of efficiency is often reflected in state telecommunications administrations by a bloated workforce, low levels of service penetration especially in rural areas, long waiting lists for telephone service, and low call completion rates (Ambrose, Hennemeyer, & Chapon, 1990).

Arguments for privatization have thus occurred in the context of improving efficiency. The eradication of inefficiency was cited by the Korean government as an objective of its privatization program, which included Korea Telecom. It is argued by some authors that even the replacement of a public monopoly by a private monopoly will increase efficiency because of reduced political interference and more effective financial constraints (Commander & Killick, 1988). The influential property rights school suggests that a change to private ownership will improve the incentives for productive efficiency performance. Other approaches contend that management, not ownership per se, is the key to efficiency of an enterprise. The Philippine Long Distance Company is certainly an exemplary case of a private but inefficient firm (Benedicto & Sussman, chapter 9, this volume). Because the question of economic efficiency remains open, one should also examine other motives for government pursuit of privatization.

Financial Resource Goals

The goals of most privatizations, including telecommunications, are to raise money.[1] Facing increasing financial constraints, there are several attractive options open to governments in the use of privatization proceeds. These include using the revenue for current expenditures, tax reductions, "social" capital expenditures, "commercial" capital expenditures including the restructuring of selected public enterprises, financing private investments, and, as noted, public debt reduction or nonincrease in public borrowing (Ramanadham, 1989).

Telecommunications Infrastructure Expansion

Many states believe that a well-developed telecommunications infrastructure is a precondition for attracting foreign investment, which is deemed necessary to increase a country's growth and competitiveness. In Jamaica, the Jamaican Labour Party (JLP) under Prime Minister Edward Seaga came to view telecommunications as a key component for encouraging foreign investment and creating a competitive exporting sector linked to the U.S. economy. The JLP sought to privatize the telecommunications

---

[1]It has been questioned, however, whether a privatization can change the long-run asset position of a government.

sector in order to increase access to the capital, technology, and foreign exchange needed to modernize the sector. Upgrading the telecommunications infrastructure is also a key concern in Hungary and in other Eastern European countries, as is the financing of this modernization effort. The telecom infrastructure is viewed similarly in much of Latin America and Asia. In fact, the cases of Hong Kong and Singapore present extreme examples of efforts by national governments to attract foreign investment, particularly in banking, services, and other information sectors, by developing and promoting their telecom and information infrastructures. These two cities compete as hubs of corporate telecom networks in Asia and also as locations for corporate headquarters, financial business, insurance, services, manufacturing, and so on.

Many governments of developing countries find themselves faced with the dilemma of replacing antiquated systems with modern equipment at a capital cost beyond their reach and are unable to borrow money abroad or to raise it domestically through government self-finance efforts. Private investors with access to commercial lending sources and expanded credit possibilities are often better able to provide the capital necessary for the development of the telecommunications infrastructure and diverse services than a government that faces competing claims on funds raised by taxation. In Chile, the decision to privatize the Compania de Telefonos de Chile (CTC), which provides domestic service, was based primarily on the premise that the only effective way to satisfy the existing demand for telephone service was through privatization, because the government could not afford to finance the required investment. Upgrading the telecommunications infrastructure was also part of the reason for the Mexican government's sale of Teléfonos de México S.A. (TELMEX), as was acquiring revenue for other development programs and debt reduction. Venezuela followed the Mexican goals and model fairly closely.

Debt Reduction and Foreign Exchange

The shortage of foreign exchange has been a compelling motive for many privatization programs. To scale back the burgeoning state sector by selling public assets to the private sector has been considered by some governments as a means of handling an acute capital shortage and acquiring funds for debt servicing or capital investments. Although the long-term goal of privatization may be economic efficiency, the short-term stimulus has often been deficit reduction (Suleiman & Waterbury, 1990). For countries whose growth of deficits and debts cannot be reversed by a continuation of the policy of state ownership, divestiture can be a means of raising revenue and reducing fiscal and credit pressures, that is, reducing the high levels of borrowing, in part, by ending subsidies. The

People's National Party under Prime Minister Michael Manley divested the government's remaining 40% of shares in Telecommunications of Jamaica as a means to acquire foreign exchange to meet financial targets, specifically international reserves tests, set by the International Monetary Fund. Debt reduction was also an objective of the widespread privatization program, which included the sale of the telecommunications monopoly, Empresa Nacional de Telecomunicaciones (ENTEL), introduced by Argentina's President Carlos Menem.

If the returns from a sale are applied strictly toward external or domestic debt, the effect will be but a one-time reduction in the government deficit, equal to the amount of the sales revenue. Profitable firms, which are more attractive candidates for sale, can yield proceeds to maintain social services and finance faster growth, and private management could increase efficiency.

PROCESS OF PRIVATIZATION

Privatization can be conducted through several means or processes. The simplest and most common is sale of stock through share offerings. In fact, many government-controlled telecommunications firms, such as Brazil's, have already sold off many shares of stock without being considered privatized, because the government retains both significant shares of stock and effective control. In this approach, stock can be sold through stock markets. Taking a more strategic approach, large blocks of stock can be negotiated with companies who then become strategic investors or partners in the privatized firm. This has been the approach with Mexico and Venezuela.

Retention of Government Shares
Versus Complete Divestiture

If the public enterprise is profitable, its privatization means that the government forfeits the future stream of income, unless it retains a large share (Cook & Kirkpatrick, 1988). In the case of Argentina, the government experienced a loss with the sale of ENTEL due to the combined effect of forfeiting future taxes and quasi-rents together with a heavy debt take-over at the time of divestiture (Abdala, 1992). The Argentine government did not retain a shareholding through a partial divestiture or denationalization effort as was done in Venezuela and Korea. Hungary has proposed that the privatization be limited to 49% of the shares, in part to avoid dependence on foreign capital. In the case of Mexico, the majority of voting shares must remain in the hands of nationals. In a complete divestiture, private individuals or firms may wholly purchase publicly owned assets, after which the government bears no further responsibility for the operation of assets and hence does not earn dividends or any other type of revenue from the enterprise. This is the case with regard to Argentina, Jamaica, and Chile's sale of its long distance and international service provider, Empresa Nacional de Telecomunicaciones (ENTEL).

Divestiture by the sale of shares has encouraged the growth of stock markets. In industrialized countries, where capital markets are well developed, a public share offering is the most common technique of privatization. There are also advantages of this method of sale for developing countries. Mexico, Chile, Jamaica, and other countries have employed public share offerings. Mexico and Chile have both witnessed dramatic growth in telecommunications stocks since the sale of their telephony companies. In Mexico, as well as in the United Kingdom and Jamaica, the government has used employee share schemes to diminish union opposition. The Venezuelan government has also sold nearly 11% of shares to employees of Compania Anonima Nacional Telefonos de Venezuela (CANTV), though there is no public market in CANTV stock. A stock market offering directed specifically at the small investor, however, may redistribute wealth in the community through broad-based share ownership. This method of divestment was advocated by the Jamaican Labour Party.

This broadening of private ownership has arguably stimulated a "people's capitalism" (Hanke, 1987). Both Korea and Chile regarded the broadening of stock ownership among the public as a goal of privatization. In fact, for South Korea, only Korean nationals may purchase stock. It may also reduce the risk of renationalization. The numerous new shareholders acquire some financial interest in the continuation of policies and governments, benefitting from the profitability of the firms that they own (Vickers & Yarrow, 1991). If, however, these small shareholders (who often fail to invest in the firm beyond the shares they purchase) sell their shares within the first year or so, a concentrated shareholding may develop. The number of shares any one individual may acquire must be limited to curtail the concentration of ownership shares by the wealthy elite, and to create a constituency for privatization as well as a ready market for the next offering (Cowan, 1990).

The narrowness of the middle-income strata in most developing countries, however, makes privatization based on the small shareholder and people's capitalism difficult. This same narrowness contributes to the "thinness" of capital markets in developing countries (Suleiman & Waterbury, 1990). In most developing countries where capital markets are rudimentary or nonexistent, denationalization is more likely to involve the sale of the enterprise as a complete entity, or, at least, controlling interest is sold to a single buyer, often a foreign buyer (Cook & Kirkpatrick, 1988). The Jamaican privatization of Telecommunications of Jamaica, in which Cable and Wireless has acquired 79% of the shares, is evidence of this situation. The privatization in relatively prosperous Korea, which is aiming to increase shareholding by domestic investors, is almost the exception that proves the rule that in most less affluent developing countries, such domestic capital is seldom sufficient.

Acquisition of Technical Expertise

In Jamaica, as well as in other developing countries, governments typically want strategic partners in the telecommunications sector with considerable experience in building and operating networks. Partners are sought for their technological and management capabilities as much as for financial commitments. In the case of Chile, the government placed no limit on foreign participation, but required that the bidders commit to injecting new equity and implementing an agreed investment program. To take advantage of privatizations, therefore, market equity investors are likely to join consortia with international carriers such as France Télécom, Telefónica de España, Cable and Wireless, or the U.S. Bell operating companies. Indeed, Southwestern Bell is a shareholder in TELMEX as is France Télécom, which also owns shares in Argentina's restructured telephony companies. Similarly, a consortium headed by GTE owns 40% of the services that CANTV is authorized to provide, and Telefónica de España owns shares in both of Chile's telephony companies, CTC and ENTEL, as well as Argentina's system.

This tendency toward increasing foreign participation in the shareholding structure of telecommunications companies has caused some governments, notably Brazil's current Itamar Franco administration, to hesitate in privatizing their telecommunications sector, which they regard as strategically important. National security is considered by some states to be a compelling reason to preserve state control over a national telecommunications system. Governments that take advantage of the system's profitability in some areas, such as the foreign exchange generated by an international carrier, to subsidize telephony service and other government operations, may also choose not to privatize their systems or at least not to engage in a complete divestiture. Those countries that have divested their telecommunication sectors, be it partially or completely, have met with mixed results, some of which may be ascribed to the various regulatory regimes instituted.

REGULATION

In principle, the design and creation of mechanisms for monitoring policy decisions about privatization should flow directly from the goals set for privatization. Given the high likelihood that imperfect competitive conditions will prevail, the creation of regulatory institutions should accompany the process of privatization. Policies and structures for government regulation should be a primary instrument for ensuring that privatized telecommunications companies follow through on commitments they make in purchasing public telecom assets to pursue public interests. In particular, because most peripheral or developing countries hope to see telecommunications expanded as part of the infrastructure for development, that goal needs to be addressed in policy and regulation. However, regulatory structures have often been tacked on almost as an afterthought, after the process of privatization is over and privatized operators are already set in motion.

There are several counterarguments against regulation and reasons why it is often neglected in the process. Assumptions about the role of the state in development have changed radically in the last decade, as a consensus about reduction in the role of the state has been promoted by the U.S. government, the Group of Seven governments, the World Bank, other development banks, and an increasing number of development experts. In a number of countries, the state is no longer seen as a benign actor capable of guiding development through active participation or even regulation of private activity. States are often seen as corrupt, inefficient, and captured by insider or clientilist private interests. Competition within the marketplace is often promoted as a replacement for regulation, even though many telecom privatizations result in companies that are private monopolies. In fact, from a regulatory point of view, some actors in countries such as Brazil are beginning to consider liberalized competition within a regulated environment more important than privatization (Siqueira, 1993).

However, even potential private investors require knowledge about the legal and regulatory regime for a privatized telecom company. Spiller (1992) noted that uncertainty over the nature of the regulatory body to enforce the Argentine privatization made some potential investors hesitant. However, a predictable regulatory regime from the point of view of an investor may be a rigid system insufficiently responsive to changing social needs, from the point of view of the government. The decision to base regulation of the Jamaican privatization in a court-supervised contract is in part a reflection of fear by the investor, Cable and Wireless, that an executive branch regulatory agency would be too responsive to changing government policy.

Sinha (chapter 13, this volume) summarizes the main issues faced in designing regulatory regimes. Those include the nature and role of the state; the impact of political institutions; the legal framework for enforcing contracts and defining property rights; the legal institutions for safeguarding legal rights (such as the judiciary) and for enforcing policy (such as regulatory agencies); the placement, autonomy, and power of regulatory agencies; the resources with which regulators must work; and the fairness and effectiveness of regulatory procedures.

The countries we analyze here give several examples of the impact of political institutions and parties on the regulatory process. In Mexico, privatization of telecom, along with many other state enterprises, was carried out from above by a powerful centralized party, the PRI, supported by almost authoritarian power exercised by the president. In Argentina, privatization was argued about by a much looser, more ideologically diverse party, the Peronists, and was only resolved by an unexpectedly strong president, Menem, who diverged from populist expectations and carried his party with him. Argentina privatized only when both President and legislature belonged to the same party, with the president strong enough to overcome diversity of opinion within the party (Sinha, chapter 13, this volume). In contrast, in Brazil, privatization has been fiercely contested by left parties, based in strong union opposition. Although Brazilian President Collor, elected in 1989, pushed for privatization, he did not have a strong political base for the idea and, after he was impeached for corruption, his Vice-President, Itamar Franco, sharply reversed course and took a more traditional, nationalist view.

The regulation of privatization takes place in a similarly varied political and institutional framework. States tend to desire a controllable, responsive regulator that can pursue their goals, so they may create regulatory bodies that are responsive, but vulnerable to politicization or corruption. Potential private operators usually favor a regulatory regime that will enforce their contract in a predictable way or perhaps an unregulated or weakly regulated system that they can capture and dominate. This seems to be the case in countries long characterized by private ownership, such as the Philippines and the Dominican Republic.

Brazil provides an example of this privatization and regulatory autonomy issue. There is need for insulation of regulation from ressure and interest groups, particularly after privatization, which may be very hard to meet in highly politicized and sometimes corrupt regimes (Sinha, chapter 13, this volume). An effective argument against privatization in Brazil is made by those who fear that both the privatization process and subsequent regulation would be dominated by powerful elite interests. They cite precedents such as the cooperation between powerful politicians, like Antonio Carlos Magalhães, current governor of Bahia, owner of a key TV Globo affiliate, and previous Minister of Communications, and Roberto Marinho, owner of TV Globo, the dominant television network, and a dominant partner in NEC Brazil, which has bid on a number of cellular licenses and would likely bid for privatized carriers. Part of the problem with regulatory autonomy rests in a fundamental lack of autonomy of state operations from overall capture by private interests.

In Jamaica, the contracted private carrier, Cable and Wireless, feared a regulator-responsive control and potential politicization of regulation by powerful executive–legislative/parliamentary coalition or combinations. To allay this concern, Jamaica resorted to judicial oversight of the privatization contract to ensure a predictable enforcement of the contract, with ultimate appeal to British courts (Spiller & Levy, 1991).

In the process of privatization, in obtaining bids, in writing the contract for the privatized operator, and in subsequent regulation of the private entity, a number of cases have shown that clarity of obligations and rules is crucial. This is especially true for obtaining and enforcing goals or targets for expansion of the telecommunication system, particularly into rural areas. For example, in Chile, the private firm stopped investing in rural expansion once the initial contract targets were met, during a fraction of its license/concession period (Sinha, chapter 13, this volume). This represents a need for ongoing regulation and renewal of social expansion targets. It also represents a very likely point of conflict between private operators, who may consider their obligations met, and regulators, who would like to require continuing investment in expanding telecom services into unprofitable areas, such as rural zones.

The regulatory structure must either be very far-sighted in its original goals or create a flexibility adequate to the anticipation of future developments, such as continuing growth targets for expansion into rural areas. Another crucial area is the eventual desire to end monopolies that might initially be given to private operators. For example, in Mexico, the initial contract specifies an end point for the monopoly after 6 years, and the regulator has a detailed set of rules regarding competition for setting up operations in long-distance service once the initial monopoly expires. This will be necessary to curtail the monopoly rights often given private operators and begin liberalized competition at some point. The Argentine and Jamaica contracts give monopoly concessions for much longer periods of time, which will limit their flexibility greatly if they wish to liberalize competition any time soon.

EFFECTS OF SECTOR REFORM ON PERFORMANCE: SOME CONCLUDING REMARKS

Ultimately, the success of any reform will have to be measured in its welfare implications. A comprehensive assessment of the reforms under review faces some serious obstacles, and a final judgment may not be possible without a more extended observation period. First, it is difficult to distinguish temporary from lasting effects of a reform. Barely anybody would deny that a change in the institutional regime after decades of continuity will have dramatic effects on the performance of a sector. In the medium and long term, it will be crucial to monitor whether more competitive structures will be sustainable. If they are not, regulation may or may not be an effective mechanism to cope with market power. Second, efficiency effects usually are accompanied by distributional effects. Moreover, local welfare improvements may be associated with spill-over effects and perhaps welfare reductions in other countries (Bauer & Straubhaar, 1994). Without a widely accepted framework for the assessment of national and transnational distributional effects, a meaningful conclusion may not be reached regarding the net welfare effect of reforms. Third, the choice of a reference scenario to assess the effects of restructuring is not straightforward. Frequently, every change occurring after a measure of reform is attributed to the reform. Such an approach ignores the fact that the established system would most likely have been subject to change even without a reform measure. Thus, the proper reference point is the trajectory of the system without reform or an external performance index that does reflect a "best practice" standard and not the status quo ante. Only in the unlikely case of an entirely static system can the status quo ante serve as a reference point.

With these caveats in mind, an assessment of the changes originating from the institutional reform processes in our sample countries is only temporary. Most analyses conclude that there are immediate efficiency improvements stemming from the reform measures. For instance, in some countries such as Argentina or Korea, the expansion of the telephone network seems to have accelerated. However, in the case of Argentina this acceleration slowed down after a few years. In other countries, such as Mexico, the rate of system expansion was not significantly different from the situation before privatization, although the slow influx of alternative providers created more diversity for the customers ("Half-way There," 1995). The liberalization of the sector has in most cases contributed to the introduction of new and innovative services. In most cases, pricing policies have been aligned with economic principles, and extra investment funds were dedicated to the industry.

At the same time, profits of the telecommunications service providers soared and the prices of newly privatized operators at the stock exchange increased dramatically, leading to significant distributional effects and windfall gains to investors in both the source and the host country. For example, Southwestern Bell stock increased by 4.8% after the announcement of the completion of the Telmex participation. This may be a sign not only of positive market expectations but also of a weak and insufficient regulatory control of the market power of the dominant carriers.

Reform processes seem to progress more rapidly if implemented by strong, centralized governments, as in the cases of Chile and Korea. The experiences of these countries also illustrate that reliance on market forces will not lead to fast deployment of services to rural and high-cost areas, and it may take a considerable time until these areas are being served. If service to these areas should be provided on a ubiquitous basis, other mechanisms such as subsidies or infrastructure side agreements must be relied upon.

The current reform processes maintain a diversity of approaches, and it will be interesting and revealing to follow the trajectories of various reform models. Accompanying research will, hopefully, improve our understanding of the intricate relations between the institutional arrangements of a sector and its performance.

REFERENCES

Abdala, M. A. (1992). *Distributional impact evaluation of divestiture in a high inflation economy: The case of ENTEL in Argentina.* Unpublished doctoral dissertation, Boston University, Boston.

Ambrose, W. W., Hennemeyer, P. R., & Chapon, J.-P. (1990). *Privatizing telecommunications systems: Business opportunities in developing countries.* Washington, DC: The World Bank and International Financial Corporation.

Bauer, J. M., & Straubhaar, J. (1994). Telecommunications in Eastern Europe. In C. Steinfield, J. M. Bauer, & L. Caby (Eds.), *Telecommunications in transition.* Thousand Oaks, CA: Sage.

Commander, S., & Killick, T. (1988). Privatisation in developing countries: A survey of the issues. In P. Cook & C. Kirkpatrick (Eds.), *Privatisation in less developed countries.* Brighton: Wheatsheaf.

Cook, P., & Kirkpatrick, C. (1988). Privatisation in less developed countries: An overview. In P. Cook & C. Kirkpatrick (Eds.), *Privatisation in less developed countries.* Brighton: Wheatsheaf.

Cowan, L. G. (1990). *Privatization in the developing world.* Westport, CT: Greenwood Press.

Half-way there. Latin American telecoms. (1995, February 4). *The Economist,* pp. 62–63.

Hanke, S. H. (1987). Towards a people's capitalism. In S. H. Hanke (Ed.), *Privatization and development* (pp. 213–221). San Francisco: Institute for Contemporary Studies.

Heller, K. (1994). Restructuring in Hungary. In B. Wellenius & P. A. Stern (Eds.), *Implementing reforms in the telecommunications sector. Lessons from experience* (pp. 375–382). Washington, DC: The World Bank.

Hemming, R., & Mansoor, A. M. (1987). *Privatization and public enterprises.* IMF Working Paper. Washington, DC: International Monetary Fund.

Kim, E., & Jin, Y.-O. (1993). The evolution of telecommunication policy in the Republic of Korea. In R. Stevenson, T. H. Oum, & H. Oniki (Eds.), *International perspectives on telecommunications policy* (International Review of Comparative Public Policy, Vol. 5, pp. 231–256). Greenwich, CT: JAI Press.

Kim, H. (1992). Financial aspects of Korean telecommunications. *Telematics and Informatics, 9,* 13–19.

Kim, J.-C., & Lee, M.-H. (1991). Universal service policies in Korea. *Telematics and Informatics, 8,* 31–40.

Kim, J.-C., & Ro, T.-S. (1993). Current policy issues in the Korean telecommunications industry. *Telecommunications Policy, 17,* 481–492.

Kiss, F. (1994). Foreign investment in Hungarian telecommunications. In E. Bohlin & O. Granstrand (Eds.), *The race to European eminence. Who are the coming tele-service multinationals?* (pp. 295–316). Amsterdam: North-Holland.

McCormick, P. K. (1993). Telecommunications privatization issues: The Jamaican experience. *Telecommunications Policy, 17,* 145–157.

Melo, J. R. (1994). Liberalization and privatization in Chile. In B. Wellenius & P. A. Stern (Eds.), *Implementing reforms in the telecommunications sector. Lessons from experience* (pp. 145–159). Washington, DC: The World Bank.

Park, R.-A. (1993). *Dramatic changes in Korean telecommunications.* Unpublished manuscript, Michigan State University.

Ramanadham, V. V. (1989). Concluding review. In V. V. Ramanadham (Ed.), *Privatisation in developing countries.* London: Routledge.

Roche, E. M. (1990). Brazilian informatics policy: The high cost of building a national industry. *The Information Society, 7,* 1–32.

Samarajiva, R. (1993). Institutional reform in Sri Lanka's telecommunication system: Regulation, corporatization, and competition. *Asian Journal of Communication, 3,* 37–63.

Siqueira, E. (1993). *Telecomunicações: Privatização ou caos?* [Telecommunications: Privatization or Chaos?]. São Paulo, Brazil: TelePress Editora.

Spiller, P. (1992). *Institutions and regulatory commitment in utilities privatization.* Washington, DC: Institute for Policy Reform.

Spiller, P., & Levy, B. (1991). *Regulations, institutions and economic efficiency.* Washington DC: The World Bank.

Suleiman, E. N., & Waterbury, J. (Eds.). (1990). *The political economy of public sector reform and privatization.* Boulder, CO: Westview Press.

Vickers, J., & Yarrow, G. (1991). Economic perspectives on privatization. *Journal of Economic Perspectives, 5,* 111–132.

Wellenius, B. (1994). Telecommunications restructuring in Latin America: An overview. In B. Wellenius & P. A. Stern (Eds.), *Implementing reforms in the telecommunications sector. Lessons from experience* (pp. 113–144). Washington, DC: The World Bank.

Whitlock, E., & Nyevrikel, E. (1992). The evolution of Hungarian telecommunications. *Telecommunications Policy, 16,* 249–258.

IV

THE ROLE OF REGULATION

11

Privatization and Developing Countries[1]

William H. Melody
Center for Tele-Information, Technical University of Denmark

The telecommunication industry is being restructured in most countries as part of a process of major economic reform. But there is substantial uncertainty as to the best way forward. The idea that telecommunication services are not just a social service and a necessary cost of doing business, but rather an enormously valuable economic resource, is only beginning to be recognized. Traditionally, telecommunication development was viewed as a natural consequence of general economic development. Thus there was no reason to promote telecommunication as a stimulus to economic development, efficiency, or productivity. More recently, telecommunication is being recognized as an integral part of economic development in both developed and developing countries and, therefore, worthy of the most serious attention from policymakers.

The general model of the telecommunication entity in need of reform has been an inefficient, government-administered PTT *in a developed country*, for example, the United Kingdom, New Zealand, or Japan. For these systems in these countries, privatization has been seen as a solution to the problems of efficiency, investment, technological upgrading, and new service development. This has led to widespread acceptance of a view that privatization is the solution to the problems of telecommunication reform in developing countries. But the conditions in developing coun-

[1]This chapter is adapted and updated from W. H. Melody, "Telecommunication Reform: Which Sectors to Privatise?" in *ITU Telecommunication Journal*, Spring 1992.

250

tries are very different. This chapter examines the relevance and applicability of privatization for the conditions that more typically prevail in developing countries (Melody, 1991).

INTERNATIONAL EXPERIENCE

The International Telecommunication Union (ITU) has examined the requirements for structural adjustment in the telecommunication sector that are arising in most countries. Its report, *The Changing Telecommunication Environment: Policy Considerations for the Members of the ITU* (ITU, 1989), draws attention to certain fundamental underlying factors that are essential to efficient telecommunication development:

1. *New technologies*: effective applications of new technologies that enable operational efficiency and new service development in national and international networks.

2. *New service development*: providing maximum opportunities for the most efficient and innovative uses and applications of telecommunication system capacity.

3. *Human resources*: (a) a restructuring of management, replacing a bureaucratic administrative culture with an entrepreneurial service-oriented management culture; (b) a restructuring of the labor force to provide incentives for output and performance; and (c) developing a workforce with a range of new skills.

4. *Financing*: attracting large amounts of capital on efficient and flexible terms on a continuing basis to support ongoing investment and system expansion.

The ITU Report concludes that in attempting to achieve effective structural adjustment, the fundamental underlying issue that must be addressed is effective *separation* of the basic functions of policymaking, operational management, and regulation. The inefficiency and unresponsiveness that have pervaded many government telecommunication administrations in both developed and developing countries derives from a mixing of macroeconomic and microeconomic, political, bureaucratic, and commercial objectives in the same decision-making structures. Whether the public telecommunication operations are used to subsidize the post, the treasury, bureaucratic inefficiency, trade union featherbedding, or the short-term needs of macroeconomic policy, the efficient development of the industry is being sacrificed to other interests. There must be a clear separation between policymaking, which provides the framework and guidelines for long-term industry development, and the operational management, which must be able to plan for effective long-term allocation and use of resources, and be held accountable for its performance.

Similarly, there must be effective separation between the ongoing regulation of the industry and both policymaking and operational management. The regulator supplies detailed expertise to interpret and enforce policy, to monitor industry developments, and to address ongoing issues and problems that arise within the overall policy framework. Independent management and independent regulation are essential to effective structural adjustment and efficient industry development (CIRCIT, 1990). Indeed, the World Bank and other international lending agencies frequently have made an effective structural separation of functions a requirement for developing countries to obtain loans (World Bank, 1988).

Within this decentralized structure, it is important for countries to develop policies in three areas: (a) for effective and efficient *integration* of the national network into the larger regional and international networks, which often entails regional system planning and service development; (b) for ensuring that *universal service* extension is a high priority, as it is essential to ensuring that the benefits of modern telecommunication are made accessible to the highest proportion of the population; and (c) for stimulating the most appropriate type and degree of competition in the industry for each particular country. Although it is clear that access to public telecommunication systems in all countries will have to be liberalized significantly, the most appropriate role for competition will depend upon the particular conditions in each country.

THE DIMENSIONS OF STRUCTURAL REFORM

A process of telecommunication reform will be most effective if close attention is paid to the order in which structural changes occur. Unless the more fundamental changes are implemented first, second-order changes will not be effective. The first-level conditions relate to the establishment of an institutional structure that clearly defines separate and distinct roles for policymaking, regulation, and management (Melody, 1990).

The essential steps are the following:

Management—To separate operational management from the government so that neither politicians nor government bureaucrats can interfere in operational decisions. The management must be accountable to a board of directors that is insulated from day-to-day interference. They may be political appointees, but for terms of significant duration and with mandates to act independently in achieving specified economic (e.g., efficiency) and social (e.g., universal service) objectives. Management must be held accountable for its performance in achieving these objectives.

In most countries, this does not require privatization (Vickers & Yarrow, 1988, 1991). Crown corporations and other equivalent structures have worked effectively in many countries. However, there are instances where new governments have concluded that the only way to break the grip of bureaucratic inefficiency, political favoritism, and apathetic service delivery is to make the greatest structural change possible. In essence, they have concluded that attempts to establish efficient management in the public sector will be unduly limited and constrained by the inherited problems. For example, this is one reason why the British government chose to privatize British Telecom and other utilities.

Regulation—To establish a regulatory agency that is independent from the telecoms operator and quasi-independent from government. The regulator should implement government policy, ensure performance accountability by the telecom operator to economic and social policy objectives, resolve disputes between competitors and between consumers and operators, monitor changing industry conditions, and advise government on developments bearing on policy.

The regulatory agency acts as a buffer between telecoms operations and government, helping to ensure the separation of functions. Whereas the operators, once separated from direct government influence, may focus too narrowly on economic objectives, the regulatory agency can ensure recognition of social and other policy objectives as well. Although regulation has been used primarily with privately owned operators, it has also been used with publicly owned operators to implement the same objectives.

Policy Development—To establish a small professional policy unit near the center of government. Policy development is directed to fundamental issues of long-term objectives and direction, not day-to-day implementation and problem solving. The policy unit is an independent, expert information and analysis support group for government policymakers. It ensures attention to long-run implications of developments and issues arising from them. It ensures that policymakers are informed and capable of addressing the need for policy change when it is required. Thus it dampens the incentive of policymakers to intervene in the affairs of the operators or the regulator except on matters of policy significance (Sappington & Stiglitz, 1987). Figure 11.1 illustrates the essential relations. The operator is separated from government by a quasi-independent regulator and board. It is accountable to the regulator, the board, and the marketplace to satisfy specified economic and social objectives.

It will be noted that private ownership is neither a necessary nor a sufficient condition to bring about these fundamental institutional

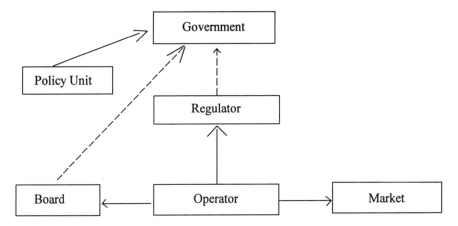

FIG. 11.1. Ideal relationship between policymaker, regulator, and operator.

changes. In certain circumstances privatization may help establish effective separation between government and operating management functions. In Fig. 11.1, private ownership will eliminate the indirect accountability of the telecoms operator to the government. But whether it is necessary or makes a significant difference depends upon the inherited circumstances and what other more important structural changes are made at the same time.

Access to Resources

The second-level issues in structural reform are concerned with access to the necessary resources for telecommunication development. The telecoms operator(s) must have access to essential resources—human, capital, and technology—if it is to be able to provide efficient telecommunication systems and services. Inasmuch as the new technologies are available globally for those who can pay the price, the resource issues fall back to labor and capital. Attention must then be paid to the structure of a country's labor, capital, and telecommunication technology markets in relation to the corresponding global markets. For developing countries the majority of the new technology must be imported.

Human Resources. Technologically advanced telecommunication systems require highly skilled human resources throughout the organizations from top management to the "shop floor." Developing countries have shortages of both skilled management and labor in this field. Few have sufficient programs of education, training, and retraining to meet the needs. Yet few countries can import massive amounts of skilled human resources in such a rapidly growing industry for long. Given the worldwide shortage of skilled labor and management resources in this sector, for developing countries this will set a limit on the speed of structural reform whether telecommunication is publicly or privately owned.

There is an immediate need to build and expand telecommunication education and training programs—technical, managerial, software design and maintenance, service development and marketing, economics and accounting, law and international relations. Efficient labor markets must be established. This will require both public and private investment in expanding the skill base, as well as in some developing countries it means establishing a labor market that is more responsive to supply and demand conditions.

Capital Resources. Telecommunication reform requires large amounts of capital in all countries. In some countries (developing and developed) the capital requirements are particularly vexatious because, at the same time, the telecommunication operations are contributing capital to the national treasury while suffering from massive underinvestment. In others, investment requirements cannot be met because of limitations on borrowing by public sector entities.

In most developing countries, capital markets are not sufficiently developed or large enough to meet the enormous telecommunication capital requirements without creating resource dislocations in other markets, raising capital costs for other purposes or denying capital to other important sectors of the economy. Moreover, a massive influx of foreign capital could disrupt domestic markets and aggravate problems of foreign debt. Finally, the necessity of foreign capital could introduce serious problems of hard currency investment in soft currency countries, with attendant implications for repayment, mounting foreign debt, and the stability of the overall economy.

Many of these capital market and financing problems are the same whether telecommunication is publicly or privately owned. Other problems are uniquely associated with public ownership and might be alleviated with a shift to private ownership. But in many cases the reason for shifting is primarily political, not economic. Under these circumstances there is no reason to believe the overall effect on the economy may not be demonstrably different.

Any reasonable forecast of needed industry growth in developing countries demonstrates that capital requirements over the next decade or two will far exceed the capital that the governments are likely to make available for telecommunications expansion. Privatization may be a solution to the immediate problem if the privatized operators are able to raise additional funds in the capital markets. But most developing countries do not have well functioning capital markets. Thus a shift from public to private capital is usually also a shift from domestic to foreign capital.

We might ask, what is the real financing problem to which privatization is proposed as the solution? Public and private capital raising is done in the same capital markets. The effect on the economy is not necessarily any different if the ownership status of the telecommunication borrower is public or private. If the government is going to permit an expansion of capital-raising by the telecommunication operator, it could raise the public sector borrowing requirement for that purpose. The economic test should rest on whether capital can be raised more efficiently (subsidies excluded) under public or private ownership. One can find examples of both. It all depends on the particular circumstances in the country.

The advantages of private ownership to attract capital are greater in those circumstances where public ownership of telecommunications has been demonstrably inefficient and where sufficient capital cannot be attracted under public ownership. In many developing countries the limitations of the capital markets are so severe as to require that capital be raised by every possible means, public and private, including subscribers, equipment suppliers and potential competitors. In others, the government is viewed as offering less security for suppliers of capital to public enterprises than to private ones. Clearly under these circumstances significant injections of new capital will only come from independent investors, private or public (e.g., other governments or international agencies), if there is greater assurance that those funds will be efficiently used and investment risks will be minimized. Under these circumstances, private ownership becomes essential to attract capital and overcome major public sector inefficiencies.

THE ROLE OF COMPETITION

Competition is often suggested as a substitute for government regulation or public ownership. Under conditions of effective competition, market forces can attract the necessary resources on efficient terms, require up-to-date production and service delivery at reasonable prices, and fully protect consumer and public interests. The market is the regulator. Neither regulation nor public ownership are needed.

But the characteristics of telecommunication markets are a long way from the idealized model of competitive markets (Mansell, 1994). Competitive market forces can be part of the equation, but they cannot be all of it. In telecommunication and most other utility industries, selected applications of competition can be an effective tool of public policy—a complement to other policies in achieving economic and social objectives—but not a substitute for them (Melody, 1986).

Competition polices can be adopted under private or public ownership, or a mixture of both. Competition between private and public ownership is a policy that some countries have deliberately adopted. Moreover there are many kinds of competition, most of which differ significantly from the classic notion of multiple suppliers offering similar products or services to consumers. In the telecommunication sector, duopoly and oligopoly are most common. Benchmark comparisons are used to attempt to stimulate quasi-monopolies to greater efficiency (OECD, 1990). Fringe competition varies significantly in different telecommunication industry markets (e.g., public network services, value-added networks [VANS], subscriber equipment, network facilities).

Competition is more readily associated with private ownership than public, leading some people to believe that real, as opposed to gentlemen's, competition is more likely to develop under private ownership than public. But experience indicates once again that it depends on the circumstances. This issue becomes especially important when one considers the interplay between ownership and competition. There is considerable evidence that competition is a much more important factor than ownership in influencing efficiency (Melody, 1989). In many proposals to introduce privatization, the case depends upon an assumption that it will stimulate more effective competition. If this is true, then the selection of public or private ownership should be made on the basis of the probability of stimulating effective competition. But it seldom is.

PRIORITIES FOR PRIVATIZATION

Resource Markets

In assessing conditions where privatization offers the greatest potential for benefit, one must distinguish among the different markets that make up the telecommunication sector. These are illustrated in Fig. 11.2. Network operators (public or private) purchase resources in equipment, labor, and capital markets. Their efficiency is highly influenced by the efficiency of these markets. In most instances, purchasing in well-functioning competitive resource markets is the most desirable market situation.

By itself privatization of facility network operations will not improve imperfections in the resource markets. However, encouraging arm's length relations between network operators and resource suppliers, and the acquisition of resources through competitive markets, is a step that could involve spinning off equipment manufacturing units, outsourcing certain specialized labor functions (e.g., engineering and legal advice), and seeking capital directly from capital markets rather than through the government. For publicly owned network operators, these are all acts of privatization. But at bottom they are simply applications of the classic make-or-buy

Resource Markets   Network Operations   Service Markets

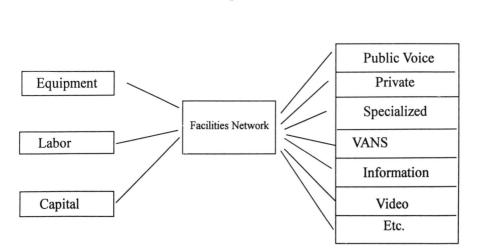

FIG. 11.2. The different markets that make up the telecommunications industries.

decisions that all organizations, public and private, must make all the time, presumably based on the criterion of long-term efficiency.

There are also occasions where private operators outsource activities to public sector agencies (e.g., gathering industry statistics). The real requirement here is that efficiency criteria and competitive opportunities be implemented seriously. Because there is ample evidence that in many areas publicly owned network operators have not applied these principles rigorously for many years, their application will result in a degree of privatization. But these problems have also been experienced with private monopolies that have had to go through the same kind of restructuring. There is no guarantee that privatizing a public network operator in itself will bring about the desired changes.

Service Markets

Until recently, monopoly network operators exercised total control over the provision of almost all services provided over the network. Liberalization of access to the network for a variety of specialized VANS and sometimes competitive telephone service providers is occurring in most countries. If the network operator is publicly owned, this generally involves a privatization of certain service development and retailing activities. This is simply the flip side of the issues raised in network operator relations in resource markets.

Again the key issue is applying efficiency criteria and establishing conditions for competitive access, not the ownership arrangements. In fact, in the past privately owned telecommunication operators established interconnection and access restrictions that were virtually identical to those established by publicly owned operators. Unsatisfactory inherited circumstances may stimulate a desire for a change of ownership. But if this is all that happens, there is no guarantee that efficient service markets will develop. In either case effective regulation will be necessary.

Network Operation

Historically, network operators have been monopolies, public or private, under the belief that such monopolies were *natural*. In recent times, we have learned that the scope of the monopoly need not be extended to cover equipment manufacturing or even equipment attached to the network. It need not extend to the provision of telecommunication-based services. And the network need not necessarily be a single, enormous monopoly. Separation by regions and by types of services (e.g., local, long distance, international) are not only possible but practiced in some places. To be sure, consumers are interested only in a single transparent global network for all their communication needs. But that does not require a monstrous monopoly; it requires coordination among the suppliers of the different parts. This question of determining the optimal structure of the network operations sector does not turn on ownership. It will be recalled that the global network involves both public and privately owned operators.

There are also important policy questions being raised in many countries about the optimal role of competition in the network operations sector. Competition is seen as a vehicle to promote efficiency and even facilitate the extension of universal service in some circumstances. In others, the risk of stimulating an inefficient duplication of network facilities is seen as a misallocation of scarce capital resources. In either case, these policy decisions are not dependent on ownership and sometimes involve competition between public and private owners.

The strongest case for privatization is where public ownership has failed and the prospects for major public sector reform are slim. But if an ownership change is all that takes place, significant improvements may still be prevented by the same institutional forces that prevented it under public ownership. Success requires a package of institutional changes, among which privatization can be a contributing part in some circumstances.

Ironically the contribution of privatization to attracting capital that could not otherwise be attracted on reasonable terms is likely to be greatest in countries where neither capital markets nor telecommunication markets are functioning very well. This inevitably requires the importation of capital, equipment, and skilled labor, with its attendant implications for hard currency foreign debt. But in these circumstances, privatization may provide an important step toward the establishment or improvment of both domestic capital and telecommunication markets.

CONCLUSION

As an agent of telecommunication reform, privatization must be carefully assessed in relation to other changes, particularly to government structure and market structure. By itself, privatization is not likely to accomplish very much in most circumstances. It is neither a necessary nor a sufficient condition for reform. Most ardent advocates of privatization assume that it will be associated with an increase in competition that will, in turn, stimulate efficiency improvements.

The keys to telecommunication reform are (a) a clear separation of operational management, regulatory and policymaking functions; and (b) an increase in competitive market forces through liberalized market entry and access to the network.

Neither requires privatization. But in some countries privatization may facilitate the implementation of the other changes. Privatization is a means of overcoming the effects of government failure under public ownership. But it introduces the possibility of market failure, that is, private monopoly behavior (Vickers & Yarrow, 1991). The strongest case for privatization is in countries where the publicly owned system is performing poorly, where prospects of reform within the public sector are slim, and where domestic capital and labor markets are not functioning well. But this often introduces a serious problem of significantly increasing hard currency foreign debt.

Before adopting privatization as a solution, policymakers would be well-advised to clarify what precisely is the problem. The effects of privatization will depend upon the related market and regulatory and institutional circumstances in which it is implemented. In appropriate circumstances privatization can provide an important plank in the platform of reform. But it is not *the* platform. And not all circumstances are appropriate. In fact, a developing country's telecommunication development requires capital investment in amounts that far exceed supply. Thus, all sources of capital will need to be tapped. If privatization simply substitutes private capital for public, there is no guarantee of benefit. If it provides significant additional sources of capital, it could be a major stimulus to system development.

REFERENCES

CIRCIT. (1990). *International telecommunications and the global economy: Policies and opportunities*. Conference Proceedings, Melbourne.

ITU. (1989). *The changing telecommunication environment: Policy considerations for the members of ITU*. Report to the Advisory Group on Telecommunication Policy, Geneva.

Mansell, R. (1994). *The new telecommunications: A political economy of network evolution*. Thousand Oaks, CA: Sage.

Melody, W. (1986). Telecommunications policy directions for the technology and information services. *Oxford Surveys in Information Technology* (Vol. 3). Oxford: Oxford University Press.

Melody, W. (1989, September). Efficiency and social policy in telecommunications: Lessons from the U.S. experience. *Journal of Economic Issues, 23*(3), 657–688.

Melody, W. (1990). *Telecommunication: Directions for Australia in the global economy* (Policy Research Paper No. 7). Melbourne: CIRCIT.

Melody, W. (1991). Telecommunications reform: Which sectors to privatize. In *Proceedings of Economic Symposium, ITU Telecom '91*. Geneva.

OECD. (1990). *Performance indicators by public telecommunications operators* (Report by Working Party on Telecommunication and Information Services Policies). Paris: Author.

Sappington, D., & Stiglitz, J. (1987). Privatization, information and incentives. *Journal of Policy Analysis and Management, 6*, 567–582.

Vickers, J., & Yarrow, G. (1988). *Privatization: An economic analysis*. Cambridge, MA: MIT Press.

Vickers, J., & Yarrow, G. (1991, Spring). Economic perspectives on privatisation. *Journal of Economic Perspectives*, 111–132.

World Bank. (1988). *Techniques of privatization of state-owned enterprises* (Tech. Papers 88, 89, & 90). Washington, DC: Author.

12

Alternatives to Private Ownership

Johannes M. Bauer
Michigan State University

Attempts by national governments, international institutions, and other stakeholders to maximize the benefits stemming from an advanced telecommunications infrastructure[1] have initiated a critical review of the established structures of the telecommunications industry in many countries. Based on the diagnosis of the inadequacy of the established models of telecommunications, governments worldwide are renewing their thrust in private enterprise and unregulated markets, by and large ignoring the fact that the failure of these arrangements several decades ago has led to stronger government involvement in the industry. Neither the assertion of the inadequacy of traditional forms of telecommunications organization nor the superiority of a more market-oriented solution are in line with a large amount of comparative empirical evidence of the performance of telecommunications systems under different institutional frameworks. No

---

[1]We interpret the term *telecommunications infrastructure* in a rather broad way encompassing the physical network layer of telecommunications, the software necessary to operate networks, as well as the service provided over those networks (Mansell, 1990). It has become widely accepted by scholars and policy makers alike that an efficient telecommunications infrastructure renders an important contribution to the economic, political, and social development of a society. Representative of the numerous studies are Parker, Hudson, Dillman, and Roscoe (1989), Cronin, Parker, Colleran, and Gold (1993) and International Telecommunications Union (ITU, 1986). However, many arguments remain largely qualitative or refer to limited examples. Overall, the positive contribution of telecommunications to economic growth is based more on shaky arguments than the fact that its absence constrains economic growth.

clear correlation between the parameters of telecommunications organization addressed by this deregulation, liberalization, and privatization debate (extent of state regulatory control, rules for market entry, ownership) can be observed. Measured with traditional aggregate indicators, such as telephone penetration rates or tariff levels, efficient national telecommunications systems can be found in more state-oriented, centralized contexts as well as in more market-oriented contexts, with several of the centralized systems outperforming market-oriented systems. In addition, telecommunications systems may perform comparatively weakly during some periods of time but then outperform alternative systems during other periods.[2]

These observations suggest several research questions. For example, as Mansell (1993) has argued, aggregate measures may be insufficient to fully capture the different dynamics of telecommunications under different institutional settings. Indeed, if more sophisticated measures are employed, for instance, by integrating spatial or distributional aspects and innovation processes into the analysis, more complex patterns, best characterized as trade-offs between different performance components, become visible.[3] Nevertheless, even more disaggregated data reveal no clear pattern between the degree of deregulation, liberalization, or privatization, and performance of the industry. Thus, a more complex framework may be needed to understand the dynamics of telecommunications network evolution and provide useful policy advice.

At the political level, the current debate over the redefinition of the boundaries between public and private domains, between state and market, is the result of a complex bundle of global and national political and economic factors that reach beyond telecommunications (McCormick, 1993; Mosco, 1990; Van de Walle, 1989). At a general socioeconomic level, two of the most important phenomena are the worldwide integration of key markets, such as the international financial system, and the transformation of *multinational* enterprise into *global* enterprise. The resulting

---

[2]For instance, Sweden has developed a highly efficient telecommunications system largely under monopoly rule. Similarly, after a long period of poor telecommunications service until the 1970s, France's highly centralized telecommunications system has since surpassed that of most other industrialized countries (Duch, 1991; Sawhney, 1993). Or, to give a last example, the performance lead of the market-oriented system of the United States to the rest of the industrialized countries has decreased during the period of accelerated deregulation in the United States.

[3]For instance, the aggregate tariff level in the United States is higher than in many other industrialized countries and resulted from a rebalancing of rates that has increased local residential rates but decreased long distance, business, and especially the rates for high speed data communications. In part the observable level of tariffs in the U.S. may also be the social cost for an environment conducive for innovation experiments and diversity (if not fragmentation) of networks and services (Nelson, 1981, provides examples for this phenomenon for other industries).

reduced autonomy for governments to pursue national policies has contributed to the decay of the Keynesian consensus of macroeconomic stabilization policy and state intervention in the economy. Instead, attention began to focus on the fiscal crisis of nation-states. These developments were enforced by a changing social stratification which in many countries fueled a general distrust in state and government activities and an emerging "culture of contentment" (Galbraith, 1992, p. 13). The dismal experience with centrally planned economies as well as examples of gross mismanagement of public enterprise and state intervention supported those attitudes.

In this chapter, we shed more light on the factors that determine the performance of infrastructure industries in general and telecommunications in particular by applying elements of institutional and evolutionary theory to the issue under investigation. Such a framework needs to go beyond the widespread "nirvana" approach of comparing ideal models of competitive markets with ideal models of the state and include "real world" characteristics of uncertainty, imperfect information, and bureaucracy, as well as various other forms of market failure and government failure. Such a generalized approach is provided by institutional theory. In an institutional perspective, the overall performance of a sector is determined by its institutional matrix, which provides a framework for the activities of individual and collective actors, an incentive system that both constrains and guides behavior. This incentive structure can be analyzed in its consequences for the performance of a sector as well as of the overall economy. The institutional matrix of a sector is not static but subject to exogenous and endogenous change. In such a perspective, the evolution of telecommunications infrastructure as well as policy reform can be analyzed as a problem of institutional design (incentive design) and change.

The following section develops an institutional model of the evolution of telecommunications networks. The third section applies the conceptual framework to briefly review the experience with exemplary models of telecommunications infrastructure development. The final section explores telecommunications policy as an institutional design problem and attempts to derive some preliminary consequences for feasible reform strategies.

TELECOMMUNICATIONS EVOLUTION:
AN INSTITUTIONAL MODEL

Institutional thinking is based on the insight that all societal activities take place in a specific institutional framework. Institutions are humanly devised constraints that structure human interaction. They are constituted by formal constraints (e.g., rules, statutory and case laws, constitutions), informal constraints (e.g., norms of behavior, conventions, codes of con-

duct, traditions, customs), and their enforcement characteristics. Within
the opportunities provided by the institutional matrix, organizations
come into existence. These institution-defined arrangements include political bodies (e.g., parties, city councils, and regulatory agencies), economic bodies (e.g., firms, trade unions, cooperatives), social bodies (e.g.,
religious associations, clubs), and educational bodies (e.g., schools, uni-
versities, vocational training centers) (Bromley, 1989; North, 1990, 1994).

The institutional matrix and the institution-defined arrangements constitute a complex incentive structure (governance structure) for a society
but also for specific sectors such as telecommunications (see Table 12.1).
From a sectoral perspective, the incentive structure is defined by features
at the societal level (e.g., the constitution of a country), the sector-specific
level (e.g., specific regulations drafted for the industry), as well as the
subsectoral, organizational level (e.g., the specific ownership form, the
financial and goal conceptions of telecommunications service providers).
In a cross-national comparison, certain ideal-typical governance structures become visible. These include the state, firms, markets, corporatist
arrangements, and networks of decision makers (Traxler & Unger, 1994).

The economic and political organization of an industry is an important
subsector of the overall institutional matrix. Important interdependent
domains constituting the economic organization of an industry are the
specific market structure as determined by the cost and demand characteristics of the industry; the governance structure of the industry which is
constituted by the specific regulatory framework of the industry as well as
more general policies towards business; and characteristics at the firm level
such as the specific assignment of ownership rights, the financial, organizational, and goal conception of a firm and so on. For practical purposes,
it is the economic and political organization of a sector that is of utmost
importance for its performance. Before we discuss the effects of economic
organization on performance, we will clarify the concept of *performance*.

Performance as a Multidimensional Concept

Performance, in a generic sense, measures the contribution of a sector to
social welfare. Neoclassical economics has developed the analytically
well-defined but narrow concept of *social surplus* (the sum of consumer
and producer rents) to measure the welfare consequences of sectoral
organization. Besides the fact that this measure is based on the rather
restrictive assumptions of orthodox economic theory, it has the fundamental disadvantage of being indifferent to the issue of distribution. For
practical purposes of policy design, a more pragmatic approach might
be justified (see, e.g., Tinbergen, 1992).

TABLE 12.1
Dimensions of the Institutional Matrix

	"Public" / "State"	"Cooperative" / "Network" <------------>	"Private" / "Market"
Institutional Framework			
• Cultural framework	Collective		Individualistic
• Political system	Centralized		Decentralized
• Economic system	Planned		Laissez-faire
Sectoral Framework			
• Sectoral public policy	Industrial policy		No explicit industrial policy
• Market access rules	Monopoly		Free market entry
• Price regulation	Strict price control		Market-based pricing
• Conduct regulation	Discretionary		Market-compatible
Organizational Framework			
• Ownership	Public		Private
• Organization/control	Government		Shareholders
• Financial conception	Public sector budget		Internal and external capital
• Goal conception	Social goals		Profit maximization

Note. The entries in the table serve to illustrate extreme attributes associated with either state or market. Most real world solutions are located somewhere along the continuum between these extremes.

265

Performance can be measured using a bundle of criteria that capture efficiency and equity dimensions. Efficiency has become a predominant concern of policy makers. In its most basic sense of Pareto-optimality, efficiency is accomplished if no alternative resource allocation exists that makes somebody better off without making somebody else worse off. Under conditions of given technology, this definition implies that both productive efficiency and allocative efficiency are accomplished. It becomes more complicated under conditions of technological change. In this case, the evolution of a sector over time has to be considered in addition to the familiar static criteria. In particular, the responsiveness of the particular form of economic organization to changing demand and supply conditions as well as its innovativeness (i.e., its ability to generate new processes, products, organizations, arrangements, etc.) need to be considered explicitly.

Economists have traditionally assumed that efficiency and equity considerations can be separated (Lane, 1985). Under competitive conditions, it has been shown that for every initial distribution of resources, a Pareto-efficient state exists and vice versa. Therefore, as long as the equity issue is solved via some political mechanisms, the policy design can focus on the efficiency dimension. To a certain extent such an approach seems to be justified, especially if mechanisms such as progressive taxes or subsidy programs can be built into the framework of the economic system to create more equal conditions. However, in a dynamic perspective one cannot assume that such an approach holds, because the overall development trajectory of a system may well depend on the distribution of purchasing power and wealth. With this caveat, we focus mainly on the efficiency dimension.

Institutional Matrix as a System of Incentives

The institutional matrix of a sector can be viewed as a system of incentives that guide and constrain the decisions and behavior of the economic agents.[4]

Governance Structure. The governance structure of an industry can be operationalized as the set of rules that constrain and guide the operations within an industry. Such rules can be part of the general legal framework of a country, such as antitrust laws or constitutional principles guaranteeing the freedom of economic activities. As such, they constitute prerequisites for the operations of a sector. Some of those rules are

---

[4]For a recent monograph on this approach, see Spulber (1989). A "classic" in this area is North (1990).

specifically designed for the sector. Examples include the specific regulations that apply to the industry, such as the conditions for market access, behavioral supervision of such parameters as prices and service quality, standards, interconnection rules, and so on. Finally, in most countries, a set of more discretionary policies, such as industrial policy, research and development (R&D) policy, or specific telecommunications infrastructure programs, are part of this governance structure of telecommunications.

This governance structure is a subset of the institutional space. It can take on various forms depending on the specific cultural and historical tradition of a country. However, it is common in the legal system of most countries that the governance structure of a sector is defined by provisions at various hierarchical levels of societal organization, such as the general constitutional level, the sectoral level, or even the firm-specific level. To a certain degree, provisions at those different levels can be substituted for each other. For instance, the close regulation of prices of a telecommunications provider might be replaced by reliance on the more general provisions of antitrust law.

Of core importance for the performance of a sector are the sector-specific forms of governance, in particular the specific forms of economic and social regulation as well as specific public policy programs designed for a sector. Economic regulation comprises all sector-specific measures that control market access and exit: economic parameters of service providers such as their prices, service quality, or investment policy. Social regulation, although not independent from economic regulation, addresses issues of equity, work safety, and related issues. The main responsibility for those forms of regulation can be vested in a separate agency such as in the United States, but it is frequently shared between the legislative and executive branches of the government as well as the main service provider, as is characteristic for most Post, Telephone and Telegraph (PTT) models.

From a normative perspective, economic regulation in industrialized countries is usually considered a substitute for competition in the presence of market failure. It is thus shaped to mimic the efficiency characteristics of competitive markets. Regulatory approaches vary in the degree of their "regulatory intensity." For instance, administrative rate of return regulation is characterized by a rather high degree of regulatory intensity in so far as it is based on close scrutiny of the business parameters of the regulated firm and provides rather little entrepreneurial freedom in the setting of prices or other business parameters. Price cap and other forms of incentive regulation, on the other hand, provide a much higher degree of entrepreneurial flexibility and are thus characterized by a lesser degree of regulatory intervention. Positive analysis has repeatedly criticized the rationale for economic regulation, and pointed to the fact that regulation was frequently welcomed by the regulated industries as a legitimate form of cartel management. The overall record of public regulation seems to suggest that the truth lies somewhere in between the public interest and the special interest theories of regulation.

Especially in the U.S. context, regulation was mainly seen as a tool to control market power. It thus focused strongly on issues of price control, or, in other words, productive and allocative efficiency under static conditions. Until recently, much less emphasis was placed on the issue of innovation and dynamic efficiency. In the framework of the PTT tradition, those tasks of economic regulation, if they were even made explicit, were usually entangled with the pursuit of other, more discretionary industrial policy goals. Based on conceptual traditions such as the theory of the socioeconomy or institutional approaches, such an approach was considered absolutely legitimate and compatible with the rationale of public ownership and control of vital infrastructure industries.

Because governance structures and, in particular, regulation can take on a large variety of forms, it is not possible to give a generic evaluation as to the efficiency consequences of those structures. The rich literature that has addressed those issues is not entirely conclusive and has shown examples of well-designed regulation that helped improve the performance of imperfect markets as well as numerous examples of ill-designed regulations that even worsened the performance of imperfect markets. Because the conditions that maximize productive efficiency given the technology and those under changing technology are not identical, it is necessary to specify the purpose of regulation as clearly as possible.

Market Structure and Competition. The most effective economic incentive system are the forces of competition in a market. The degree of competitiveness of a market is dependent on the entry and exit conditions of this market as well as the characteristics of technology and demand, which determine the equilibrium market structure. Most real world markets do not show the characteristics of perfectly competitive markets or their ideal property to enforce productive and allocative efficiency. As Schumpeter (1950) and others have argued, it is even questionable whether such perfectly competitive markets lead to an optimal allocation of resources for innovation purposes and, hence, it can be challenged whether such market structures should serve as a policy guideline at all.

Under real world characteristics, it is more relevant to understand the processes of rivalry in markets with oligopolistic or monopolistically competitive structures. Such market structures usually do not fully meet the conditions of productive and allocative efficiency, although, under certain conditions, they might get sufficiently close. Due to the strategic interdependence of firms, markets with only a few suppliers do not allow for simple analysis and numerous models of oligopolistic behavior have been studied. In markets with less than four rivals, strong incentives for open or tacit collusion exist. Based more on empirical pragmatism than on rigorous analysis, the theory of "workable" competition has suggested that in oligopolistic markets the forces of rivalry are sufficient to discipline firms to achieve productive and allocative efficiency, if there are four or more suppliers that do not engage in any form of collusion in the market and the market share of the dominant firm is less than 40%.

The effectiveness of oligopolistic rivalry to generate innovation also needs critical scrutiny. Based on Schumpeter's arguments, various writers have studied the ability of oligopolistic firms to generate sufficient funds for R&D and to appropriate temporary extra profits from their innovation activities, both factors commonly considered conducive to innovation (Kamien & Schwartz, 1982). These arguments focus too strongly on the market structure and do not explicitly model the innovation process. In an evolutionary perspective, innovation can be modeled as a search process under conditions of uncertainty (Nelson & Winter, 1982, p. 275). Possible avenues of technology and applications are only gradually unraveled. Because different people and organizations have different views as to the most promising directions of research, diversity and pluralism in R&D is desirable from a societal point of view. Neither perfectly competitive nor monopoly markets create this diversity. However, it is a well-known phenomenon that oligopolistic markets do not have this characteristic either, because companies engage in "defensive" research to fend off possible threats from their competitors (Nelson, 1981).

Under several circumstances, notably the condition of a "natural" monopoly or the prevalence of significant externalities, market forces fail to achieve the desired coordination and efficiency incentives or they convey misleading signals to the economic actors. Although no conclusive empirical evidence is available, it has been a widely shared belief that telecommunications is indeed an example of a natural monopoly and, thus, should deliberately be exempted form the exposure to market forces. Despite its popularity, the natural monopoly argument is relatively weak as a guideline for policy decisions. First, if technology is endogenous to the structure of a market, then we would expect the implementation of decreasing cost technologies by monopoly suppliers. Second, the argument is a static one that does not properly take account of changes in technology or demand. Third, decreasing average cost at the sectoral level could be the result of the aggregation of decreasing cost technologies at a much lower, possibly firm scale of telecommunications service provision and thus not necessarily call for a national service provider (Antonelli, 1994).

The externality argument is evenly unexplored. Indeed, if significant regional, intertemporal, and network externalities prevail, market forces would send misleading forces to investors and consumers alike and result in underinvestment or overinvestment (and only by chance in the correct investment level) into telecommunications. Although frequently referred to, no conclusive evidence as to the extent of such possible externalities has been presented so far.[5]

A last point deserves mentioning. Even if the technological and demand characteristics of a formerly monopolized market would allow entry into selected parts of telecommunications and indeed attract new market entrants, this should not be confused with the creation of a competitive market. For one, the incumbent supplier usually commands significant market power and enjoys large advantages due to the existence of an installed base. Second, the transition zone from a monopolized market to a workable competitive one can be significant (Shepherd, 1984). Finally, market entry might occur because of inefficient pricing strategies of the incumbent firm due to a policy of cross-subsidization and, through such a process of cream skimming, even deteriorate the overall efficiency of the sector.

Organizational Factors at the Firm Level. At the micro level of the firm, variables such as the organizational form of an undertaking, its financial and goal conception, and the assignment of ownership rights are important factors influencing the performance of an industry. Much of the privatization debate focuses on the ownership dimension. The specification of ownership rights in an undertaking can be of importance for its efficiency because of the necessity to monitor entrepreneurial processes. In an owner-managed firm, there might be a coincidence between the self-interest of the owner and the performance of a company and, hence, a strong motivation to achieve a high performance level. In investor-owned joint stock companies, this link between self-interest and performance is mediated through management and the interests of the investors might not fully coincide with the interests of the organization.

In both cases, but especially in the latter case of the separation of ownership and control, the principals (owners) do not have complete information about the internal performance of the company. Therefore, monitoring becomes necessary. Monitoring of intrafirm processes is less costly and may even be replaced with the comparison of firm data to

---

[5]First attempts to quantify the externality dimension have been made by Antonelli (1991) and researchers at Bellcore, although those approaches need further development. A priori, one can assume that the extent of such externalities will be more significant in poorly developed than in saturated telecommunications networks.

observations derived from other firms if the market structure is competitive. However, if such data is not readily available, monitoring may
become a costly undertaking if the degree of information asymmetry
between the owners (principals) and the agents (management) is high. It
has been shown that in this latter case ownership per se becomes a rather
weak mechanism to induce high performance due to free-rider behavior
of the owners, which leads to suboptimal low levels of monitoring (Vickers & Yarrow, 1988).

Public ownership may, in theory, avoid the problems associated with
the free-rider behavior of private owners but possibly faces other severe
limitations. Frequently, the monitoring function is not performed effectively because the principal is not defined clearly (Aharoni, 1982) or the
function of the principal might change frequently alongside election cycles
or other political events (Backhaus, 1989). Moreover, whereas the ultimate
goals of private enterprise in most cases seem sufficiently clear as the
maximization of long-run profits, this cannot be said of many public
enterprises, especially if the goals conception of a public entity incorpo-
rates industrial and other public policy goals in its operations. Losses and
other forms of inefficiencies are frequently legitimized by referring to
such public interest obligations, thus curtailing the effectiveness of monitoring an enterprise's efficiency.

Although ownership per se does not give clear advantages to either
publicly or privately owned companies, such advantages might rest in
other parameters of organization at the firm level. In addition to the
incentives exerted through ownership rights, public and private companies are subject to other disciplining forces, such as the control derived
from financial markets, take-over threats, or the "market for managers"
(Vickers & Yarrow, 1988, p. 7, provide an overview). Whereas the effectiveness of those mechanisms to induce efficiency should not be over-
stated, their functioning differs widely between the typical private and
public undertaking. Public enterprises are, by definition, not subject to
the disciplining forces of capital markets and are only weakly exposed
to the forces of credit markets. Their financial basis might frequently be
determined on political grounds rather than on business terms. Nor are
public undertakings exposed to take-over threats. Because the recruitment
of management in public firms follows different practices than the recruitment of management for private companies, this incentive might not
be very effective either.

A final determinant of efficiency at the firm level is the organizational
form of an undertaking. Because of the specific requirements of the public
sector, many public telecommunications operators were set up as public
administrations and, as such, make up part of the ministerial bureaucracy.
Public administrations, due to the requirements of administrative law,

usually are characterized by slow and lengthy decision-making processes as well as relatively weak reward structures. Thus, they tend to perform poorly with respect to static efficiency and, in particular, with respect to dynamic efficiency. Only in rare cases, perhaps in organizations run by highly enthusiastic and committed officials, can we expect to observe efficient behavior. To a certain extent, these disadvantages can be avoided by a "corporatization" of public undertakings, especially if their management is given autonomy with respect to personnel and financial decisions. In addition, the reward system for employees and management can be structured to provide stronger performance incentives.

Ownership and the related incentive mechanisms cannot be analyzed in complete isolation from the competitive structure of an industry. Much of the literature on the performance of public versus private enterprise has brought forward similar conclusions. In competitive markets, both public and private enterprise face strong disciplining forces to behave efficiently. In imperfect and especially monopolistic markets, public enterprises tend to perform better with respect to allocative efficiency whereas private enterprises tend to realize a higher degree of productive efficiency. Thus, the overall results remain somewhat ambiguous. Most of these studies remain incomplete, however, because they do not address the dynamic efficiency aspects, such as the responsiveness of an organization to changing market conditions or the innovativeness of organizations. The scattered evidence makes us believe, though, that the mentioned general conclusions also hold for the dimensions of dynamic efficiency.

The Institutional Design Problem

As the discussion so far has illustrated, the overall performance of a sector is determined by a bundle of incentives related to the governance structure of an industry and its competitive structure, as well as micro-organizational factors. Within the framework and according to the specific rules of a society, these incentive structures can be modified to improve the performance of a sector. Ownership and the openness of market access are only two of the parameters in this overall institutional matrix, and their effectiveness in improving the performance of a sector such as telecommunications is dependent on the particular structure and combination of the other incentives. A change in one parameter of the institutional matrix might need to be accompanied by other measures. For instance, the privatization of telecommunications in the former Eastern European countries might not improve the efficiency of operations significantly as long as a general institutional framework for capital markets, credit markets, and other preconditions of private ownership in productive resources are missing.

At this point it is important to see that some of the different components of the institutional matrix are at least partial substitutes for each other, whereas others are complements to each other. For instance, planning of telecommunications infrastructure development via a public operator might be a substitute for the use of private capital in the build-up of this infrastructure, perhaps under regulatory oversight. A priori it is hard to say which solution will perform its task more efficiently. Planning might outperform private initiative or vice versa. An answer will have to be based on close observation of the conditions of both solutions. Other dimensions of the institutional matrix show a complimentary relation. For instance, the liberalization of market access to more network providers will, in general, need arrangements to secure interconnection between networks.

The institutional matrix opens a wide variety of different solutions to organize the provision of telecommunications services. Historically, the various parameters appear bundled in specific ways, such as in the PTT model, the regulated competition model, or the unregulated competition model of telecommunications. Under conditions of perfect information, all those solutions would lead to equally performing, efficient solutions. Under real world conditions of imperfect information, the relative performance of the various options is not uniquely defined. This is due to the partial substitutability of the various components of the institutional matrix. Thus, public administrations may achieve a high degree of performance, if run by "ideal" administrators. Private enterprise may achieve a high rate of infrastructure development, if properly governed by a framework to cope for distributional concerns or forms of market failure. On the other hand, public undertakings or private undertakings can show extremely poor performance, if the institutional matrix is improperly designed.

In all, it is not possible to rank typical institutional solutions in a unique order according to their efficiency properties. We can expect to observe some degree of "institutional equivalence" between different models of infrastructure development. In a dynamic context, different models might produce characteristic "trajectories" of infrastructure development. For the purposes of telecommunications reform, this implies that a variety of options need to be considered by policy makers that may suit the specific needs of a country equally well. In general, it is easier to design gradual policy changes than radical ones. However, not all imaginable solutions are feasible, due to the particular political, economic, and international situation. Before we look more closely into such options, we briefly review the experience with some typical models of telecommunications infrastructure development.

TELECOMMUNICATIONS INFRASTRUCTURE
IN INDUSTRIALIZED COUNTRIES

Due to the number of variables involved and the resulting problem of institutional equivalence, it is very difficult, perhaps impossible, to fully test the relationship between the institutional framework of a sector and its performance empirically. Nevertheless, we can assume that observations not visible in the aggregate might remain invisible with more sophisticated methods of analysis. Underneath the surface of the PTT model and the approach of regulated private enterprise in the United States and Canada, industrialized countries have developed a variety of specific institutional approaches to organize telecommunications that may help us shed more light onto the relation between the institutional framework and performance.

Within the industrialized world, national approaches differ in various dimensions of the institutional matrix. Some countries, such as the United States, Italy, Denmark, or Finland, have licensed more than one supplier of telecommunications services. Nevertheless, they have maintained a policy of exclusive geographical or service-specific licenses and have not, until perhaps recently in the United States, introduced immediate competition between the carriers. National approaches also differ with respect to the ownership model adopted. After the defeat of proposals to nationalize telecommunications operators, the United States relied mostly on regulated private companies with only a minuscule number of municipal and cooperative providers. Many countries relied on the "traditional" public monopoly model. Finland or Denmark, on the other hand, have used the municipal and cooperative models to a much larger extent.

Countries have also pursued various strategies with respect to the organizational structure they created for their providers. In the private enterprise environment of the United States, the organizational form of the main providers was determined endogenously by the logic of private corporate business. Countries have accepted various models to regulate telecommunications within the spectrum between the "internal" regulation through public enterprise and the "external" regulation through expert commissions. Very general oversight of business activities by the parliament is contrasted by the rather explicit rules of price control as visible in the U.S. regulatory model. In countries with public telecommunications operators, the choice of the organizational form becomes itself part of the political process and the process of institutional design. Again, a variety of approaches is visible, ranging from the traditional model of a public administration (e.g., in Portugal, Germany, or France) to models of public corporations with much more degree of autonomy of the public sector (e.g., in Sweden). Last but not least, national approaches differ with respect to the degree of discretionary public policy directed towards telecommunications. Denmark or Sweden have pursued a rather open policy, countries such as France or Germany have adopted a relatively high degree of public planning of telecommunications infrastructure development, and the U.S. is characterized by a fairly high degree of reluctance to engage in any open industrial policy for telecommunications.

A comparative analysis of aggregate performance figures for Denmark, Finland, France, Portugal, and the United States shows some interesting results. In the 25-year period between 1965 and 1990, all five countries were able to expand significantly their telecommunications infrastructure, measured as the penetration rate of telephone service. However, the relative starting positions also differed significantly, with the United States and Denmark starting from a relatively well-developed telecommunications infrastructure and Finland, France, and Portugal starting from a relatively poor one (see Table 12.2). The four European nations were able to expand their telephone networks faster than real economic growth, whereas the United States expanded its network at a lower rate than its economic expansion. In addition, Denmark was able to accomplish this expansion at tariff levels that are below the U.S. level.

The five selected countries illustrate some of the more theoretical points. Beginning in the late 1950s, the United States started to introduce a higher degree of competition into the provision of telecommunications infrastructure. Gradually, the comprehensive monopoly privileges of the integrated Bell System, which also acted as a leader for the business

TABLE 12.2
Growth Rates of Telephone Penetration and
Gross Domestic Product (1965–1990)

	Main Lines per 100 Population (CAGR 1965–1990)[1]	GDP per Capita (CAGR 1965–1990)[2]	Telecom Ownership	Degree of Competition
Denmark	4.0	2.1	Public	Yardstick[3]
Finland	5.6	3.2	Public	Yardstick[3]
France	8.9	2.7	Public	No
Portugal	5.9	4.6	Public	No
U.S.	2.2	2.5	Private	Partial

[1]CAGR = compound annual growth rate.
[2]GDP = gross domestic product in 1985 U.S.$.
[3]Yardstick competition existed between the various local and regional monopolies (see Shleifer, 1985, for the concept of yardstick competition).
Sources: ITU, *Yearbook of common carrier telecommunications statistics*, various years, Geneva: ITU. Own research.

strategies of about 1,600 "independent" telephone companies, were reduced to core areas such as the provision of local telephone services. More open market access was first introduced in the area of terminal equipment, which was basically fully deregulated by 1975. Beginning in 1959 with the "Above 890" decision of the FCC, the provision of specialized telecommunications services, both facilities based and on a leased line basis, were opened to competitive market access. Finally, in 1980, long distance telephone was opened to competitive provision.

In the case of the United States, the simple data of Table 12.2 tell only part of the story. First, a significant part of infrastructure development took place before 1965. The United States had achieved almost a 30% telephone service penetration rate by then, which it had nearly 90 years to build. There is considerable controversy as to the major causes of telephone penetration growth. After Alexander Graham Bell received his patent for the telephone, the Bell System grew for 13 years protected by its patent rights. During this period, the Bell System developed, partly in a strategic fashion, partly following the logic of private business, telephone service in the major urban centers and, by and large, neglected rural areas. After the expiration of the patents between 1893 and 1896, numerous new companies started to serve the hitherto neglected areas, thus generating a burst in telephone penetration.

However, it would be misleading to interpret this phenomenon as a proof that competitive provision of telecommunications necessarily outperforms a monopolistic system. For one, the Bell System was not subject to any substantive regulation and followed the strategies of a monopolist. By not serving rural areas, it had unintendedly created opportunities for the entry of other providers, and soon the first boom of independent companies started to reintegrate the system. One has to be careful not to confuse the short-term effect of open market entry with the medium- and long-term effects of such a policy.

It is necessary to recall that in the wake of an antitrust lawsuit in the 1910s, the Bell System actively sought regulatory oversight and promised, in turn, the introduction of universal telephone service. During the stable phase of regulated monopoly, basically 1930–1960, telecommunications infrastructure development was accomplished in part through a mechanism of cross-subsidies based on the averaging of prices throughout the network. This was more of a side effect of the regulatory system of price control than a deliberate master plan to improve telecommunication service. Such an explicit policy was introduced in 1949, when Congress expanded the authority of the Rural Electrification Administration (REA) to make funds available to rural telephone companies. Under this program, rural telephone companies received long-term, low interest loans to expand and improve telephone service in rural areas. As Fuhr (1990) discusses, in 1950, 17 telephone companies with a total of 29,100 subscribers borrowed funds at an interest rate of 2%. In 1986, about 4.2% of all people with telephone service were served by telephone companies under the REA program. The average subsidy per subscriber was about $0.63 per month or $7.56 a year. Given the low price elasticity of demand for access to telephone service, the overall effect of the REA program on achieving universal service has to be judged as rather low. The REA program did not solve another problem, the unequal penetration of telephone service in different income strata of the population. In 1988, household penetration ratios varied between 72.3% for households with incomes below $5,000 per year to a penetration of 99.6% for households with annual incomes above $60,000. To target this problem, individual subsidies, introduced after the divestiture of AT&T in the form of lifeline programs and the universal service fund, are more appropriate. Thus, although the REA program has some serious shortcomings, the program illustrates the use of subsidy programs to improve service penetration.

A last point deserves mentioning and illustrates the effects of the opening of market access in telecommunications markets. As a result of competitive strategies such as product differentiation, a multivendor market structure tends to create a higher degree of diversity than a monopolistic market structure. This might contribute to better matching of customer needs, especially for the business sector with its highly differentiated demand for telecommunications services. On the other hand, in network industries, diversity creates costs of coordination between "networks of networks." Moreover, if network externalities prevail, it is possible that open market access creates a suboptimal high degree of diversity. In any case, there are social costs related to diversity; this factor may explain the fact that some telephone tariffs are higher in the United States than in other countries.

France has achieved a remarkable upgrade of its telecommunications infrastructure within a rather monopolistic and planned framework, especially during the late 1970s and 1980s. In the preceding decades, the French telecommunications system was known for its notorious poor quality. Alarmed by the conclusions of the Nora and Minc (1980) study, ways were sought to rapidly upgrade the French telecommunications system. The ambitious *plan cable* in the early 1980s to deploy a broadband infrastructure throughout France using cable TV as a driver ended unsuccessfully due to bureaucratic rigidities and poorly defined goals and strategies. The most significant upgrade of the telecommunications network was accomplished by the French monopoly provider of telephone service, France Télécom. Due to its monopoly position, France Télécom could charge relatively high prices to generate sufficient internal funds to embark on a program of rapid telecommunications investment. To a certain degree, France was able to benefit from a "second mover" advantage as it leap-frogged an entire generation of telephone technology and realized its network upgrade in digital technology. Thus, France can serve as an example of a rigid, bureaucratized behemoth while showing how national determination and a monopoly position can contribute to a successful modernization and expansion of a poor telecommunications infrastructure.

Both Denmark and Finland are interesting examples that licensed more than one provider of telecommunications services, although Denmark re-integrated regional providers in a holding company in 1990 (TeleDenmark) to deal with increasing international competitive pressure. Nevertheless, the infrastructure build-up in the past decades has been accomplished within the more diverse structure. Until the reorganization in 1990, telecommunications services in Denmark were provided by four regional companies that were responsible for local and trunk service within their territories and one company under the control of the Danish Post and Telegraph office that was responsible for national and international long distance service.[6] Until the mid-1980s, the state, based on an agreement with the regional companies (the concordat), acted as a mediator on issues of prices, cost sharing, traffic planning, and standardization (Olsen, 1989). This arrangement was replaced by the Telecommunications Council in 1986 and later by an assignment of regulatory tasks to the Ministry of Public Works.

The Finnish telephone system is operated by 56 local companies, either private firms, municipal enterprises, or subscriber cooperatives that are directly accountable to their subscribers, and a state-owned PT&T that also offers local service and, exclusively, domestic and international long-distance service. Until 1987, when the Ministry of Transport and Communications took over regulatory responsibilities, the PT&T was also the main regulatory body for Finnish telecommunications.

The Danish and especially Finnish operators are of relatively small size and might thus forgive possible economies of scale. It is interesting, however, that in both countries the systems are run very efficiently and that tariffs are among the lowest in the world, especially in Denmark. This leads one to question the traditional rationale of telecommunications as a nationwide natural monopoly. Cost advantages for the entire system might also be achieved with a multitude of carriers. In both countries, the regulatory authorities took advantage of the multitude of carriers by comparing their relative performance and basing their regulatory policy

---

[6]The regional companies serve Copenhagen (KTAS), Jutland (JTAS), Funen (FKT), and South Jutland (Tele Sonderjylland). The PTT (now Telecom Denmark) provided national and international long-distance service.

on this information. Thus, each company acted as a yardstick for the others and the fragmented structure created information that would not have been available otherwise. Moreover, the vicinity of control of the companies to their customers seems to have induced a responsiveness to the local needs and a commitment to dedicate funds to expand and upgrade the network and services.

In our brief comparison, Portugal can serve as an example of telecommunications development in a relatively "poor" industrialized country. Portuguese telecommunications services are provided by three companies, Correios e Telecommunicacoes de Portugal (CTT), Telefones de Lisboa e Porto (TLP), and Compania Portugesa Radio Marconi (CPRM). All three report to the Ministry of Public Works, Transportation, and Communications. Although telecommunications received increased attention from the government in the 1980s, the investment per line was among the lowest within the group of OECD countries until it increased significantly in 1988. Portugal was able to expand its telephone network faster than its overall real growth. However, the poor initial conditions and the scarcity of resources even in a moderately rich country have kept telephone penetration in Portugal below the level of the other European countries. A 1990 telecommunications law integrated the three companies in a holding company, Telecom Portugal (TLP), and allowed up to 49% private investment in the holding in an attempt to increase funding for the expansion of the telecommunications network. Although results remain to be seen, this move by the conservative center-right government illustrates the decision-making problem faced by many poor countries.

TELECOMMUNICATIONS REFORM: AN INSTITUTIONAL DESIGN PROBLEM

Telecommunications as Infrastructure

Telecommunications is different from most other industries in so far as it is a network-based industry in which the various components need to interoperate in a seamless way to provide efficient service. Its infrastructure character derives from the pervasive use of telecommunications services throughout all sectors of the economy as well as in the area of final consumption. Although partial substitutes for electronic telecommunications services, such as transportation or mail service, exist, the absence of an adequate telecommunications infrastructure can become a major obstacle to economic development (Trebing, 1994).

Any decision to invest in the telecommunications infrastructure needs to consider factors that go beyond the industry. Given scarce resources, it is not sufficient to prove a social rate of return for telecommunications investment, but it is necessary to compare telecommunications investment with other, alternative uses of the funds under consideration. Therefore, a proper decision-making process has to evaluate the net social rate of return of telecommunications investment, equal to the rate of return minus the opportunity cost of sacrificed alternative investment projects.

The network character of telecommunications adds additional complexity to the solution of this question. Ideally, funds would be committed until the net social rate of return on telecommunications infrastructure investment approaches zero. This requires the equation of the cost and benefits of telecommunications investment. The measurement of the benefits of telecommunications investment is, however, a complicated problem that, so far, has not been properly solved (Bauer, in press). The benefits of investment in infrastructure tend to be widely dispersed across regions as well as time and, hence, hard to quantify.

Investment in telecommunications networks exhibits a high degree of idiosyncrasy (i.e., the facilities usually do not have any other use). Given the relatively high amount of fixed investment needed to run a telecommunications network, the sector is characterized by a fairly high share of sunk costs in its overall cost structure. In periods of rapid technological change this can render the calculation of the cost of telecommunications investment as complicated as the assessment of the benefits. In all, the design of an optimal investment strategy may be very hard to accomplish.

What, if anything, can be learned from these examples? Although case studies should be used very cautiously when conclusions are drawn for other cultural, political, and economic contexts, the previous analysis can provide the basis for an overall assessment of policy option to improve the performance of the telecommunications industry, especially in developing countries and the former Communist countries in transition, which are currently looking for refreshed approaches to telecommunications. Some of the basic differences in the starting conditions of countries need to be considered explicitly. First, income per capita varies tremendously between countries, with significant gaps between the industrialized and developing nations. Second, many developing countries face rates of population growth that pose an additional challenge for any policy designed to improve the telecommunications penetration rate. Third, unlike most industrialized countries, which due to the limited diversity of telecommunications technology up to the 1960s could invest in one basic infrastructure for various groups of users, developing countries and economies in transition face the problem of basic infrastructure build-up under conditions of extremely diverse demand from residential and business users.

Institutional Prerequisites for Market-Oriented Reform

Private enterprise market systems are embedded in a complex web of institutions (legal framework, established financial sector, individual attitudes to life, etc.) that is a prerequisite for the functioning of markets. The potential effectiveness of privatization and liberalization in improving the performance of telecommunications relative to the status quo is largely dependent on the existence of such an institutional framework. Because markets are socially constructed institutions and must be politically maintained (Kenis, 1992) privatization within an established structure of political influence and power (such as in Latin American countries) might not lead to the overall desired effects, although in the medium and long run an endogenous dynamics toward more market-like institutions may be initiated.

As our earlier discussion has illustrated, the overall performance of an industry, especially its dynamic performance, might be improved by policies of liberalization and privatization, especially if the status quo ante is characterized by rather restrictive rules. However, although short-term effects of an opening of market entry might seemingly improve conditions as compared to the status quo ante, core parts of the telecommunications industry are characterized by noncompetitive market structures and, hence, prone to possible allocative inefficiencies (Shepherd, 1984). Public regulation could, ideally, cope with these problems. But like the private enterprise system as such, public regulation needs a proper institutional framework as well as expertise (and regulatory experience and tradition) to be effective. If policy networks in existence prior to liberalization and/or privatization measures are maintained, the overall effects of such measures may at least be doubted.

The Spectrum of Possible Reforms

In assessing the performance of telecommunications, it is important to choose the appropriate yardstick (e.g., a "best practice" model). Depending on the specific status quo ante, improvements in the performance of the telecommunications sector can be accomplished in more gradual as well as in more radical ways.[7] One major decision is the degree of liber-

---

[7]See World Bank (1994), International Telecommunications Union (ITU, 1994). Noam (1994a, 1994b) suggests a model of telecommunications network evolution in which the integrated network is replaced by a more pluralistic network as the network expands and matures. In this approach, the design options of policy are constrained by this endogenous dynamic of network evolution. Although the model has found widespread acceptance, its general applicability to the problems faced by developing countries needs more careful review.

alization of telecommunications. Although most of the discussion in industrialized countries framed this debate by looking at the market structure in different part of telecommunications, a more proper perspective should include the dynamic aspects of more liberal conditions of market access: The dynamic performance of companies with substantial market power might exceed the efficiency loss due to this market power.

However, it is important to understand the dynamics of investment in profit-seeking companies. Especially under conditions of substantial excess demand, users with the highest purchasing power and willingness to pay will be served first. This strategy may conflict with the goals of more equal infrastructure development and also not fully realize possible externalities involved in telecommunications investment. Proper regulation (e.g., through investment prerequisites to channel the excess revenues into reinvestment, a value-added infrastructure tax) can reduce this problem, but the major suppliers may have a strong strategic position to undermine such a policy.

Even if core parts of telecommunications are not considered to be liberalized, various ways to improve the performance of established operators exist: The performance of a public administration can be significantly improved by providing it with a more flexible corporate structure; significant performance improvements might be accomplished by using more effective yardsticks to evaluate the past performance of telecommunications provider; a decentralization of decision making as well as financing may generate improved performance; last but not least, regulation can be substantially improved by switching to more efficient forms of "incentive" regulation.

An important issue is whether foreign investors should be sought. Although in most cases this would open up access to otherwise possibly closed sources of funds, it also provides a new source of vulnerability for the national government if international investment strategies are not in line with the national welfare (Bauer, in press). As some of the new trade literature shows, liberalization of trade may not improve national welfare (Krugman, 1990). To improve monitoring and control, it is helpful to establish clear policy objectives that include an infrastructure development plan and proper incentives to the telecommunications organization to implement such a plan. To develop technologies suited to the specific needs of developing countries, a coordinated R&D policy should prove beneficial. Finally, to prevent "blurring" of various goals, a separation of performance and distributional or industrial policy goals might be useful.

Not all theoretically possible reforms are feasible given the economic, political, and institutional status quo. For instance, there are limits in the possible degree of organizational transformation. Foreign funds are in many cases only accessible through privatization and/or liberalization. The specific political conditions (social structure, majority requirements to reach decisions, international pressures) frequently render reforms impossible.

Last but not least, it is important to recognize that the various possible reform patterns show certain tradeoffs. Liberalization of market access may improve performance but have a negative impact on distributional goals if no corrections are established. By the same token, the liberalization of telecommunications might improve national performance of the telecommunications industry, but, at the same time, deteriorate employment in national industries, possibly leaving the net welfare effects unclear. In all, privatization and liberalization are no panacea for improved performance of the telecommunications industry, but under the specific economic, political, and foreign policy constraints of many developing countries, no better feasible solution may be possible.

CONCLUSION: QUESTIONS FOR FURTHER RESEARCH

As the previous theoretical and empirical arguments have illustrated, no proper mix of public and private activities in telecommunications holds for every environment. Similar levels of overall performance might be achieved under diverse institutional settings with various roles for the state and private sector. However, depending on the specific model adopted, different trajectories of telecommunications infrastructure development might result and, thus, some models might be more appropriate for the setting of developing or industrialized countries than others. For instance, models with free market access for telecommunications service providers seem to show a higher degree of diversity of services but not necessarily at the lowest possible overall social cost because diversity exacts a price in the form of duplication of investment, resources spent for R&D, or expenses for customer recruitment and retention. Models with more closed markets for basic infrastructure development tend to generate a lower degree of diversity but possibly at a lower overall social cost.

Comparative research into the dynamics of telecommunications infrastructure development is at best at its beginning. The diverging solutions to telecommunications as we observe them in the United States, Europe, and the Pacific Rim should provide ample empirical evidence to continue a research program in this area that will help contribute to our understanding of the working of network industries. Areas of particular interest are the relative dynamics of innovation processes under different institutional matrixes, implications of different institutional solutions for various groups of telecommunications users, or the dynamic interrelationships between institutional frameworks and sectoral performance as telecommunications systems evolve.

ACKNOWLEDGMENTS

The author wishes to thank Jill Hills, Peter Lewis, and Rohan Samarajiva for helpful comments on an earlier draft. The usual disclaimer applies.

REFERENCES

Aharoni, Y. (1982). *The evolution and management of state-owned enterprises.* Cambridge, MA: Ballinger.

Antonelli, C. (Ed.). (1991). *The economics of information networks.* New York: Elsevier Science.

Antonelli, C. (1994). Increasing returns: Networks versus natural monopoly. The case of telecommunications. In G. Pogorel (Ed.), *Global telecommunications strategies and technological changes* (pp. 113–134). Amsterdam: North-Holland.

Backhaus, J. (1989). Privatization and nationalization. A suggested approach. *Annals of Public and Cooperative Economy, 60,* 307–328.

Bauer, J. M. (in press). The anatomy and regulatory repercussions of global telecommunications strategies. In G. W. Brock (Ed.), *Current issues in telecommunications policy.* Proceedings of the 22nd Telecommunications Policy Research Conference. Hillsdale, NJ: Lawrence Erlbaum Associates.

Bromley, D. W. (1989). Institutional change and economic efficiency. *Journal of Economic Issues, 23,* 735–759.

Cronin, F. J., Parker, E. B., Colleran, E. K., & Gold, M. A. (1993). Telecommunications infrastructure investment and economic development. *Telecommunications Policy, 17,* 415–430.

Duch, R. (1991). *Privatizing the economy.* Ann Arbor, MI: University of Michigan Press.

Fuhr, J. P., Jr. (1990). The subsidization of rural telephone service in the U.S. *Telecommunications Policy, 14,* 183–188.

Galbraith, J. K. (1992). *The culture of contentment.* Boston: Houghton-Mifflin.

International Telecommunications Union. (1986). *Information telecommunications and development.* Geneva: Author.

International Telecommunications Union. (1994). *World telecommunication development report.* Geneva: Author.

Kamien, M. I., & Schwartz, N. L. (1982). *Market structure and innovation.* Cambridge, MA: Cambridge University Press.

Kenis, P. (1992). *The social construction of an industry: A world of chemical fibres.* Boulder, CO: Westview Press.

Krugman, P. R. (1990). *Rethinking international trade.* Cambridge, MA: MIT Press.

Lane, J.-E. (1985). Introduction: Public policy or markets? The demarcation problem. In J.-E. Lane (Ed.), *State and market: The politics of the public and the private* (Modern Politics Series Vol. 9, pp. 3–52). London: Sage.

Mansell, R. (1990). Rethinking the telecommunication infrastructure: The new "black box." *Research Policy, 19,* 501–515.

Mansell, R. (1993). *The new telecommunications. A political economy of network evolution.* London: Sage.

McCormick, P. K. (1993). Telecommunications privatization issues: The Jamaican experience. *Telecommunications Policy, 17,* 145–157.

Mosco, V. (1990). The mythodology of telecommunications deregulation. *Journal of Communication, 40,* 36–49.

Nelson, R. R. (1981). Assessing private enterprise: An exegesis of tangled doctrine. *Bell Journal of Economics, 12,* 93–111.

Nelson, R. R., & Winter, S. G. (1982). *An evolutionary theory of economic change.* Cambridge, MA: Belknap Press.

Noam, E. M. (1994a). Beyond liberalization. From the network of networks to the system of systems. *Telecommunications Policy, 18,* 286–294.

Noam, E. M. (1994b). Beyond liberalization II: The impending doom of common carriage. *Telecommunications Policy, 18,* 435–452.

Nora, S., & Minc, A. (1980). *The computerization of society. A report to the President of France.* Cambridge, MA: MIT Press.

North, D. C. (1990). *Institutions, institutional change, and economic performance.* Cambridge, MA: Cambridge University Press.

North, D. C. (1994). Economic performance through time. *American Economic Review, 84,* 359–368.

Olsen, O. J. (1989). Deregulation and reorganization—the case of Danish telecommunications. *Annals of Public and Cooperative Economy, 56,* 323–342.

Parker, E. B., Hudson, H. E., Dillman, D. A., & Roscoe, A. (1989). *Rural America in the information age: Telecommunications policy and rural development.* Lanham, MD: University Press of America.

Sawhney, H. (1993). Circumventing the centre. The realities of creating a telecommunications infrastructure in the USA. *Telecommunications Policy, 17,* 504–516.

Schumpeter, J. A. (1950). *Capitalism, socialism, and democracy* (3rd ed.). New York: Harper.

Shepherd, W. G. (1984). Contestability vs. competition. *American Economic Review, 74,* 572–587.

Shleifer, A. (1985). A theory of yardstick competition. *Rand Journal of Economics, 16,* 319–327.

Spulber, D. F. (1989). *Regulation and markets.* Cambridge, MA: MIT Press.

Tinbergen, J. (1992). The measurement of welfare. *Journal of Econometrics, 54,* 1–12.

Traxler, F., & Unger, B. (1994). Governance, economic restructuring, and international competitiveness. *Journal of Economic Issues, 28,* 1–23.

Trebing, H. M. (1994). The network as infrastructure—The reestablishment of market power. *Journal of Economic Issues, 28,* 379–389.

Van de Walle, N. (1989). Privatization in developing countries: A review of the issues. *World Development, 17,* 601–615.

Vickers, J., & Yarrow, G. (1988). *Privatization: An economic analysis.* Cambridge, MA: MIT Press.

World Bank. (1994). *World development report 1994. Infrastructure for development.* Washington, DC: Author.

13

Regulatory Reform:
An Institutional Perspective

Nikhil Sinha
University of Texas at Austin

The restructuring of telecommunications sectors now under way across the developing world has created an urgent need for the development of effective and credible regulatory structures and processes. The success of reform in a number of countries will hinge significantly on the effectiveness of regulatory oversight of the restructured markets. A well-structured regulatory system could significantly enhance sector development by ensuring that reform takes place in accord with public goals and that governmental decision making does not unnecessarily constrain the innovations and efficiencies of private enterprise. On the other hand, a poorly structured regulatory system can vitiate the goals of reform by eroding the confidence of private investors or by permitting private companies to engage in monopolistic or predatory practices.

Privatization and liberalization, embraced by governments as mechanisms to overcome chronic deficiencies in investment in telecommunications and to respond to the challenge of rapidly changing technology, may generate problems of their own. The main dilemma being faced by governments is how to respond to the growing technological and economic pressures to liberalize, while safeguarding important social and political goals and objectives (Nulty & Schneidewind, 1989). This problem of developing mechanisms to reconcile operational efficiency, rational management, investment generation, competitive discipline with national security and social equity is a relatively new one for governments. "Under the traditional organization of telecommunications systems, these many contradictions were reconciled directly with the single political body that both operated the telecommunications system and made social, national and economic policy: the government" (p. 30). Privatization sunders this unified regime and in turn requires the development of alternative credible and effective regulatory structures to reconcile these often divergent economic, technical, social, and political objectives.

Countries will inevitable devise their own unique regulatory system based on their political and economic environment, the legal system, and the objectives that are sought to be achieved through reform and regulation. However, there are common elements, particularly in the key areas of structural, jurisdictional, and procedural issues, that need to be resolved in each situation. In this chapter, I examine some of the core issues in the development of regulatory structures and procedures in developing countries. Instead of focusing on the techniques of regulation, or what Galal, Tandon, and Vogelsang (1992) called "regulatory engineering" (p. 32), this essay adopts an institutional perspective, arguing that the success or failure to develop credible and effective regulatory mechanisms in developing countries, and by extension the success of the reform process as a whole, hinges on the institutional context of regulatory reform.

Although economic analyses of regulatory instruments such as the efficiency of rate of return versus price-cap tariff mechanisms are useful, they cannot, in isolation, provide much insight into why countries adopt different instruments or why countries with similar instruments witness such varied payoffs. North (1993) argued that institutions determine the payoffs. Institutions are the structures that constrain economic activity and, together with other constraints (capital, technology, etc.), determine the choices that organizations make that shape the performance of a sector. "Institutions consist of formal rules, informal constraints (norms of behavior, conventions, and self-imposed codes of conduct), and the enforcement characteristics of both" (p. 36).

Economic change, or at least changes in the form of economic organization and governance, involves changes in institutional rules and their enforcement characteristics. Changes in formal rules may occur as a result of legislative changes such as the passage of a new statute, of judicial changes stemming from court decisions, of regulator rule changes enacted by regulatory agencies, and of constitutional rule changes that alter the rules by which other rules are made (North, 1993). Informal changes may result from a transformation of the dominant ideas and ideologies guiding economic processes in a society. These in turn may be engendered through the rise of new economic and political interests or as a response to the failure of existing ideologies to motivate economic progress. Both sets of changes will play a large part in determining the nature, effectiveness, and credibility of regulatory processes.

"If institutions are the rules of the game, organizations are the players" (North, 1993, p. 36). The constraints imposed by the institutional framework, together with other constraints (e.g., technology, the availability of finance), define the opportunity set and therefore the kind of organizations that will come into existence (North, 1993). Regulatory structures play a dual role in the telecommunications arena. They are at once organizations subject to the institutional rules of the game and at the same time they can be effective instruments of institutional change itself.

Institutional regimes governing economic activity can take three generic forms: *markets, hierarchies* (or bureaucracies), and *hybrids* that combine the characteristics of markets and hierarchies. The reform of telecommunication sectors now under way is a move from *hierarchies* to *markets*. But the first step in this process, and indeed some argue the only viable and realistic structure for the sector, is a *hybrid* form characterized by market forces in the form of firms, investors and competitors, and hierarchies in the form of regulatory and other safeguarding institutions.

Two sets of problems make a *hybrid* governance regime inevitable and desirable for telecommunication sectors in the Third World. The first stems from the market power of the dominant provider. Telecommunication services in the developing world have been traditionally provided by government monopolies. Privatization of these companies without liberalization or effective regulatory oversight will simply lead to greater opportunity for the private firm to exploit its monopoly power. "In this case the purposes of reforming telecommunication entities—efficiency, innovation and flexibility—may be aborted" (Nulty & Schneidewind, 1989, p. 29). Alternatively, if privatization is introduced with unregulated competition, the dominant provider may destroy the competition through predatory pricing or control over key resources, resulting once again in a monopoly situation. The introduction of unregulated competition could also lead to "cream-skimming" with the competing firms investing only in profitable areas and services. "This situation will lead to uneven development, will undermine economies of scale available in an integrated system, and will prevent growth of a nationwide structure" (p. 29).

Second, there are problems related to national security and the protection of public goals like social equity. Telecommunications are vital for national security, both military and civilian, and are essential infrastructure for economic and social development. In the absence of strong regulatory oversight both of these objectives may be jeopardized, leading Nulty and Schneidewind (1989) to conclude that a "telecommunications system that relies solely on private, competitive firms will tend to be both inadequate overall and too unevenly distributed to meet the needs of society as a whole" (p. 29).

The institutional matrix of regulatory systems may be, therefore, the key element in successfully initiating the reform process and for ensuring its continuing success. Although there is no correct blueprint for what the make-up of a successful institutional matrix may be, there are certain themes and characteristics that will determine the effectiveness and credibility of the regulatory system. First, the reform of the telecommunications sectors as a whole and the development of sector policies, in particular, is a political process, and the political system and within it the nature of the state, the relationship between political institutions, and the dominant political ideology are key factors in the reform process. Second, formal institutional rules, particularly those setting the legal framework, that is, the nature of the contractual system and the property rights regime, are critical to providing stability to the reform process. Finally, the nature and effectiveness of safeguarding institutions like the judiciary and, most importantly, the regulatory agencies and instruments developed to oversee and enforce the reform process, are crucial to determining the effectiveness of the management of reform. The remainder of this chapter is devoted to a detailed analysis of these three broad institutional determinants of regulatory reform.

THE POLITICAL SYSTEM

The nature of the political system within which telecommunication reforms are being introduced may be the single most important determinant of the effectiveness of the regulatory process. The polity and the economy are inextricable linked in any understanding of the performance of an economy:

> A set of institutional constraints and consequent organizations defines the exchange relationships between the two and therefore determines the way a political/economic system works. Not only do polities specify and enforce property rights that shape the basic incentive structure of an economy: in the modern world the share of gross national product going through government and the ubiquitous and ever-changing regulations imposed by it are the keys to economic performance. (North, 1993, p. 45)

Noll (1986) argued that the "the performance of the communication sector—indeed, of any part of the economy—depends on the political and economic environment in which it operates" (p. 43). The political and economic environment is taken to mean the kinds of organizations that make decisions affecting process, technology, service, and other important variables; the relations among these organizations; the rules governing these relationships; and the nature of the process for changing these rules. The most important of these organizations is the state.

The State

Katz (1988) maintained that the growth of "telecommunications technology [in developing countries] is strongly determined by government policy" (p. 58). Similarly, Jonscher (1983) argued that improvements in the information sector "may not be realizable without active intervention by governments and other policymakers" (p. 27). In short, states may be the active drivers of growth in the telecommunications, apart from and/or in addition to economic variables like market structure, tariff structures, and conditions of entry.

In its most obvious manifestation, the restructuring now under way is being initiated and directed by states, and the privatization and liberalization process is often referred to as "state reform programs" (Petrazzini, 1993). This emphasis on the role of the state should not undermine the clear and strong influence of external forces like transnational corporations, multilateral development agencies, and the collapse of Communism in erstwhile Soviet-bloc countries, in triggering the current wave of reforms. However, there is little doubt that Third World governments have directed, if not initiated, the changes that are now under way in the telecommunications sector. As Skocpol (1985) argued, states "necessarily stand at the intersections between domestic sociopolitical orders and the transnational relations within which they must maneuver for survival and advantage" (p. 8). As such, they are the filter through which international forces impact on the development of domestic telecommunications. Therefore, the extent of state commitment to the reform of a sector may be critical in determining the extent its success.

The success of divestiture depends heavily on the actions of the state. The nature of divestiture, the way that it is structured, the regulatory regime introduced, all these factors will influence the eventual outcome, and all these factors are determined by the government. "Thus the ability of the government to plan and implement a divestiture program will play an important role in determining the success of the program" (Galal et al., 1992, p. 34).

The role of the state in economic development is neither new nor limited to developing countries. Since the depression of the 1930s there has been a steady growth of the state's role in countries across the globe. In the industrialized democracies, the emergence of the managerial state to combat the crises and resolve the contradictions of capitalism has been one of the significant developments of the present century (Dahrendorf, 1959, 1977). In addition, growing state regulative and welfare functions since World War II have contributed to an enormous expansion of the state apparatus and corresponding state activities in the industrial and related sectors. This expansion is even clearer in the structuring of eco-

292   SINHA

nomic and social systems in Third World countries in general (Skocpol,
1985) and in the process of industrialization in particular (Evans, 1985;
Stepan, 1978).

For instance, Evans (1985) argued that the state may act as a relatively
autonomous actor that helps shape the development of local productive
forces. Indeed, the importance of the state in the development of the
industrial sector in Asian countries like India, Pakistan, Malaysia, and Sri
Lanka (Pattnayak, 1990) and in Latin America (Cardoso & Faletto, 1978)
cannot be overestimated. Moran (1974) studied the evolutions of the
state's role in the extractive industries in Peru; he argued that even modest
attempts by the government to shift resources to the sector resulted in
rapid growth and the formation of a pool of trained workers. Similarly,
in South Korea, an aggressive state helped rapidly increase industrial
production (Frieden, 1987; Sen, 1981).

In fact, if we compare the performance of the industrial sector in Brazil
and South Korea to that of Argentina and Chile, the importance of state
commitment becomes all the more apparent. During the 1970s, highly
active governments committed to stepping up industrial production in
Brazil and South Korea fared better than those in Argentina and Chile
where the intent was more to allow the unhindered operation of market
forces (Diamond, 1983). Diamond argued that the productivity of the
industrial sector in developing countries depends considerably upon government behavior. This behavior is manifest not just through the amount
of resources it allocates to different sectors, but, even more importantly,
through sector policies.

It is obvious that the ability of the state to direct economic activity is
greatest "if the connection between elected officials and the industry's
decision makers is direct and does not need to be mediated through
another organization" (Noll, 1986, p. 43). The best example of these are,
of course, state-owned or public enterprises. Rueschemeyer and Evans
(1985) argued that through public enterprises the state becomes an active
participant in production and market exchange and partially supersedes
the way in which markets combine information, incentives, and economic
power. Through such enterprises the state itself becomes an agent of
capital investment. This is justified by the need to overcome impediments
to private investment created by high sunk costs, long gestation periods,
and large externalities.

As Jones and Mason (1980) pointed out, the fact that state-owned
enterprises tend to be located in sectors with high capital requirements and
longer payback periods suggests that the disciplines and incentives of
competition cannot be counted upon to produce optimal behavior on the
part of private capital. The effectiveness of state participation in the market
is further enhanced if such sectors have important forward or backward

linkages. Typically, these sectors are usually part of the economic infrastructure like power, health, education, transportation, and telecommunications.

Finally, state enterprises also permit states to most effectively control the growth and development of sectors, by enabling them to control the allocation of resources. In addition, it also enables them to pursue distributional outcomes by directly influencing the allocation priorities within the sector. As Rueschemeyer and Evans (1985) pointed out, states engaged in redistribution efforts cannot rely on the mechanism of the market but must seek the same results through administrative means. One way to overcome the problem is through the establishment of public-sector enterprises.

The role of states in the industrialization process can be multilayered, and states can influence economic activity without necessarily being involved directly in the production of goods and services. In its most rudimentary form, states set the legal and institutional environment for the operation of the rules of property and commerce. In its most usual form, states regulate markets since the effective operation of markets in developing countries often requires the presence of strong interventionist states. Another example is corporatism as practiced in some European countries and in some Japanese and South Korean industries. Here, government ministries possess the means to control or at least strongly influence private decisions (such as in controlling credit or allocation of research) and implement their plans through direct contact with officials of industry (Noll, 1986).

The role of the state in economic activity does not necessarily diminish with privatization, it changes, because what matters is not its right to intervene but how it intervenes. Stepan (1978) defines the state as the "the continuous administrative, legal, bureaucratic and coercive systems that attempt not only to structure relationships between civil society and public authority but also to structure many critical relations within civil society as well" (p. xii). Mosco (1988) argued that the relationship between the state and telecommunications policy must be understood in the context of power relations. Mosco identified four ways that capitalist societies legitimately settle social claims. These include representation (representatives exercising political power), the market (defined as monetary or exchange power), social control (power built into the daily mechanisms of society), and expertise (power derived from the possession of skills and information). All of these modes of settling social claims involve state intervention in one form or another, for instance, regulation combines representation with the private market and hence constitutes a political tool for settling social claims.

The effectiveness of this tool will depend on a number of characteristics of the nature of the state and political process within which regulation is introduced. Many institutionalists, like Theodore Veblen, have taken it as virtually axiomatic that the state heavily reflects the dominant economic and class interests in society. Others have viewed the state as at least partially independent of economic interests and thereby capable of independent and quasi-independent programs action (Skocpol, 1985). In any case, evidence from a number of developing countries engaged in the process of telecommunications reform suggest that the effectiveness of regulatory mechanisms depends considerably on the configuration of power in the state apparatus (Galal, 1993; Petrazzini, 1993).

Petrazzini (1993) argued that the differential success of the restructuring of telecommunications in Argentina, Mexico, Thailand, and Malaysia can be largely understood through an examination of the "relative strength and autonomy of the State vis-à-vis domestic interest group coalitions; and the degree of power concentration within the State apparatus" (p. 6). His analysis of the four countries shows that Mexico and Malaysia, with higher degrees of power concentration and with political systems that are relatively more effective in insulating themselves from interest groups, were more successful in pushing through privatization programs than Argentina and Thailand.

However, this analysis falls short of explaining the performance of the sector and the effectiveness of regulatory oversight after the reforms. In such situations the relationship between successful regulation and the concentration of power and the extent of insulation from interest groups becomes more complex. Although a strong executive body may be effective in initiating reform, as Petrazzini accurately argues, it may vitiate the objectives of reform by diluting the power and effectiveness of the regulatory process. In the case of Chile, Galal traces how military control allowed the government to introduce privatization legislation in the early 1980s. But after democracy was restored, the fact that the presidency and the Parliament were controlled by two different parties restricted the ability of both to arbitrarily change the legal regimes governing telecommunications.

Similarly, in Argentina, the Menem government was able to privatize ENTEL only after the Peronist party gained control over both Parliament and the presidency, the first time since 1976. This unusual extent of political control permitted Menem to concentrate the power to direct economic activity in the presidency and push through the passage of legislation necessary for privatization (the Economic Emergency Law and the Public Sector Reform Law) with minor opposition from the legislature.

However, the regulatory agency established to oversee the privatized duopoly that replaced the government monopoly was not sufficiently insulated from the presidency, from the other ministries, from Congress, or from the STET/France and Cable et Radio and Telefonica de Espana, the two multinational consortia that bought the northern and southern interests in ENTEL. The failure of a strong and independent regulatory structure is one of the principal reasons why the privatization process in Argentina failed to achieve the main objectives of reform. Thus while the concentration of power within the executive branch may be an important element in introducing reforms, the stability of reforms may well depend on an effective balance of power between the executive and the legislature.

Political Institutions

Attention must also be focused, therefore, on the internal composition of the political system, particularly the nature of the relationship between political institutions. For example, Spiller (1992) attempted to explain the nature of Jamaica's regulatory arrangement by focusing on the nature of the country's political system. He argued that a license-based regulatory system was largely inevitable because a legislation-based regulatory system (e.g., the U.S. system) would set up an incomplete contract between the government and the private telecommunications company (Telco). The establishment of independent regulatory agencies is difficult in Jamaica because the two-party parliamentary system gives the party in power full control over both the executive and the legislature. Consequently, regulatory laws will not usually serve as ex-ante constraints on regulatory agencies. For instance, a court ruling that a particular decision violates the statute can be overturned by appropriate legislation during the same administration. Licenses function as contracts between the government and the Telco, and while the government can amend laws it cannot unilaterally alter the terms of the contract. Because of the nature of the country's courts, independent, with long-lasting tenure and with a final appeal to the Privy Council in London, they can be called upon to determine alleged violations of the contract by either party.

The effectiveness of the regulatory regime also depends in part on the ability of the reform constituencies to sufficiently insulate it from interest groups. If the power of competing groups is in flux, it may become necessary to provide greater independence to regulatory agencies and have safeguarding institutions that maintain the stability of the regulatory system. As Noll (1986) suggested,

> regional representation, single non-transferable votes, party list systems, all have very different implications to how sensitive the composition of the legislative body is to minor changes in constituencies' interests, and also to the role that parties, and party machines, may play in the polity. (p. 14)

The inability of the Choonhavan government in Thailand to overcome the opposition of labor unions, the military, and the management to its privatization program stemmed in part from the "atomized and diverging interests of the legislature, and fluctuating support of parties within the governing coalition" (Petrazzini, 1993, p. 13), which left the government with little political leverage.

Privatization ground to a halt with the overthrow of the Choonhavan administration by a military coup in 1991. Government stability and the continuation of core policies despite changes in the government, goes a long way in reducing the need for safeguarding institutions. The stability of political parties may be an important element in determining the extent to which they can exercise control over their policies over time. Comparing the populist PRI in Mexico, Petrazzini (1993) argued that the long tradition of party discipline and hierarchical power relations that characterized the PRI allowed President Salinas to push through reforms with guaranteed support from his party. President Menem struggled to elicit his party's support because the Peronists have long been a loose conglomerate of groups with conflicting interests and ideologies.

The ability of countries to successfully initiate and regulate reform is clearly dependent on the nature of its political institutional structures, particularly the effectiveness of the state, the balance of power within political institutions, the legislative process, and the nature of interest group coalitions. Political institutions set the rule by which conflicting interests in telecommunications restructuring are mediated. Since these rules vary, the outcome of these struggles will also vary. This is one of the major reasons why privatization, liberalization, and regulation have been successfully implemented in some countries and not in others.

LEGAL FRAMEWORK

The role of political institutions in the development of regulatory systems for restructured telecommunication sectors will be constrained, in part, by the legal and constitutional framework of each country. The ability to allocate authority and options for structuring the regulatory system varies according to the constitutionally determined horizontal and vertical distribution of jurisdiction and authority in a country. Horizontally, the constitutional framework will determine the formal relationships between political institutions like the executive, legislature, and judiciary. It will also determine whether the country's legal system is one based on common law, civil law, or dictatorial fiat. Vertically, the constitutional framework will delineate the power and jurisdictions of central, regional, state, and local authorities and the mechanisms by which disputes between them can be resolved.

Enforcing Contracts

In all cases, effective regulation of privatized and/or liberalized telecommunication sectors in developing countries requires the establishment of a comprehensive framework of laws, rules, and regulations that clearly identify the contractual obligations and property rights of both governments and private actors. This framework should be both transparent and enforceable. Spiller and Levy (1991) argued that a main reason for the development of telecommunications in the public sector has been the difficulty in resolving the contracting problems between governments and firms. Successful reform must involve the resolution of these contractual problems (Galal, 1993).

Contracting problems between governments and firms arise when governments are unable to enforce investment and performance standards and firms are uncertain about the security of their investments either because of the fear of administrative expropriation or the lack of autonomy in setting prices. In the absence of a strong regulatory system, firms tend to invest less than the potential amount and/or invest only in high-profit areas. On the other hand, governments may try to administratively set prices below replacement costs or force the firms to make unprofitable investments. For instance, investments in the Jamaica telephone system were suspended for about five years from 1962 during a license renegotiation dispute between the government and the Jamaican Telephone Co. (JTC), a period characterized by the absence of an effective regulatory system (Spiller & Sampson, 1992).

A major purpose of instituting a strong regulatory structure is to overcome what Spiller and Sampson (1992) called "the license end-game problem," which is an outgrowth of the incompleteness of the contract between firms and governments. However, even regulatory structures may not be able to ensure the flow of investments by themselves. In Jamaica, investments once again stalled in the early 1970s during a pricing dispute between JTC and the Public Utility Commission that had been established in the late 1960s to regulate the country's utilities. Some system of resolving disputes between firms and regulatory agencies will also have to be instituted if countries are to successfully overcome the contracting problems between governments and firms, an issue that is taken up later in the discussion of safeguarding institutions.

In contrast to Jamaica, Chile has always had the capacity to write credible contracts and enforce them because of the relative stability of the country's political institutions and independence of its judiciary (Galal, 1993). But even so, the contractual problems between the government and the private telecommunications provider could not be easily resolved. Tracing the early history of the country's telecommunications system Galal (1993) also argued that even though the 1925 law granted CTC (Compania de Telefono de Chile) strong concessions on services, limited the government's ability to expropriate its assets, and relied on the Supreme Court for resolving disputes the sector failed to expand after CTC fulfilled its initial investment commitment because the law did not guarantee the company a specific rate of return and left room for the government to intervene in the company's operation "under vaguely defined circumstances."

But the law also failed to obligate the firm sufficiently. Even though the concession was to last for 50 years, the company's obligations for investment in infrastructure development, equipment, plant, and extension of services were laid down for only a fraction of that time, and in no case extended beyond 15 years. Once the company fulfilled these commitments, it slowed investments, particularly in extending service to rural areas and integrating the various regions of the country. In effect, the law provided neither incentives for CTC nor stipulated obligations for the company to increase investment beyond a fraction of the concession period. Subagreements in 1958 and 1967 resolved some but not all of these problems. Thus, the sector grew only modestly from 1925 to 1971 since the contractual problem between the government and CTC was only partially resolved by the law.

Property Rights

The state's capacity to institute and redefine property rights is an important tool in the process of privatization not only because it enables policymakers to transfer ownership from the state to private firms but also because it enables them to influence the degree to which economic production and exchange is regulated by markets, cartels, joint ventures, administrative agencies, courts, or other organizational forms (Campbell, 1993). Institutional studies of the development of capitalism have recognized that the state has regulated economic activity in the advanced industrialized countries through the manipulation of property rights (Lindberg & Campbell, 1991; North, 1981, 1990). Although some attention has been paid to the potential for East European countries to use property rights to regulate economic activity (Campbell, 1993), similar attention has not yet been directed toward Third World countries. Developing countries have relied heavily on the provision and allocation of credit, subsidies, and budgetary and plan allocations to regulate the growth of their telecommunication sectors. Privatization, liberalization, and the influx of capital and technology from the developed world are reducing the options available to them to govern telecommunications activity through such direct allocations.

Privatization necessitates the introduction of private property rights into sectors that have traditionally been governed under legal regimes for public assets. The performance of these newly privatized sectors will depend largely on the nature of the property regimes introduced to govern and enforce them. According to Furubotn and Pejovich (1974) three elements needed to be defined: "(a) the right to use the asset . . .

(b) the right to appropriate returns from the asset . . . and (c) the right to change the asset's form and/or substance" (p. 4).

As in much of the privatization-liberalization process, regulatory regimes must reflect a compromise between the "ideal" property rights of privatized firms and the restraints that may be imposed on them. The right to "use the asset" may bring into conflict the twin aims of privatization and liberalization. The stronger the right, the more private investors, both domestic and international, will be likely to invest with confidence. But the introduction and promotion of competition may require that the rights of the privatized Telco be curtailed. Real competition may depend on the rights of competitors to use the PSTN under conditions restrictive to the dominant provider. In introducing competition in value-added services, Mexico placed several regulatory restrictions on the newly privatized Telmex including forcing it to share network information and to develop the network in a manner designed to provide easy interconnection for competitors (Petrazzini, 1993).

The second right, to appropriate returns from the asset, may also be subject to restriction. Regulators may have to enforce investment targets, particularly for growth of the network in rural areas, to fulfill important social objectives. As the evidence of the importance of telecommunications in the development process as a whole and in rural areas in particular, begins to mount, regulatory regimes must ensure that the privatized Telcos remain committed to expanding the reach of telecommunication services. The most obvious restriction will be to the right to change the form or substance of the asset, with specific restrictions on resale of all or part of the company or otherwise disposal of assets. Experiences from privatization in Eastern Europe provide unfortunate examples of what can happen when the capital assets of privatized companies are not protected (Etzioni, 1993).

As important as the form of the property regimes is the issue of the security of the rights. Williamson (1991) identified two types of problems: expropriation by the government and expropriation by other economic actors. The stability of private investment in the sector will be greatly strengthened if governments make "credible commitments" (Williamson, 1983, p. 520) and private investors have "security of expectations" (p. 520) with regard to property rights. If property rights are subject to reassignment or suspension or if adequate compensation is not paid for expropriation, then investments will not flow into acquiring privatized companies (Williamson, 1991). Property rights may also be attenuated through devaluation of the privatized assets by governments. By changing the tariff setting mechanism, Argentina seriously reduced the profitability of the privatized companies leading to reduced interest by investors in the divestiture.

Antitrust Laws

Finally, the role of the regulator is also related to the existence and effectiveness of antitrust laws. In Venezuela, where there are no applicable antitrust laws, substantial attention was paid to the creation of an effective regulatory entity and system at the outset of the privatization process. In contrast, Chile underpowered its regulatory agency at the time of privatization, placing substantial reliance on the then existing antitrust laws to ensure opportunities for competitive entry. However, in recognition of the fact that antitrust laws do not substitute for a forward-looking policy (and because of the slew of costly law suits that have resulted), Chile is now reexamining the issue of regulatory functions.

SAFEGUARDING INSTITUTIONS

Even if the legal framework provides formal rules, regulations, and rights that can be used to regulate telecommunications, the mechanisms to enforce these laws may not be well developed. Galal (1993) argued that the problem of expropriation by the government or default by firms can be largely overcome if the state has the capacity to write and enforce credible contracts. This is at least partly achieved through the stability of political institutions and the existence of a rigorous legal system. But equally important may be the development of safeguarding institutions.

According to Spiller (1993), safeguarding institutions play two important roles in the privatization process. They help lower the real and perceived probability of the state expropriating assets of private companies; that is, of violating contractual obligations. Second, safeguarding institutions serve as mechanisms to limit the possibility of deviation from the intended objective of privatization and liberalization by both governments and firms.

The Judiciary

A well-functioning judicial system that respects property rights and enforces contracts can serve as an important safeguarding institution. An added bonus would be a judiciary that has a tradition of reviewing administrative agencies. But such judicial systems are not very prevalent in the developing world. Most of them do not have administrative tribunals, and where these do exist they are more geared to reviewing procurement decisions than reviewing regulatory procedures and instruments. The courts too may not have a tradition of restraining administrative decisions. Where such traditions do exist, the backlog of cases and the slow pace of the judicial review process may undermine the effectiveness of the system. However, even though courts do not have a tradition of restraining government decisions, judicial enforcement of contracts may provide a safeguarding environment.

Jamaica and India are examples of two countries with relatively effective judicial systems. Both derive their legal roots in British common law and both have parliamentary systems of government that derive from the British system. In India the relationship between the judiciary and Parliament is constitutionally governed in a form that more closely resembles the separation of powers in the U.S. Constitution than it does the sovereignty of the British Parliament. It also provides for the judicial review of statutes, executive orders, and administrative procedures. The Indian judicial system, therefore, can exercise judicial review of government decisions and also enforce contracts between firms as well as between firms and the government. However, the effectiveness of the Indian judicial system is severely hampered by long delays in the disposal of cases. For instance, the inability of the courts to rapidly decide cases has delayed the restructuring of the country's cellular markets.

In contrast, the Jamaican judicial system more closely resembles the British system of parliamentary sovereignty, and judicial review or restraint of government decisions is virtually nonexistent. Judicial reversals of government decisions can be overturned through the passage of legislation. But the courts' respect of contract rights ensures that agreements between the government and private firms will be protected and enforced. Therefore, the regulation of private utilities in Jamaica is done through licenses. The license, which cannot be changed unilaterally by the government or by the firm, specifies the way regulation is to be undertaken (Spiller, 1993). These licenses are enforced as contracts by the courts, thus providing a safeguard from both unilateral government actions and noncompliance by firms.

Hence, countries with well-developed and functioning judicial systems like India, Jamaica, Chile, and Costa Rica can rely on the judiciary to oversee the process of liberalization and privatization, to balance the interests of private firms, and to monitor the objectives of government. However, in countries where the judicial system is not reliable or effectively independent of the government (like Brazil, Argentina, or a number of countries in sub-Saharan Africa), other safeguarding mechanisms may have to be sought.

Regulatory Agencies

Independent regulatory agencies, not subject to manipulation by the government or capture by private industry, are crucial to the success of reforms. The effectiveness, credibility, and independence of regulatory agencies depends on a variety of factors, including: (a) the organizational and hierarchical location of the agency, (b) the financial and technical resources of the agency, and (c) the nature of regulatory procedures and regulatory instruments.

Organizational Structure and Hierarchical Location of the Agency. The nature of the reform being initiated is, of course, an important determinant of the structure of the regulatory agency. In countries where the PTT is being reorganized as a State Owned Enterprise (SOE), it is important that the regulatory function be removed from the newly formed company. The most frequent placement for it is the Ministry of Communications. If the regulatory function is to remain within the executive branch, one alternative may be to separate telecommunications from the postal services and other public services (e.g., transportation) with which it has usually been combined and to create a new ministry or department committed solely to telecommunications. Though this provides a high-level and centralized locus for the issuance of telecom-munication policies, it does not sufficiently insulate the regulatory agency from political interference.

The organizational structure and location of the regulatory agency may be particularly important if the reforms involve partial or complete dives-titure. Although both independent regulatory commissions and executive departments have advantages and disadvantages, Spiller (1993) pointed out that if both are subject to stringent judicial review, it may not make much of a difference in terms of functioning. However, in countries where the administrative system requires different standards of review, the choice of the nature of the regulatory body may become critical.

Spiller (1993) maintained that the little interest that the Argentinean privatization process raised among international investors can be traced, at least in part, to the nature of the regulatory regime that was instituted to regulate the private duopoly. The licenses issued by the administration spelled out the basis of the regulatory system. However, at the time of the bidding, there was not a clear idea of how, and by whom, the terms of the licenses would be enforced. The National Telecommunications Commission (CNT) was formed by presidential decree after the bids were invited and only days before they were to be adjudicated. "The fact that the body was going to implement these regulations did not exist nor the rules and regulations under which it would operate were spelled out couldn't but help increase the uncertainty of the entire reform process" (Spiller, 1993, p. 22). CNT was initially created as part of the Ministry of Public Works, a quite obscure and not very influential ministry. As a consequence, the potential for manipulation of the commission by both the executive and legislature was quite high. Spiller (1993) also stated that the decisions by the CNT were not subject to judicial review, but rather could be appealed only to the executive. "Thus the organization of the regulatory commission did not provide further reassurances of regulatory objectivity" (p. 22).

Galal et al. (1992) argued that part of the explanation of why divestiture of the telecommunication carrier increased domestic consumer welfare in Chile but not in Mexico can be found in the nature of their respective regulatory institutions. Chile introduced comprehensive regulatory legislation and created a new department to oversee the implementation of telecommunication regulation years before privatization (even so there were weaknesses in the regulatory agency that will be discussed later). On the other hand, Mexico merely changed the way in which the secretariat of communications set tariffs. The ability of the secretariat to effectively oversee and enforce the terms of the contract between Telmex and the government is still uncertain.

The hierarchical position of the regulatory agency can also be crucial. Galal noted that during the period of public ownership of Chile's two main telecommunication carriers CTC and ENTEL (1974 to 1982 under the military regime) the two carriers were legally regulated by an independent agency, SUBTEL. But since SUBTEL was headed by a colonel and CTC and ENTEL by generals, the two companies were left virtually unregulated. Even after widespread changes introduced in the Telecommunications Law of 1982, key regulatory issues were left unresolved. Most importantly, SUBTEL was underpowered and its role in the regulatory process was left undefined. Consequently, in key areas like tariff setting and granting of concessions, SUBTEL was largely ineffective. The government remedied this problem in 1985 by empowering SUBTEL with the authority to supervise legal, procedural, and technical issues. However, SUBTEL remained a relatively weak body since it was not provided the financial and administrative independence other Chilean regulatory agencies, like those in the electricity and banking centers, enjoy. It was located within the Ministry of Transportation and Telecommunications and its staff operated according to the general civil service code. Once again, the government is moving to strengthen the agency.

Chile provides the example of a country with an explicit legal framework for the sector and a strong judiciary to enforce it that still suffered because of a relatively weak regulatory agency. A number of costly lawsuits have had to be resolved by the courts, including the Supreme Court, because of what Galal (1993) called "excessive reliance on market forces to define market boundaries, the drive for profit maximization following privatization" and the strong incentive for cream skimming (p. 25). Even in a country like Chile, with strong political institutions, a well-developed civil legal system, and an independent judiciary, a stronger regulatory agency has proved to be necessary.

Financial and Technical Resources of the Agency. Who runs the regulatory organization, how are they appointed and what is the security of their tenure may all play a part in determining the extent of independence of the agency. Understaffed organizations are obviously going to be dependent on outsiders (political or industry) for directions. Chile assigned SUBTEL the primary responsibility of regulating the industry and empowered its board with ministerial level members and staffed it with qualified personnel. Its Mexican counterpart, the secretariat of communication "still lacks this kind of expertise and therefore may be at a disadvantage when it comes to negotiations with Telemex" (Galal et al., 1992, p. 30). Lack of expertise may open up the regulatory agency to manipulation by the regulated firms, particularly in the case of a private monopoly. Along with technical and personnel resources, dedicated financial resources may also strengthen the agency's independence.

However, the possibility of introducing truly independent regulatory agencies in developing countries is remote indeed. The institutional mechanisms for implementing policy will also reflect the political forces giving rise to policy. (Fiorina, 1985; McCubbins, 1985). Financial independence does not guarantee insulation from executive manipulation. The Argentine CNT, for instance, gets .5% of all telecommunications revenues, but is still tightly controlled and directed by the government. While financial independence may insulate the regulatory body from the legislature in theory, in practice it makes the political affiliation an important element of the regulatory mix. In the Argentine case that affiliation allowed for close control of regulatory activities by both the executive and the legislature, undermining the independent functioning of the agency.

Regulatory Procedures and Instruments. The procedures with which regulatory agencies engage in the process of oversight and decision making are important aspects of the make-up of the regulatory institution. A major concern in establishing regulatory stability is being able to limit the discretionary power of the regulators. One approach is to make the process of amending regulations as transparent as possible. The U.S. system, which is based on public hearings, ascertainment, sunshine rules, and so on, limits the discretionary power of the FCC by forcing it to take into account diverse views and by making its rulemaking process subject to public scrunity.

Most developing countries do not have the well-developed political culture of interest groups politics essential to make such a system work. An alternative is to limit the discretionary power of the regulatory agency, at least in the crucial area of tariff policy, by imposing some sort of rate of return or price cap system. Chile and Mexico have both adopted price cap systems that allow their telcos to raise prices in line with inflation,

less X percent for technological change and other cost reduction procedures. The Argentine system, which was initially devised as a two-step
system with an interim rate of return leading to an eventual price cap
system, was later amended under opposition from Parliament and a
complete price cap system was introduced. In other words, such mechanisms will also fail unless the agency's rulemaking and changing process
is sufficiently insulated from the political system. In countries where this
is not realistic or that have strong legal systems, regulation can be imposed
through license agreements, such as in Jamaica.

License agreements determine the form, powers, and limits of the
regulatory regime. Embedding the regulatory system in the license limits
the power of the government to unilaterally alter the regulatory instruments. Specification of regulatory procedures in licenses can also limit
the discretionary power of the regulators. Virtually the entire Jamaican
telecommunications regulatory system is specified in the 25-year monopoly license given to the Telecommunications of Jamaica Co. The license
stipulates a rate of return method with guaranteed profitability of between 17.5% and 20.0% (Spiller, 1993). The government may not disqualify investment or challenge costs, and even arbitration to settle disputes
between the government and the TJC must ensure that profitability is
maintained in the specified range.

Although such an arrangement provides good short-run investment
incentives, its long-term consequences may not be quite so felicitous. The
system provides safeguards, but it also reduces regulatory flexibility since
licenses cannot be changed without the agreement of the private firm.
This limits the ability of the regulators to impose investment and expansion incentives, and it limits their ability to force the company to respond
to new technologies or provide new products. Spiller (1993) maintained
that in the Jamaican case, with a long history of strife among the Jamaican
Regulatory Commission, the government, and the Telco, this may have
been the only way to attract private capital. In any case, there is little
doubt that in this situation the balance between national goals and company incentives tilted clearly in favor of the Telco.

Licenses, moreover, cannot be fully contingent and will "necessarily
contain ex-ante rigidities and inefficiencies" (Spiller, 1993, p. 26). One of
these rigidities stems from the fact, that differing from most legislation,
these contracts tend to have a specific finite term. As a consequence,
periodic "end-games" will develop. Because the government has the
power not to grant a new license, the companies will take actions to
protect their specific assets, while the government may take action to
"soften" the company to accept different contract terms. "Thus politics
(including the ability of the companies to use the political process) as
well as the nature of the long term contract will be key determinants of

these renegotiation costs" (Spiller, 1993, p. 26). Spiller further argued that in Jamaica's case, the very high assured rate of return coupled with a very long monopoly license may provide some future government with an incentive to renegotiate the license.

CONCLUSIONS

Swept along by the wave of deregulation and liberalization that is dramatically changing telecommunication sectors in the industrialized countries, Third World countries are currently engaged in the widespread reform and restructuring of their telecommunication sectors. These reforms have involved various degrees of corporatization, privatization, or liberalization.

This movement is in contrast to the historical trend, which, in the past few decades, was away from commercial provision of telecommunications and toward governmental control. The reason was a perception that private telecommunication companies exploited their monopoly position by providing inadequate service at excessive prices and ignored important social concerns.

Nulty and Schneidewind (1989) argued that "if telecommunications services are to be successfully moved back into a more market-oriented mode of operation that is also stable and viable, the problems that led to nationalization in the first place must be addressed" (p. 30). To do so, effective and credible regulatory mechanisms will have to be developed to balance commercial interests against political or social concerns, "otherwise reform and restructuring of the telecommunications sector will not be stable or sustainable" (p. 30).

Clearly there is no one correct blueprint for such a mechanism, and each country must find its own solution depending on its political dynamics, the nature of its legal system, and the independence of the judiciary and the regulatory agency. In other words, regulatory systems are specific to a combination of institutional constraints that may vary over time and across countries. These systems are highly sensitive to changes in institutional constraints, such as those imposed by political actors, the nature of the property rights regime, or the regulatory structure, which together determine the margins at which organization within the regulatory system operate. Third World regulatory systems will fail if the institutional constraints define a set of political and economic payoffs that do not encourage efficient and balanced development of the sector. Therefore, reform of telecommunications in developing countries must include ways to restructure the institutional rules so as to redirect incentives, both political and economic, in a manner that will encourage organizations to engage in regulatory and productive activities that lead to the efficient development of the sector.

For instance, although privatization may eliminate the management and the organizational shortcomings of public enterprises, the potential for regulatory failure can arise because firms "capture" regulatory agencies influencing and controlling the regulatory process. Capture aside, regulatory failure may be caused by information asymmetry between regulators and firms, unclear tariff formulas, weak or no conflict resolution mechanisms, and limited institutional capacity to enforce regulatory rules. In short, trading a public monopolist for its private equivalent is not exactly guaranteed to enhance efficiency. Nor does it constitute sufficient conditions for improving domestic welfare, as is evident from the cases of Mexico and Argentina (Galal et al., 1992).

Regulatory systems will be the key mechanisms to bridge the divide between public objectives and private incentives for the development of telecommunications in Third World countries. Countries should seriously consider the introduction of regulation and the institution of appropriate regulatory bodies prior to reform. What should be done will vary according to the objectives and nature of reform, the development of market forces, and the political climate of the country. What can be done will further constrained by political, legal, and safeguarding institutions.

REFERENCES

Campbell, J. E. (1993). Property rights and governance transformations in Eastern Europe and the United States. In S. Sjostrand (Ed.), *Institutional change: Theory and empirical findings.* Armonk, NY: M. E. Sharpe.

Cardoso, E., & Faletto, E. (1978). *Class and class conflict in industrial society.* Stanford, CA: Stanford University Press.

Dahrendorf, R. (1959). *Class and class conflict in industrial society.* Stanford, CA: Stanford University Press.

Dahrendorf, R. (1977). Scientific-technological revolution: Social aspects. Beverly Hills, CA: Sage.

Diamond, M. (1983). Overcoming Argentina's stop and go economic cycles. In J. Hartlyn & S. Morely (Eds.), *Latin American political economy.* Boulder, CO: Westview Press.

Etzioni, A. (1993). A socio-economic perspective on friction. In S. Sjostrand (Ed.), *Institutional change: Theory and empirical findings.* Armonk, NY: M. E. Sharpe.

Evans, P. (1985). After dependence: Recent studies of class, state and industrialization. *Latin American Research Review, 20,* 149–160.

Fiorina, M. P. (1985). Group concentration and the delegation of regulatory authority. In R. Noll (Ed.), *Regulatory policy and the social sciences.* Berkeley, CA: University of California Press.

Fligstein, N. (1990). *The transformation of corporate control.* Cambridge, MA: Harvard University Press.

Frieden, J. (1987). *International political economy: Perspectives on global power and wealth.* New York: St. Martin's Press.

Furubotn, E., & Pejovich, S. (1974). *The economics of property rights.* Cambridge, MA: Ballinger.

Galal, A. (1993). *Regulation, commitment and development of telecommunications in Chile.* Washington, DC: The World Bank.

Galal, A., Tandon, P., & Vogelsang, I. (1992). *Welfare consequences of selling public enterprises.* Washington, DC: The World Bank.

Jones, L., & Mason, M. (1980). *The role of economic factors in determining the size and structure of public enterprises in mixed economy LDCs.* Paper presented at the 2nd Annual Boston Area Public Enterprise Group Conference, Boston.

Jonscher, C. (1983). Information resources and economic productivity. *Information Economics and Policy, 1,* 13–35.

Katz, R. (1988). *The information society: An international perspective.* New York: Praeger.

Lindberg, L. N., & Campbell, J. L. (1991). The state and the organization of economic activity. In J. L. Campbell, J. R. Hollingsworth, & L. N. Lindberg (Eds.), *Governance of the American economy.* New York: Cambridge University Press.

McCubbins, M. (1985). Regulating the regulators: A theory of legislative delegation. *American Journal of Political Science, 29,* 721–748.

Moran, T. (1974). *Multinational corporations and the politics of dependence.* Princeton, NJ: Princeton University Press.

Mosco, V. (1988). Toward a theory of the state and telecommunications policy. *Journal of Communication, 38,* 107–125.

Noll, R. (1986). The political and institutional context of communications policy. In M. Snow (Ed.), *Marketplace for telecommunications: Regulation and deregulation in developing countries.* White Plains, NY: Longman.

North, D. (1981). *Structure and change in economic history.* New York: W. W. Norton.

North, D. (1990). *Institutions, institutional change and economic performance.* New York: Cambridge University Press.

North, D. (1993). Institutional change: A framework of analysis. In S. Sjostrand (Ed.), *Institutional change: Theory and empirical findings.* Armonk, NY: M. E. Sharpe.

Nulty, T., & Schneidewind, E. (1989). Regulatory policy for telecommunications. In B. Wellenius, P. Stern, T. Nulty, & R. Stern (Eds.), *Restructuring and managing the telecommunications sector.* Washington, DC: The World Bank.

Pattnayak, S. (1990). *Explaining differential levels of industrial growth.* Unpublished doctoral dissertation, Vanderbilt University, Nashville.

Petrazzini, B. A. (1993). The politics of telecommunications reform in developing countries. *Pacific Telecommunications Review, 14*(3), 4–23.

Rueschemeyer, D., & Evans, P. (1985). The state and economic transformation: Toward an analysis of the conditions underlying effective intervention. In P. Evans, D. Rueschemeyer, & T. Skocpol (Eds.), *Bringing the state back in.* Cambridge: Cambridge University Press.

Sen, A. (1981). Public action and the quality of life in developing countries. *Oxford Bulletin of Economics and Statistics, 43,* 287–319.

Skocpol, T. (1985). Bringing the state back in: Current research. In P. Evans, D. Rueschemeyer, & T. Skocpol (Eds.), *Bringing the state back in.* Cambridge: Cambridge University Press.

Spiller, P. T., & Levy, B. (1991). *Regulations, institutions and economic efficiency.* Washington, DC: The World Bank.

Spiller, P. T. (1993). *Institutions and regulatory commitment in utilities privatization.* Washington, DC: Institute for Policy Reform.

Spiller, P. T., & Sampson, C. (1992). *Regulation, institutions and commitment: The privatization of Jamaican telecommunications.* Berkeley: University of California Press.

Stepan, A. (1978). *The state and society: Peru in comparative perspective.* Princeton, NJ: Princeton University Press.

Williamson, O. E. (1983). Credible commitments: Using hostages to support exchange. *American Economic Review, 73,* 519–540.

Williamson, O. E. (1991). Comparative economic organization: The analysis of discrete structural alternatives. *Administrative Science Quarterly, 36*(2), 269–296.

14

Privatization and the Public Interest: Is Reconciliation Through Regulation Possible?

Harry M. Trebing
Michigan State University

The movement toward the privatization of public utilities in both industrialized and developing nations began in the mid-1980s and has continued to grow. Whether measured in number of transactions or value of assets transferred from the public to the private sector, the magnitude of the change has been impressive. The greatest growth in terms of the number of transactions has occurred in Latin America rather than in Europe or Asia, whereas privatization in Africa has barely begun. However, numerical comparisons based on transactions alone can be misleading, because the privatization of a major telecommunications carrier such as Nippon Telegraph and Telephone (NTT) or British Telecom (BT) can far exceed the dollar value of transactions in developing nations for a comparable year. But regardless of the measure, privatization continues to grow, with no visible end in sight (Sader, 1993; Schwartz & Lopes, 1993).

A review of privatization in the public utility field reveals little or no general pattern with respect to the transfer of assets, attendant industry restructuring, or the type of regulatory oversight that has been put in place. In some cases privatization has been accomplished by the sale or distribution of shares to the general public, whereas in other cases utility properties have been sold to foreign investors through bidding or auctioning arrangements. Restructuring may or may not have taken place. The privatization of Japanese National Railways (JNR) and Britain's Central Electric Generating Board was carried out through extensive industry-wide restructuring. Telecommunications systems, on the other hand, appear to have been transferred intact except for liberalized entry provisions designed to encourage competition. BT and Telmex serve as examples. The regulatory role assumed by government covers the spectrum from nonintervention to comprehensive oversight of pricing and performance. New Zealand appears to be an example of nonintervention, where legislation adopted in 1987 leaves all matters pertaining to interconnection, pricing, and service largely to the market with no agency set up to exercise regulatory responsibility. In contrast, separate regulatory agencies were created for each privatized industry in Great Britain (OFTEL for telecommunications, OFGAS for gas, and OFFER for electricity).

It is probably too early to draw any definitive conclusions regarding the success or failure of the privatization movement. Several review articles written by persons sympathetic to privatization claim significant improvements (Beesley & Littlechild, 1989; Caves, 1990). Similarly, anecdotal evidence suggests instances of success. For example, Chile has doubled its phone capacity in the four years following the sale of its telephone properties, and Mexico has reduced its per-unit labor costs with telephone privatization. However, there is other evidence that is far less sanguine. Complaints continue about BT's high profits and poor service. This has prompted OFTEL to attempt to exert a downward pressure on telecommunications prices by successive increases in the imputed level of productivity gain.[1] Similarly, Great Britain's two privatized power-generating entities have been pressured to sell off 10% to 15% of their capacity to increase arms-length bargaining with the transmission network, while at the same time generating prices have been reduced by 7% by regulatory action (Smith, 1994). In Japan, a survey of the privatized rail system reveals evidence of a deterioration in quality of service and safety standards (Abe, 1991).

In developing nations, World Bank studies have also expressed a fear that privatization per se, without the necessary concomitant regulation will not result in the establishment of competitive markets. The World Bank's studies note that privatization alone may be conducive to anti-

---

[1]OFTEL originally set the productivity offset under price cap regulation at 3%. The Retail Price Index minus the productivity offset would determine a permissible net price increase. The offset was kept at 3% from 1984 to 1989, but complaints about poor service and excessive earnings prompted successive increases to 4.5% between 1989 and 1991, 6.25% between 1991 and 1993, and 7.5% between 1993 and 1997. Still, the rate of return remains high (16%–18%), and quality improvements have been minimal. BT's desire to become a global player in international communications markets may have been a factor affecting domestic performance. Whether entry into domestic markets by British and American cable carriers together with the regional Bell holding companies will improve performance and lower prices remains to be seen.

competitive behavior that will serve to entrench privatized monopolies (Wellenius et al., 1993). Continuing, the World Bank notes that privatization can create firms that are interested only in serving highly profitable niche markets such as cellular and international communications, rather than in the establishment of core communications services for the general populace.

This chapter presents the privatization movement within a broader holistic context that considers the interrelationship among three areas: (a) the pressures that led to privatization, (b) fundamental structural features of the telecommunications industry within which competition will have to be established, and (c) the regulatory options available to government if public interest goals are to be achieved. Hopefully, an analysis of these issues will contribute toward putting a regulatory framework in place that will address the World Bank's concern that "the single most troubling issue in recent reforms (privatization) is the slow progress in developing regulatory capabilities" (Wellenius et al., 1993, p. 10).

PRESSURES FOR PRIVATIZATION

There are at least five significant factors that have served to promote the privatization movement since the mid-1980s. The first was the widespread disenchantment with the performance of public utility industries that were placed in the public sector in the years following World War II. In developing nations, communications systems were initially installed by foreign firms (such as ITT) or colonial governments but were nationalized in the 1960s. These systems appear to have been plagued by a common set of problems. Second, the level of investment in a new plant was much lower than that needed to meet demand. Nationalized industries had little or no access to foreign capital, and local governments did not have sufficient funds to meet the requirements for expansion. Third, governments routinely appropriated any telecommunications surplus that might have appeared for redeployment elsewhere in the economy, thereby worsening the investment problem. Fourth, telephone systems were required to provide subsidized service for various sectors of the economy to allay public discontent. At the same time, the government resisted price increases for telecommunications, further aggravating the problem of maintaining adequate service. Fifth, employment in the industry became highly politicized, with the result that telephone systems were often burdened with a redundant work force. A common solution to this revenue-capital shortfall was to increase foreign borrowing, with the result that much of the rapidly growing debt in developing nations was attributable to poor performance in the nationalized public utility industries.

In the industrialized nations, publicly owned postal telephone and telegraph systems (PTTs) had been in place for a much longer period. These PTTs were also more fortunate in that they did not have to deal with many of the labor problems that beset the state-owned railways or the burden of misdirected investment in nuclear power that plagued many government-owned electric utility systems. As a result, the telecommunications systems did not contribute as much to the ballooning need for subsidies as did JNR nor did they demonstrate the substantial operational inefficiencies that characterized Great Britain's electricity network.[2] For the most part, the PTTs appear to have been slow to innovate and inclined toward conservative management practices, and they were typically burdened with the need to subsidize postal systems. Some attempts were made in the 1960s and 1970s to improve the performance of government enterprise, particularly in Great Britain, but without much success.

In both developing and industrialized nations, privatization was assumed to constitute a significant step toward industry self-sufficiency, more effective cost control, greater innovation, efficient pricing, greater responsiveness to consumer needs, and the replacement of chronic deficit borrowing with operating profits and the attraction of foreign and private capital. This transformation would be accomplished through the introduction of property rights, profit incentives, and private management. Little thought apparently was given to the type of competitive market that might emerge or to the accompanying regulatory intervention that would be required.

A second factor promoting privatization was the growing belief that economic efficiency was the primary goal of public policy and that an earlier emphasis on equity, fairness, and income distribution was misplaced. Achieving economic efficiency would reduce prices and raise the standard of living, whereas governmental efforts to promote equity and

---

[2]It is appropriate to urge caution before condemning all publicly owned enterprises as inefficient and inept, while praising private ownership as inherently efficient and innovative. Electricity supply serves as an important example. In the 1960s and 1970s, both publicly owned British and privately owned U.S. electric utilities erred on the side of building large coal and nuclear generating units that were subsequently plagued with poor reliability records and massive cost overruns. But the U.S. experience was no better than the British and in some cases was even worse. Furthermore, publicly owned systems such as the Tennessee Valley Authority, Ontario Hydro, and Bonneville Power Administration pioneered in matters of network integration and high-voltage transmission long before the privately owned electric utilities. It should also be noted that France's publicly owned electricity system had one of the best records for low-cost nuclear power plant construction and efficient operation. It also pioneered in major pricing reforms such as the French Green Tariff, introduced in 1957. Utility-type enterprises may fail to perform over time, but the causal factors are much more complex than simply the form of ownership.

fairness would culminate in the protection of special interest groups, ultimately lowering efficiency and raising prices. There appears to have been insufficient recognition of the fact that attaining efficiency goals can have distributional consequences. For example, the accelerated writeoff of redundant telephone plants in order to modernize telecommunications networks would probably be charged against those consumers whose demand functions are most inelastic and therefore most vulnerable to price increases. It also could be argued that these core market customers would be those least apt to benefit from modernization. Moreover, market efficiency is not synonymous with societal efficiency. The latter is concerned with social values and social costs. If one assumes that these values and costs can be quantified, then they must be considered as an integral part of any improvement in efficiency.

Third, there was a growing recognition that telecommunications, like electricity, water supply, pipelines, highways, railways, and airways, are an integral part of infrastructure. It has also been shown that infrastructure investment is closely correlated with a growth in productivity, and productivity, in turn, is a major determinant of the level of real income (Peterson, 1994). The network as infrastructure provides a platform for growth by expanding markets and providing greater opportunity for all types of businesses and services to utilize network services more cheaply and efficiently. The economic characteristics of networks will be discussed later, but it is sufficient at this point to indicate that the attainment of all of the economies inherent in a network will tend to lead to high levels of market concentration. That is, an efficiently functioning telecommunications network will require a minimum efficient market share (MEMS) to achieve these economies and this, in turn, will result in facility-based carriers garnering a large share of the market. Thus, a relationship emerges in which network economies are conducive to concentration, and concentration is conducive to market dominance, high levels of profitability, and opportunities to employ pricing and related strategies that will perpetuate such dominance and circumscribe competitive market forces (Trebing, 1994). Accordingly, although the desire to achieve network economies has been a major force for privatization, the inherent characteristics of an efficient network will tend to have an adverse affect on the sustainability of competition over time.

A fourth pressure for privatization has come from a variety of special interests that seek to benefit from the transfer of ownership from public to private enterprise. State enterprises, potential foreign entrants, and large business customers capable of exercising monopsonistic buying power are major players in the privatization movement, particularly in advanced industrialized nations where the impact of globalization is apt to be most dramatic. There is little question that British Telecom and NTT perceived privatization as an opportunity for greater freedom to expand their scope of activities, while at the same time opening up new sources for attracting capital. Privatization would permit them to serve new, lucrative markets made possible by the computer revolution and rapid changes in the multimedia field. Privatization would also permit entry in global telecommunications markets and create a potential for participating in these markets through joint ventures, cross-industry ownership patterns, and formal alliances. Examples include BT's acquisition of a 20% interest in MCI for $4.3 billion; France Telecom's and German Telekom's proposed investment of $4.2 billion in Sprint, and the comprehensive alliances formed between Singapore Telecom, KDD, and AT&T, which have subsequently been expanded to include the telecommunications systems in Sweden, Switzerland, and the Netherlands.

Concurrently, potential foreign entrants into national telecommunications markets have viewed privatization as an opportunity to enter markets previously closed to them. Entry by direct foreign investment has proven to be particularly attractive to the regional Bell holding companies (RBHCs) that are currently restricted by the Modified Final Judgment and the Cable Act of 1984 in their efforts to diversify within domestic markets. It is not surprising that the nine largest U.S. telephone carriers (excluding AT&T and MCI) have 265 programs in 52 foreign countries (Wasden, 1993).

Finally, there are the large multinational corporations who wish to exercise their bargaining power to extract price concessions from the telecommunications carriers. This will clearly be easier if the supplier is privatized and given a high degree of latitude for setting prices for particular classes of service, as opposed to dealing with a lethargic public enterprise that is obligated to make price concessions to basic service customers as a matter of public policy. Now the concessions will flow to the multinational corporation. The misallocation inherent in designing networks and making concessions to multinationals was summarized by Mansell (1993).

A fifth pressure for privatization has come from the change in lending policies of the World Bank. This, of course, applies primarily to developing countries. In the 1960s and 1970s, the World Bank pursued lending programs that focused on funding modernization and organizational improvements for state-owned enterprises. By the mid-1980s, the Bank had apparently concluded that governments were unable to allocate sufficient funds to promote the telecommunications improvements needed if their nations were to participate in global markets. As a result, the Bank shifted its emphasis to support the privatization of state enterprises. Bank working papers expressed great faith in the creation of competitive markets as a stimulus for efficiency and a source for new capital. However, it is interesting to note that the World Bank also took a strong stand in favor of putting both competitive markets and regulation in place. This juxtaposition of competition and regulation stands in sharp contrast to much of the deregulation literature in the United States. The World Bank argued that there is a need for regulation and that the "no regulation option" would impose significant costs by limiting entry and permitting anticompetitive behavior to become entrenched (Wellenius, 1993). Regrettably, the Bank did not move to the next stage and set forth a specific set of regulatory proposals together with a detailed critique of the strengths and weaknesses of each option. Instead, it called for transparency (decisions made in open forum), less government micromanagement, and greater managerial autonomy. In its working papers, the Bank apparently endorsed regulatory licensing criteria that include price caps, interconnection, promotion of basic service, and so on but with little or no exploration of the ramifications associated with reliance on such measures in industries that have unique structural characteristics.

INTERACTION OF PRIVATIZATION, NETWORK CHARACTERISTICS, AND IMPERFECT MARKETS

Whether these disparate forces promoting privatization will yield an outcome that promotes social and market efficiency, economic growth, and an acceptable distribution of income will depend on the roles played by markets and regulatory intervention. The adequacy of market constraints and regulatory techniques will depend on how successfully they can accommodate the structural and behavioral characteristics of the telecommunications industry. In effect, this structure and behavior will prescribe the competitive nature of emerging markets and the strengths and weaknesses of different tools available for regulation.

Because the United States has had the greatest experience with telecommunications deregulation and increased reliance on market forces, its patterns of change provide an important insight into what can be expected from privatization on a global basis. Individual adjustments will have to be made for industrialized and developing nations, but the U.S. pattern could serve as a starting point for analyzing both markets and public policy options.

Networks require a significant minimum threshold investment because of the need to interconnect all customers in a given service territory. At the same time, a network has the capability of providing multiple services at a lower cost than if each service were to pay its stand-alone cost. This is because of economies of scope, economies of joint production, and economies of scale. Networks also have the capability of minimizing reserve margins because routing alternatives and pooled reserves permit increased reliability at minimum cost. Closely connected to this feature is the ability of a comprehensive network to reduce the level of capacity needed to meet coincident system peaks. Another cost characteristic of a network is its ability to add additional services at a small incremental cost as the size of the overall network increases. Achieving these inherent economies typically requires a large share of the market (MEMS) and a long gestation period. For example, it took MCI 10 years to achieve financial viability. Both of these factors constitute a significant barrier to new entry.

At the same time, telecommunications networks serve a wide variety of different customer classes with differing demand elasticities. An ability to differentiate customers along these lines is a prerequisite for price discrimination,[3] cross-subsidization,[4] and limit-entry pricing.[5] The management of a telecommunications network will be under pressure for two reasons to employ these pricing strategies. First, achieving full utilization requires that rate design exploit the demand characteristics of individual markets in such a fashion as to achieve high-load, diversity, and capacity factors. Second, telecommunications management faces major challenges because rapid technological change has created new markets and new delivery systems. However, future markets are ill-defined, consumer preferences are unknown, and appropriate pricing and marketing strategies are highly conjectural. Faced with high levels of uncertainty, potential market volatility, and heavy capital investment needed for modernization, it follows that the firm will seek to employ those pricing strategies that minimize overall risk, ensure stable or increasing revenues, and limit

---

[3]Price discrimination may be defined as selling the same product or service at two or more different prices when conditions of supply are comparable. Essentially, this means that costs of production are comparable but prices vary between differentiated markets. Overall efficiency may or may not be adversely affected, but income distribution (between those paying higher and those paying lower prices) will clearly be affected.

[4]Cross-subsidization may be defined as selling a specific product or service at a price that does not yield sufficient revenue to cover its cost of production, yet the overall revenue of the firm from the sale of all services is sufficient to cover its total cost (including its cost of capital). There is general agreement that the incremental cost constitutes one standard for judging the cost of production for a specific service but less agreement on the use of fully distributed cost as a measure of the cost of production. Both efficiency and income distribution will be adversely affected by charging a class of customers a price that is greater or less than an acceptable cost standard for serving that class.

[5]Limit-entry pricing involves setting a price for a specific service sufficiently low to foreclose entry. The motivation may not be predation, but rather the discouragement of entry. Predatory pricing involves setting a price to drive rivals out of the market or prevent the entry of efficient rivals. This involves setting a price below short-run marginal cost, with the result that long-term earnings will be reduced if the price is not increased at a later date. Limit-entry pricing would seem to be the preferred strategy of public enterprises facing competition.

potential entrants. Furthermore, the need to employ such strategies will be reinforced when management perceives that neither government per se nor regulation affords sufficient protection against technological change or the actions of potential rivals.

The pricing problem for the telecommunications firm is complicated even more by the fact that there appears to be a high concentration of business sales in the hands of relatively few large monopsonistic buyers. One estimate indicates that 20% of business consumers account for more than 80% of business revenues (Trebing, 1989). Another estimate indicates that there are approximately 2,000 large multinational corporations potentially capable of exercising such power on a global basis (Andrews, 1994). A possible response for network management will be to form global alliances or joint ventures of the type previously discussed. This would control the options available to multinationals seeking global outlets.

Another dimension of network management behavior is deep-pocket diversification, a means for ensuring that the firm will not be foreclosed from new markets, whether cable TV, wireless communications, or multimedia services. It also provides a hedge against miscalculations of future market growth, thereby reducing overall risk and uncertainty. The disadvantage of diversification is that management's efforts to redeploy resources into areas of greatest potential profit may result in a denigration of basic service. Unfortunately, the neoclassical answer for controlling disinvestment, that is, raising the rate of return on basic services to match the opportunity cost of foregoing diversification, is not readily applicable to a carrier seeking to cherry pick among nonregulated activities and foreign countries of varying degrees of political vulnerability.

Focusing on the U.S. experience since AT&T divestiture in 1984, it is interesting to explore whether the industry displays characteristics usually associated with workable competition, notably, an absence of market dominance (largest firm has no more than 35% to 40% of the relevant market), no excessive profits over time, and no sustained evidence of exploitative pricing practices or price leadership.

High levels of concentration continue to persist in telecommunications. Although AT&T's share has declined, it still retains about 60%–65% of the interexchange/interLATA market and about 69% of the overseas market. It is also important to note that the rate of decline in AT&T's share of the total domestic market has flattened out in the last several years. At the local exchange level, dominance still persists despite the appearance of competitive access providers (CAPs) that link big business customers to long-distance carriers.

Of course, rapid technological change can produce bypass delivery systems (e.g., personal communication systems) and head-to-head confrontation. Huber, Kellogg, and Thorne (1993) argued that local exchanges are geodesic—that is, highly interconnected networks with multiple routings so that no monopoly control is possible. However, Selwyn and Hatfield (1994) argued that telecommunications is structured along hierarchical lines from the local office to the tandem switch and the long-distance carriers' point of presence. Further, local networks are becoming more concentrated because of high-capacity fiber and digital switching. Regardless of which side one takes, concentration levels remain high and the prospect at best is for duopoly (between cable and local exchange carriers) or tight oligopoly.

Special attention must be given to the theory of contestable markets that holds that the threat of potential entry will prevent the incumbent firm from earning excessive profits or practicing exploitative pricing. In this situation, concentration is no longer a measure of market power. Hit-and-run entry by a facility-based carrier with high sunk costs is unlikely, but what about brokers and resellers? These firms exist in larger numbers, but they buy capacity from the carrier and typically compete with the same carrier in retail markets. Accordingly, they are vulnerable to a vertical price squeeze. At best, brokers and resellers sell in a market characterized by bounded contestability.

Existing market structures have also given little evidence that they control profit levels. The traditional measure of an acceptable level of profits under regulation was a market-to-book ratio of 1.05 to 1.10. AT&T's market-to-book ratio averaged 1.05 between 1969 and 1978. However, for the 12 months ending December 1993, the market-to-book ratio for AT&T had increased to 5.02. For MCI and Sprint, the ratio was 3.03, and for all telecommunications it was 2.7. These high levels of profitability would have instigated rate case proceedings under stringent rate base–rate of return regulation.

A third result that casts doubt on the ability of markets to control behavior is the chronic persistence of evidence of exploitative pricing and price leadership. These include the shifting of accelerated depreciation charges to basic service customers, the recovery of non-traffic-sensitive costs through subscriber line charges, and the recovery of local transport charges through a residual interconnection charge. An intriguing pattern of pricing has also appeared in long-distance telecommunications. The 1993 Huber Report states that AT&T has a strong incentive for price leadership to maintain profits and a veneer of competition. The resulting umbrella pricing supposedly permits MCI and Sprint to coexist. Huber argues that if AT&T were to move aggressively to fully exploit its network economies through price reductions, the result would be detrimental to profits and would restore AT&T to its former monopoly position (Huber et al., 1993). AT&T would then face either reregulation or new antitrust action. The consequence of umbrella pricing is a failure to exploit network

economies and a shifting of the burden of inefficiency forward to consumers. Whatever the merits of the Huber argument, there is empirical
evidence that discernible patterns of price leadership exist in portions of
the long-distance market. This undoubtedly stems from the pressure to
avoid price competition to maintain profits whenever market conditions
permit. For example, between 1988 and 1993, AT&T, MCI, and Sprint
charged identical prices for message toll telephone service in the inter-
LATA intrastate market in Virginia. In August 1993, AT&T raised prices
for medium- and long-haul MTS. MCI and Sprint followed with identical
increases (Virginia State Corporation Commission, 1994). On the other
hand, there does appear to be intense price rivalry for the large business
market. But since these prices are essentially private negotiations, there
is no empirical basis for testing pricing behavior.

On the basis of the U.S. experience in the 10 years since AT&T divestiture, it would appear that nations endorsing privatization in telecommunications can ultimately expect to face (a) a highly concentrated industry characterized by tight oligopoly; (b) firms that will seek to
maximize profits through price discrimination, cross-subsidization, and
price leadership whenever possible; and (c) investment patterns and
interconnection policies designed to maintain the long-run viability of
the firm rather than social goals such as universal service. This means
that public policy must move to promote competition whenever possible
but also intervene directly through regulation to ensure that the benefits
inherent in telecommunications network infrastructure are maximized
for all segments. This chapter will next consider regulatory options for
achieving these goals.

A CRITIQUE OF REGULATORY OPTIONS

A regulatory program may proceed in two different directions, depending
on the underlying assumptions. If one assumes that all markets are
potentially competitive and network economies are minimal, then it will
be sufficient to carry out the regulatory program through "light" regulatory intervention. This could be achieved by mandating open access
through unbundling the local exchange function to permit competitive
entry, the encouragement of bypass of the local exchange by specialized
carriers serving large business customers, the application of simple price
caps tied to the rate of inflation, and incentive allowances designed to
stimulate technological innovation and new service offerings. However,
if one assumes that telecommunications market structures will remain
highly concentrated, that potential entry among facility-based carriers is
limited, and that tight oligopoly rather than workable competition will

be the norm, then a more comprehensive form of regulatory oversight is required. The analysis of industry structure and corporate behavior suggests that the latter alternative must be addressed. Regulatory options will be critiqued in that context.

At least three major areas of regulatory intervention must be considered. First, there must be surveillance of network coverage and quality of service. This includes the development of empirical tests for determining universal service and the proper level of investment in the network. It also includes an assessment of the benefits from modernization as distributed among major user groups. Given the rapid rate of technological change in telecommunications, it is important to determine who will get the "technology dividend." This dividend is measured by the difference between the average embedded cost of the existing plant and the incremental or stand-alone cost of a new plant utilizing the most modern technology. If the burden of obsolescence is to be assigned to residential customers, then this should be a clear matter for public policy rather than corporate strategy. Similarly, the need to ensure universal service to all classes of customers may require that a user charge be imposed on all classes of customers, whether business or residential.

Second, technological advance permits the creation of alternative delivery systems that may be based upon landline, wireless, or satellite technologies. Regulatory policy must determine whether alternative technologies will be supplied by a single carrier or by independent systems acting autonomously. If the latter option is chosen, then interconnection must be far more comprehensive than simple open-network architecture. Full interconnection would permit the consumer to choose between different delivery systems based on quality of service, reliability, and price.

Third, there remains the question of developing adequate guidelines for constraining the adverse consequences of varying pricing strategies. There appear to be two major options. The first involves the establishment of pricing criteria or guidelines for individual service categories. The second involves structural separations. Each will be considered separately, but there is no reason why they cannot be employed jointly in a real-world situation.

Pricing Guidelines

The primary objective of pricing guidelines is to protect captive–basic service customers and promote economic efficiency. The promotion of other social policy objectives would be a secondary goal for pricing.

There are at least six different sets of pricing guidelines. The first is fully distributed cost (FDC). This involves the calculation of directly assignable costs for each service and the allocation of remaining overhead, common, and joint costs between services or between regulated and nonregulated activities on the basis of some usage criterion. The FCC's Part 64 Cost Allocation Manual is an example of the use of FDC to apportion costs between regulated and nonregulated services. The principal criticism of FDC involves the allocation of overhead, common, and joint costs. There is no generally accepted method for making such allocations, and each option poses problems. For example, an allocation on the basis of historic usage may distort future prices, whereas a forecasted allocation is subject to obvious abuses. Also, the cost-causation rationale for such allocations can be weak or entirely missing. An FDC allocation based on time of use (i.e., peak vs. off-peak usage) probably has fewer problems, but it does give rise to distributional issues when sunk costs are assessed against basic service markets in order to facilitate entry into new markets. The strength of FDC is that it is reconcilable with the books of account and the gross revenue requirements of the regulated firm. In an environment moving toward competitive markets, regulated firms will bitterly resist the imposition of FDC allocations because they would be viewed as an impediment to management's pricing flexibility.

The second guideline is coverage of incremental cost. Incremental cost is defined as the additional cost for adding an entire service over a future period of time. It is more vague than the economist's concept of long-run marginal cost. The rationale of incremental cost coverage is that any new service must cover the incremental resource cost (including rate of return on additional investment) in order to avoid cross-subsidization. The so-called net revenue test is an application of the incremental cost concept. It states that an additional service is compensatory when the additional revenues of a new service equal or exceed the incremental cost of that service. The FCC adopted the net revenue test in Docket 84-1235 for optional calling plans.

There are substantial problems with incremental cost. The first, of course, is the problem of incremental cost measurement for the complex network that consists of alternative routings and continuous advances in technology. A second problem is the susceptibility of incremental cost to manipulation and, as a consequence, to risk and cost shifting. As typically applied, residual pricing provides such an example. Under this approach the incremental revenues for each new service must cover the incremental cost for that service, and these revenues, in turn, are deducted from the gross revenue requirements of the firm. The entire residual balance is then assigned to basic service customers. This creates a strong incentive for the firm to underestimate the incremental cost of serving competitive markets and thereby shift as much of the cost burden as possible to basic service markets. This practice of cost and risk shifting can be further aggravated by the introduction of average incremental cost (which averages a range of incremental costs for a specific service). When average incremental cost is narrowly defined to include only those costs directly identified with the provision of additional capacity and associated operating costs for that service, it is possible to argue that a general upgrading or modernization of switches, transmission, and subscriber loops for the entire network should be excluded from the average incremental cost calculation if this upgrading is considered to be a part of the basic plant rather than the incremental cost of a new service.

The principal advantages of the incremental cost floor are that it gives management greater pricing flexibility and, if properly calculated, it may approximate economic costs as defined by neoclassical economists.

The third guideline is the neoclassical economist's maximum/minimum price standard for judging cross-subsidization. Neoclassicists argue that cross-subsidy cannot take place if the maximum price for a service does not exceed its stand-alone cost and the minimum price is not less than its incremental cost. Assume that A is the monopoly service and that B is the competitive service. Then, as long as A's price does not exceed its stand-alone cost, it cannot be argued that it is subsidizing B. If the price of A exceeds its stand-alone cost, then profits will increase above a normal level (sufficient to attract capital) or prices will fall below incremental cost for B. This supposedly avoids the need to allocate common costs between Services A and B. The stand-alone cost of A would be equal to the cost of providing that service alone, based primarily on the cost that a new entrant would incur to provide A (Baumol & Sidak, 1994).

There are serious shortcomings associated with this approach. There is an inherent tendency to deny economies of joint production to monopoly Market A. Instead, these gains will be assigned to competitive Market B. This, of course, ignores the fact that most services of the network are typically developed on a joint basis in the real world. In addition, this approach is still beset by the problems of estimating incremental cost and the need to exercise continuing oversight over maximum prices, minimum prices, and profit levels.

Furthermore, this approach fails to establish a clearly defined stand-alone cost for A except in cases where the incumbent firm makes its own estimate or when a potential entrant is a single-product firm producing only A. As soon as the stand-alone cost must include different combinations of output for both the incumbent and the potential entrant, the task becomes much more complex. Moreover, if one assumes that the market structure is composed of two or three large multiproduct firms serving differentiated markets through networks that incorporate large common and joint costs, the stand-alone concept loses much of its meaning when employed as a standard for establishing the ceiling price for A. Each rival's ability to exploit its network economies within the relevant structure of markets will serve to determine both its ability to serve A and the actual ceiling price for A. Under these circumstances, common and joint costs will enter into the determination of ceiling prices for individual markets.

A fourth guideline involves the imposition of price ceilings through a series of price caps placed on individual market-basket groupings of services. Price increases are tied to some measure of inflation, but an offsetting productivity adjustment based on an expected level of productivity gain is deducted from the inflation rate to determine the net price change. For example, if the annual rate of inflation was 5% and the offsetting productivity adjustment was 3%, then there would be a net increase in price of 2% for all services grouped in this basket. The rationale for price caps is premised on the belief that they can constrain price increases in captive markets (thereby controlling cross-subsidization), while at the same time introducing an incentive for greater efficiency by not directly controlling profits or the rate of return. Price caps have been particularly popular in telecommunications, having been applied in France, Holland, Germany, Sweden, Great Britain, Australia, and Argentina. They have also been extensively applied in the fields of electricity and natural gas. In passing, a distinction could be made between "pure" price caps and modified price caps. In the case of the former, any review of earnings is strictly off limits; in the case of the latter, the productivity offset may be adjusted to compensate for poor service or excessive earnings.

Despite their popularity, price caps suffer from a number of major deficiencies. First, price caps will not constrain price discrimination and cross-subsidization when there is at least one monopoly market in a particular market basket and prices increase faster than the increase in the cost of service for that basket. Second, there is no incentive to reduce prices for monopoly–captive markets, particularly if the demand function for the service is highly inelastic. Third, the compound effect of index-driven prices can exaggerate the spread between the ceiling prices charged consumers in captive markets and the actual cost of service or the cost that would have prevailed had these markets been highly competitive. Fourth, there are major computational problems involved in the selection of a proper inflation adjustment and productivity offset.

An argument can be made that the inflation adjustment should not be based on a broad index such as the Consumer Price Index, but rather on the actual price increases associated with inputs purchased by the telecommunications industry. Given the rapid decline in the cost of telecommunications equipment, the spread between these two indices could be considerable. The same argument can be made with respect to the productivity offset. Should it be based on some general estimate such as 3.5%

per year or should it reflect the actual productivity gain in the telecommunications and computer industries?

There is a further complication with price caps, as demonstrated in the case of British Telecom, when OFTEL has had to successively raise the productivity offset over the past decade to correct for excessive earnings and poor quality of service. One is inclined to agree with Veljanovski (1993) that a broader regulatory perspective should be introduced to replace the simplistic price-cap structure. Of course, it should also be noted that price caps do nothing to control patterns of price leadership and conscious parallelism.

A fifth pricing guideline would involve the creation of an interservice sliding scale. Under such an arrangement, price decreases (or increases) in competitive markets would trigger proportionate price decreases (or increases) in monopoly markets. This would diffuse the benefits of network economies and new technology to all classes of users. The difficulty, of course, focuses on how to set up such a sliding scale that would accurately reflect the relative cost of serving different classes of customers or markets. The task would be further complicated by the need for substantial reporting requirements for both regulated and nonregulated sectors. This approach assumes that cross-subsidization and price discrimination can be controlled by spreading price reductions or price increases across the board.

A sixth pricing guideline could be based on the Glaeser model (Trebing, 1989). Essentially this is an approach to controlling cross-subsidization by establishing a revenue contribution by class of service that reflects both the direct costs of serving that class and an allocation of common and joint costs made in proportion to the benefits received by each customer class from the use of the network. It also imputes costs to those who are causally responsible for placing demands upon the network. In its simplest form, the Glaeser model subtracts the direct cost of each service from the total cost of the project. The resultant common–joint costs are allocated on the basis of the relative benefit derived. The benefit is equal to the difference between the stand-alone cost of the service and its direct cost when provided over the network.

This approach has a number of distinct advantages. It recognizes that a service may not be offered unless it realizes economies of joint development and scope from utilizing the network. When the spread between the stand-alone cost and the direct cost is great for a service, then that service will bear a proportionally greater share of the common/joint costs. When the spread is small, the reverse is true. When stand-alone and direct costs are identical, then that service would bear none of the common/joint costs. Application of the Glaeser model would prevent uneconomic bypass of the network. If the relative allocation to particular classes of service is considered as an intermediate step between the calculation of total revenue requirements and the pricing of individual services, then a significant shortcoming of FDC pricing would have been overcome. The revenue contribution by class of service would reflect cost proportionality. The final pricing for that class of service would permit recognition of peak–off-peak usage and various demand factors.

An additional note on the Glaeser model is appropriate. The model provides a strong stimulus for networks to add more and more services. To the extent that additional services result in the allocation of a smaller amount of common/joint costs to basic service customers, management could be given the freedom to expand the system, and the benefits of this expansion would accrue, in part, to monopoly core customers.

Structural Separations: The Creation of Separate Subsidiaries

The FCC's Computer II decision (1980) took a nonprice approach to the control of cross-subsidization. The commission stated that customer premises equipment and enhanced services (largely deregulated) should be supplied through a separate subsidiary. The FCC's Computer III decision (1986) dropped this separate subsidiary requirement for nonregulated services.

There are a number of shortcomings with the FCC's employment of the separate subsidiary concept: It is difficult to separate common costs between the parent and the separate subsidiary, to ensure an arms-length relationship between the parent and the separate subsidiary, and to isolate the cost of capital effects of diversification. In the event that the separate subsidiary becomes a financial failure, it will raise the cost of capital to the parent and therefore to all those who use the network, including basic service customers.

A more promising approach to structural separations has been incorporated in the so-called Rochester Plan (New York Public Service Commission, 1994). Under this plan the network would be separated from an unregulated affiliate. The network would remain under regulatory surveillance with full reporting requirements, monitoring for quality of service, and price regulation. It would sell comparable service on equal terms to all parties on a wholesale basis, and would provide basic service to all customers requesting that service. The unregulated affiliate, on the other hand, would be set up to buy at wholesale and sell at retail, in competition with other brokers and resellers. It would be free to provide electronic shopping, video on demand, call waiting, and private-line service and to engage in overseas as well as domestic diversification projects. The network would also be able to promote packages of service that incorporate reselling basic service. The unregulated affiliate would receive all gains and bear all losses from such activities.

The network would have separate debt financing, separate management, and a separate board of directors to insulate it from the activities of the unregulated affiliate.

A major advantage of the Rochester Plan is that it would create a competitive pressure on the network to modernize and be efficient because the unregulated affiliate and all other brokers, resellers, CAPs, and IXCs (interexchange carriers) would be free to buy from the network or from any other service provider. Yet, because it was maintained as an integrated whole, the network would be free to maximize all inherent network economies and translate these into wholesale prices. Aside from the obligation to provide basic service, the network would not be compelled to undertake any unique or special welfare-type programs, thereby avoiding the burden that was traditionally placed on publicly owned utilities.

However, the Rochester Plan still leaves a number of issues to be resolved. For example, where are the boundaries of the network? Should the network include landline and broadband, as well as wireless services? Also, should the network assume an active or passive role in promoting new services? Expressed somewhat differently, should the network be passive insofar as modernization and expansion are concerned and wait for customers to make demands upon it, or should it take the initiative in promoting the telecommunications revolution?

CONCLUSION

The World Bank is very wise in suggesting that regulation has a crucial role to play in the future of telecommunications under privatization. If the industry is characterized by high concentration, persistent profits, and deep-pocket diversification, as well as collaborative strategies designed to maintain market dominance, then the challenge to regulators in both industrialized and developing nations will be enormous. To the extent that these strategies transcend national boundaries, regulators will find themselves in the same jurisdictional predicament that plagues state commissions in the United States when they are confronted by regional or nationwide holding companies. Furthermore, in an industry of tight oligopoly, matters more closely identified with industrial organization studies (e.g., anticompetitive behavior, definitions of workable competition, and the impact of market dynamics) will occupy much more of a regulatory agency's time than traditional price and earnings controls. To handle these issues, regulators must have more than transparency and independence. They will need improved reporting requirements, a national and global perspective on issues, and a capability to refine and apply new concepts that maximize the interactive contribution of regulation and market pressures. Toward that end, the concepts outlined in this chapter are only a primitive first step.

REFERENCES

Abe, S. (1991). Privatization of Japanese national railways and its consequences. *Business Review, 41*(5/6), 111–130.

Andrews, E. L. (1994, June 24). AT&T teams up with 3 Europe partners. *New York Times.*

Baumol, W. J., & Sidak, J. G. (1994). *Toward competition in local telephony.* Cambridge, MA: MIT Press.

Beesley, M. E., & Littlechild, S. C. (1989). The regulation of privatized monopolies in the United Kingdom. *Rand Journal of Economics, 20*(3), 454–472.

Caves, R. E. (1990). Lessons from privatization in Britain: State enterprise behavior, public choice, and corporate governance. *Journal of Economic Behavior & Organization, 13*(2), 145–169.

Huber, P. W., Kellogg, M. K., & Thorne, J. (1993). *The geodesic network II. 1993 report on competition in the telephone industry.* Washington, DC: The Geodesic Company.

Mansell, R. (1993). From telecommunications infrastructure to the network economy: Realigning the control structure. In J. Wasko, V. Mosco, & M. Pendakur (Eds.), *Illuminating the blindspots* (pp.181–195). Norwood, NJ: Ablex.

New York Public Service Commission (1994, May 13). *Petition of Rochester Telephone for approval of proposed restructuring plan—Joint stipulation and agreement.* Case No. 93-C-0103.

Peterson, W. C. (1994). *The silent depression.* New York: W.W. Norton & Co.

Sader, F. (1993). *Privatization and foreign investment in the developing world, 1988–92* (Policy research working paper). Washington, DC: The World Bank.

Schwartz, G., & Lopes, P. (1993, June). Privatization: Expectations, trade-offs, and results. *Finance & Development*, 14–17.

Selwyn, L. L., & Hatfield, D. N. (1994). *The enduring local bottleneck—Monopoly power and the local exchange carriers.* Boston, MA: Economics & Technology, Inc.

Smith, M. (1994, February 12). Generators in deal to sell plant and reduce prices. *Financial Times*, p. 1.

Trebing, H. M. (1989). Telecommunications regulation—the continuing dilemma. In K. Nowotny, D. B. Smith, & H. M. Trebing (Eds.), *Public utility regulation* (pp. 93–130). Boston: Kluwer.

Trebing, H. M. (1994, June). The networks as infrastructure—the reestablishment of market power, *Journal of Economic Issues, 28*(2), 379–389.

Veljanovski, C. (1993, January). The future of industry regulation in the UK. *European Policy Forum.*

Virginia State Corporation Commission (1994). *The interlata market in Virginia, 4th quarter, 1993* (Division of Economics and Finance report).

Wasden, C. D. (1993). *A descriptive compendium of the international activities of major US-based utility holding companies.* Columbus: Ohio State University/National Regulatory Research Institute.

Wellenius, B. et al. (1993). *Telecommunications: World Bank experience and strategy.* (Discussion Paper No. 192, pp. 10–14). Washington, DC: The World Bank.

Author Index

Subject Index

C

Competition, 255–256
Corporatization, 24–26

D

Debt crisis, 136–140
Domestic capital
  ASEAN countries, 165–178
  capital circuit, 154–156
  Latin America, 135–163
  social capital, 159

F

Foreign direct investment, *see also* Global
    capital
  attributes of the Regional Bell Operating
    Companies (RBOCs), 101
  costs and benefits, 89–91
  modes of international participation,
    102–103
  services offered, 101–102
  firm-specific advantages (RBOCs), 101
  firm-specific determinants (RBOCs)
    Ameritech, 92–94

Bell Atlantic, 94–95
BellSouth, 95–97
NYNEX, 98
Pacific Telesis, 98
Southwestern Bell, 97
US West, 99
Dunning's eclectic theory, 87–89
location-specific determinants, 103–105
regional examples, 105–106
possible impacts, 106–109

G

Global capital
  financial integration
    capital markets, 69–72
    defining and measuring financial
      integration, 66–67
    distributional effects of capital flows,
      73–74
    foreign direct investment, 72–73
    integration of developing countries, 67
    intersection of capital markets and
      telecommunications, 54–58
  fixed capital, 62–65
    independent versus enclosed forms of
      fixed capital, 65

335